Reporting
Technical
Information   FOURTH EDITION

Kenneth W. Houp
Associate Professor of English Composition
The Pennsylvania State University

Thomas E. Pearsall
Professor and Head, Department of Rhetoric
University of Minnesota

Macmillan Publishing Co., Inc.
New York
Collier Macmillan Publishers
London

For Frederic Sanford Cushing

Glencoe Publishing Co., Inc.
17337 Ventura Boulevard
Encino, California 91316
Collier Macmillan Canada, Ltd.

Library of Congress Catalog Card Number: 78–71738

        5 6 7 8 9 10   83 82

ISBN 0–02–475630–X

Contents

15 Proposals   342

16 Progress Reports   366

17 Physical Research Reports   381

18 Feasibility Reports   393

19 Oral Reports and Group Conferences   417

Preface

Shortly before our revision for this fourth edition ended, we came into possession of a memorandum that delighted us. It was from Robert A. Lauer, the training officer for the Saint Paul Companies, a large insurance firm. The memorandum was directed to company department heads and outlined the objectives for a proposed report-writing class to be offered to company employees. The objectives were drawn from comments made by the department heads themselves and were as follows:

Report Writing Trainees should be able to

1. Identify the target audience(s) of each of their reports by functional area and job level.
2. **Write a one-sentence statement detailing the purpose of the report, that is, what the information is to be used for.**
3. List the significant findings of the report in descending order of importance to the target audience.
4. Construct a framework (outline) for a report that is appropriate to both the target audience and stated purpose.
5. Compose (from the outline) a report that is
   A. Informative and concise
   B. Logically organized
   C. Free of departmental jargon (foreign to target audience)
   D. Free of mechanical errors
   E. Readable
   F. An accurate depiction, where appropriate, of statistical data

We were delighted for several reasons, not the least being that Mr. Lauer's memorandum shows that some business people—an increasing number, we find—appreciate effective writing and know how to obtain it. For another reason, this message from the business world closely mirrors our own objectives for the users of this book—that they should be able to analyze an audience correctly; to find and organize material appropriate to audience, purpose, and situation; to design a functional report or letter that answers the purposes of both writer and readers; and to write that report or letter correctly, clearly, and persuasively.

Finally, Mr. Lauer's memorandum supports writing courses where this text, or others like it, is used—courses that in an objective, rational, structured way teach students how to do the writing that all professionals need to do once they are on the job.

What will you find in this book?

Part One is basic and introductory. It covers library research, information gathering, audience analysis, the use of the rhetorical modes, and achieving clear style.

Part Two includes instruction on the elements of a well-organized professional report, including graphics and prose elements such as abstracts, introductions, conclusions, tables of contents, and headings. It concludes with a chapter on planning, writing, and revising the professional report.

Part Three covers advanced and extended applications of the basic principles. Here you are shown how to write correspondence, proposals, progress reports, feasibility reports, and the like.

Part Four is a handbook for ready reference when questions of grammatical usage, punctuation, and mechanics arise.

In the appendixes you will find an extended student report, guides to library research and computerized information retrieval systems, metric information, and a bibliography of books that can help you to learn more about report writing.

What changes have we made for this fourth edition?

- A new chapter, "The Professional Report," is based upon our belief that no really well-defined categories called "formal reports" and "informal reports" exist. Rather, there is a continuum from simple memorandums to complex reports. Formal report elements such as abstracts and tables of contents should be used to satisfy functional needs created by increasing complexity and not to satisfy some traditional notion of formality.

- Our chapter on correspondence has a greatly expanded section on resumes and employment letters. We provide a list of resources helpful to the job seeker and a modern resume modeled on the one used at the Harvard University Graduate School of Business Administration. We have also added a section on memorandums to this chapter.

- In our oral reports chapter we have added a new section on the use of visual aids in speaking.

- In our mechanical elements chapter we have revised the section on documentation to make it compatible with the guidelines set in the University of Chicago Press *Manual of Style*.

- Our exercises have been modified to call for even more realistic writing and also to get students writing earlier in the course. We also suggest (with suitable cross-reference to the correspondence chapter) that many early reports could be written as memorandums.

- We have added a section on using computerized information retrieval systems to our library chapter.

- We have widened both the number and scope of our examples drawing upon the natural and social sciences, engineering, and numerous other technical and professional fields.

Many of these changes and numerous other smaller ones reflect the growth of technical writing classes. More students from more disciplines than ever before fill increasing sections of technical writing. Two of the major journals that now inform the technical writing teacher—*The Journal of Technical Writing and Communication* and *The Technical Writing Teacher*—did not even exist when the first edition of this text was published in 1968. Articles on technical writing are published much more frequently than in the past in such journals as *College Composition and Communication* and *College English*. *Technical Communication* continues to publish useful articles. As a result, we now have more and better sources than ever before to draw upon when we revise our text. We are grateful for these important sources of ideas and information and happy to acknowledge them. We also happily acknowledge the help of all the following:

John Muller, Air Force Institute of Technology; Frances Blosser Maguire, Tarrant County Junior College; and W. Keats Sparrow, East Carolina University, who read our book and made many constructive comments.

Those who used earlier editions of *Reporting Technical Information* and provided suggestions and insights for improving this fourth edition, including Eleanor Bergholz, Strayer College, Virginia Murray, University of Houston, Beth Waggenspack, Central Ohio Technical College, Alben C. Johnson, California State Polytechnic University at Pomona, Ted Valvoda, Lakeland Community College, Douglas

Wixson, University of Missouri-Rolla, Mead R. Johnson, California Polytechnic State University at San Luis Obispo, Freda F. Stohrer, Old Dominion University, Daniel Riordan, University of Wisconsin-Stout, James B. Steele, Metropolitan State College, Louise Garrison, Oregon State University, Thomas L. Warren, Oklahoma State University, M. A. Stugrin, University of Pittsburgh, Thomas M. Sawyer, Northern Montana College, and Alan N. Steen and Mark Larson, Humboldt State University.

Our many friends in the business world and the Society for Technical Communication, in particular, J. Paul Blakely, Oak Ridge National Laboratory, Mary Fran Buehler, Jet Propulsion Laboratory, California Institute of Technology, Karen Bunting, The Trane Company, Gerald Cohen, International Business Machines Corporation, and Charles R. Pearsall, Deere and Company, who sent us example materials.

Donald J. Barrett, Chief Reference Librarian, United States Air Force Academy, who once again has revised Appendix B, "Technical Reference Guides."

And our wives, Lois and Anne, who know all too well the time-consuming effect of our touchstone phrase, "All writing is subject to infinite improvement."

Kenneth W. Houp
Thomas E. Pearsall

PART 1

Process of Technical Reporting

We have designed Part 1 to serve as an introduction to technical report writing for the student. Part 1 is particularly suited to the scientific, engineering, or technical student who must occasionally write reports. But the student who thinks of becoming a full-time technical writer will also find it valuable.

In Part 1 we lay a foundation that will enable a student to write simple technical reports. We cover three stages of report preparation: investigating, planning, and organizing. In addition, chapters on audience analysis, rhetorical methods, and style are included.

CHAPTER 1

An Overall View of Technical Reporting

This first chapter is purely introductory. It is intended to give you the broadest possible view of report writing. Beginning with Chapter 2 we will go into details. But for the details to be most meaningful, they have to be seen against the background given here.

Some Matters of Definition

What does the term "report writing" mean to you? Does the term "technical reporting" convey any clearer notion? In this book these terms are often used interchangeably. Therefore we are compelled at the outset to provide you with a working definition, which we will then expand and refine in later chapters.

This need for definition always arises when a novel and sometimes complex term is introduced. Suppose you were to learn from a bulletin board announcement that a lecture is to be given tomorrow evening on "operations research." If you didn't know the meaning of the term, would you attend the lecture anyway? Probably not.

Operations Research: analysis, usually mathematical, to determine the effectiveness of a process, operation, or the like, to increase efficiency.

This definition, brief as it is, might very well enable you to make up your mind.

Now let us turn specifically to the title and subject matter of this book: *Reporting Technical Information.*

reporting: providing an account or description of what has been learned by experience, observation, or investigation.

technical: peculiar to or characteristic of a particular art, science, trade, or profession.

information: a body of knowledge gained from experience, observation, or investigation.

The reporting of technical information is thus seen to involve three elements at one or more stages of the process:

1. A problem or subject matter that is not popular knowledge but, rather, is specialized in that it belongs to art, science, medicine, engineering, or the like.
2. Study, observation, analysis, experimentation, and measurement to obtain accurate and precise information about the problem or subject matter.
3. Organizing and presenting the information thus gained so that it will be clear and meaningful to the person or persons for whom it is intended.

Next, but still by way of introduction, we want to elaborate upon this opening definition or set of definitions. We believe that we can best do this by treating the problem under these seven headings:

Various Writing Styles
Where Technical Writers Work
A Day with Two Writers
Who Reads Technical Reports
Goals of Technical Writers
What Makes a Good Report
Reports Compared with Writing in General

Various Writing Styles

Example	*Commentary*
. . . the very nice plant my mother had on her table in the front hall.	Everyday, homey diction; much depends on the reader's imagination.
. . . a properly potted and displayed specimen of the family Begoniacae.	Abstract, general, formal; open to interpretation.
. . . in a shaft of yellow sunlight, a white-flowering begonia in a red clay pot.	Pictorial, vivid, sensory; "shows" rather than "tells about."
. . . a twelve-inch begonia propagated from a three-inch cutting; age, 42 days.	Specific, "technical"; factually informative.

As a writer, whether part-time or full-time, you may have to use all of these "languages," for your job will be to convey the important truth to your intended readers. By playing the right tune with these languages in different combinations, and by adding other writing skills in generous measure, you can produce leaflets, brochures, and sales literature; reports to stockholders; a great variety of letters; and articles for magazines and journals.

When you write as a technician, scientist, or engineer or for technicians, scientists, or engineers, however, you will usually have to follow the closing begonia example in both diction and point of view. The diction is highly specific, so the meaning is clear and unequivocal. By relying on this factual language, you can produce operating manuals, inspection and test reports, specifications, progress reports, and similar materials, some listed in the table of contents of this book.

We do not pretend that all technical writing is exactly alike. It is not. The substance being communicated, the writer's immediate intentions, and the interests and capacity of the chosen audience should and undoubtedly will influence the prose. At the risk of oversimplification, however, we suggest that many varieties of technical writing reveal these attributes:

1. The writing is characterized by a no-nonsense approach to the subject it treats. It is single-minded and earnest. Interesting points are seldom introduced for their interest value alone; they must also be pertinent.

2. The purpose of the article or paper is usually spelled out in the opening paragraph or two. All included information bears upon the accomplishment of the stated purpose. For example, a technical paper on smoke detectors may set forth only one major objective: to determine the relative effectiveness of photoelectric and ionization chamber types in detecting smoldering fires, flaming fires, and high temperatures. Other major topics would be reserved for other papers.

3. The vocabulary tends to be specialized. Some of the terms may not appear in small dictionaries for general use. Often the specialized terms are not defined within the text, on the assumption that members of the profession the writer is addressing (e.g., forestry, optics, nursing) will be familiar with them.

4. The sentences are tightly packed with information, for the intended readers are highly motivated and will not tire when faced with an array of facts.

5. When appropriate to the material, numbers and dimensions are numerous. These are usually in Arabic form and are exact rather than rounded out to the nearest whole number.

6. Signs, symbols, and formulas may pepper the prose. The terms may be listed and defined in accompanying glossaries (mini-dictionaries).

7. Graphs and tables may substitute for prose or reinforce and expand upon the surrounding prose.
8. Documentation and credits appear in footnotes and bibliographic supplements.

Of course, this list of characteristics of technical prose could be greatly extended. (You may be able to think of additional points.) In any case, excessive technicality in writing is never a virtue. Be technical if you must, but only if you *must*.

Where Technical Writers Work

Research is one of our largest industries, and this research produces information that must be reported. Let us look at a few sources of research-based reports and other varieties of technical paperwork.

1. Many government agencies, scientific laboratories, and commercial companies make research their principal business. They may undertake this research to satisfy their internal needs or the needs of related organizations. The persons who do the research may include chemists, physicists, mathematicians, psychologists—the whole array of professional specialists. They record and transmit much of this research via reports.
2. Many companies and laboratories do research for outsiders on a commercial basis. The "outsiders" may be agencies of government, or other institutions that are inadequately equipped to do their own research. Reports may be the only products of commercial research companies and laboratories.
3. Many companies do research to meet their own business needs—developing new products, determining market needs, predicting demand, and so on. Research units within these companies transmit their findings to executives and other company units via reports.
4. Many individuals have a personal or professional curiosity that entices them into research. If they believe that their findings are important, they publicize the information in various ways—books, journal articles, papers for professional societies.
5. Students are often assigned research problems to further their academic training. What they have done and learned may be presented in a laboratory report, monograph, or thesis.

From myriad sources such as these comes a great flood of paperwork. Some of it is only of passing interest; some of it makes history. All of it is prepared by report writers: sometimes those who have made a part-time or full-time career of technical writing, but more often researchers who are reporting on their own investigations.

A Day with Two Writers

Let us recreate two representative writers, whom we shall identify as Marie Enderson and Ted Freedman.

Marie Enderson: Junior Engineer and Occasional Technical Writer

Marie Enderson works as a junior engineer in the design and development department of a small electronics firm employing some 700 persons. She still has eighteen credits to earn before obtaining her B.S. in Electrical Engineering from State University, and is working off this requirement at the rate of three credit hours per term.

Enderson has been with the company about nine months. Since her early teens she has been recognized in her neighborhood as a whiz at building and repairing electronic gadgets. Therefore she has practical experience not possessed by employees several years her senior. Her specialty is that of designing and "bread-boarding" novel electronic circuits. Her current project is to develop a solid-state frequency multiplier.

Marie is especially valued for her originality and drive. She is also great with schematics, pliers, and soldering iron.

Her first project with the company concerned bow-tie antennas. On the Monday following completion of the engineering work, Marie was reminded of the requirement to submit a full report on the six-month project. She had a ghastly time the next two weeks. She found, as do many novices at writing, that she knew *what* she wanted to say, but not *where* or *how* to say it. But the twenty-five page report did somehow get written, and after a thorough overhaul by Ted Freedman (whom we'll meet next) went to publications.

Her first experience with on-the-job reporting taught Marie four unforgettable things: (1) even a junior engineer is not simply a fix-it person whose only product is a gadget that works; (2) things that go on in your head and hands are lost unless they are recorded; (3) reporting what you have thought and done is a recurring necessity; and (4) reporting, strange and difficult as it may seem at first, is something that can be learned by anyone of reasonable intelligence and perseverance.

Marie Enderson is now earning $6.00 per hour as a junior engineer, and she will be eligible for promotion the first of July or January after obtaining her B.S. in Electrical Engineering. At the moment, her biggest worry is whether her report on the frequency multiplier will get by Mr. Freedman.

Ted Freedman: Technical Writer and Company Editor

Ted Freedman was hired as a technical writer/editor. Freedman is twenty-seven. He holds a B.S. degree in Technical Communication from a university in the Midwest.

His office is a sparsely furnished cubicle down the hall from the publications and mailing departments. His office possessions include a fairly new typewriter; a three-foot shelf of dictionaries, reference manuals, a company style manual, and military specifications manuals; and a nondescript floor fan for July and August.

At 8:45 this morning Ted is scheduled for a project review session in the company auditorium. He arrives at the auditorium with five minutes to spare. For the next hour he studies flip charts, slide projections on the huge screen, chalk-and-blackboard plans for company reorganization (minor), and staffing proposals for three new projects totaling $278,400. From the platform, Chief Scientist Muldoon requests Freedman to develop research timetables and to preview reporting needs.

At 10:20 he meets with a commercial printer to examine the artwork and layout for a plush report his company is preparing for the Tri-State Commission. The work looks good, mighty good, but a little lacking in typographical variety, he suggests.

At 12:55, back from lunch in the company cafeteria, he glances over the memos that collected on his desk during the morning—nothing urgent. Then he opens the manila folder lying in his mail rack. It is the manuscript on Enderson's frequency multiplier project.

At 1:30 he gets Enderson on the telephone and arranges for a meeting at 3 so that they can run through the manuscript together. But he feels that he should do some work on it before the meeting to demonstrate the variety of things needing to be done.

He comes upon a disjointed, mixed-up paragraph. Turning to the typewriter, he makes a complete rewrite of it. He detects a "hole" or discontinuity in the second chapter. Another telephone call to Enderson confirms that a page somehow got lost and will have to be reinserted.

However, it is the two-page set of conclusions that really stirs him into action. The conclusions are a hodgepodge of preliminary intentions, procedures employed during the research, reservations about the quality of the work, and some unsupported speculation. Therefore, he pulls this section and attaches a covering critique for Enderson's guidance. Thus the afternoon wears on.

At 3 Enderson appears and the two confer, make changes, and plan later alterations in the manuscript. As always, they work amicably together. To relieve tension they intersperse their writing and editing with an occasional trip to the coffee urn, a chat with a department head, and a trip to the library to consult a specialized reference work.

For this kind of work Ted is paid $18,200 a year, with two weeks of paid vacation and ten days of sick leave. He is good at his work and considered to have a great future.

Enderson and Freedman are roughly representative of many thousands of technical writers/editors, most of them, like Enderson, part-time as the need arises. Of course, to gain a more rounded understanding of their duties and behavior we would have to pay them many additional visits. It is evident, though, that much of the time they are not writing at all, in the popular sense. Some of the time they are simply listening hard to what people are saying to one another —trying to clarify, simplify, and translate into other terms. A generous portion of their time is spent on tasks that have little direct connection with writing, but eventually provide grist for the writing mill.

At their best, they are adaptable persons, because they have to be. At Freedman's level, especially, they must be willing and able to cut into the communication process at any point along the line: from the moment when a theory or a new piece of hardware is only a gleam in the engineer's eye until the last *i* is dotted in the final report.

Who Reads Technical Reports

When you write a report, you need to know as precisely as you can who will read it, for a report must suit the needs, abilities, and interests of its principal users. Because the adaptation of reports to readers is a many-sided problem, we will not go into details here. Rather, we will simply try to make you aware of the many kinds of people your reports may have to satisfy. *How* you satisfy them, we will save for later chapters (see particularly Chapters 4, 9, 10, and 12).

Some reports—such as interdepartmental letters, technical notes, and memorandums—are intended only for use within the company. Some of these go from superiors to subordinates; some from subordinates to superiors; others to colleagues at the same level. If they move in more than one direction, they may have to be drafted in more than one version. Company policy, tact, and need-to-know are important considerations for intracompany paperwork.

Many examples come to mind: the boss has to collect information from subordinates for a report to the board of directors; existing departments have to be informed by management that they are about to be reorganized; new regulations concerning vacations, sick leave, and holidays are about to become effective. In fact, the outsider cannot conceive of the amount and variety of paperwork a company must generate simply to keep its internal affairs in order.

Some reports are prepared for outside customers, including commercial enterprises and agencies of government. Let us cite a few of the many possibilities: (1) a tax-free organization has to support its privileged status in a report to the Internal Revenue Service; (2) a company may have to demonstrate that it has the proper facilities in order to be awarded government contracts; (3) a research laboratory has to report on its findings at the close of a year-long contract with the Department of the Army; (4) a subsidiary may have to disclose and justify its operations in a report to the parent company.

Some reports are prepared for public consumption. A state department of forests and waters is entrusted not only with conserving our natural resources but also with making the public aware of these resources. Pamphlets, posted notices, and radio and television announcements are commonly used to spread this kind of information. Profit-earning companies must create and improve their public image and also attract customers and applicants for employment. Airlines, railroads, distributors of goods and services, all have to keep in the public view.

As you see, the possibilities are endless, and it would be tedious to suggest more than a few of them.

Goals of Technical Writers

To write clear and effective reports, you have to build upon the natural talents you have in communicating ideas to others. But how can you build successfully? What skills and attitudes are valued most in the report writer? What characteristics are bound up in the person we are glad to certify as a "good report writer"? Let us see.

1. You have to be reasonably methodical and painstaking. Plan your work for the day and for the rest of the week. Look up from time to time to take stock of what you and the others are doing, so that you do not squander your time and energy on minor tasks that should be put off or dispensed with altogether. File your corrrespondence. Keep at your desk the supplies you need to do your work. Keep a clear head about ways and means for accomplishing your purpose.
2. Be objective. Try not to get emotionally attached to anything you have written: be ready to chuck any or all of it into the wastebasket. While reading your own prose or that of your colleagues, ask not whether you or they are to be pleased, but whether the intended audience will be pleased, informed, and satisfied.
3. As a technician, engineer, or scientist, keep in mind that most of what you do will eventually have to be presented in writing. Do your work so that it will be honestly and effectively reportable. Keep a notebook or a deck of note cards. Record what you do and learn.
4. Remind yourself every morning and afternoon that *clarity* is the most important single objective of writing. Until the sense of a piece of writing is made indisputably clear, nothing else can profitably be done with it.
5. As a technician, engineer, or scientist—who must sometimes write— understand that writing is something that must be learned, even as chemistry, physics, and mathematics must be. The rules and formulas of writing are not as exact, perhaps, as those of science, but they can never be thrown overboard if you are to bring your substance home to your reader.

What Makes a Good Report

We have no patented list of attributes to offer you. The salient attributes vary somewhat from report to report, but for the moment we will propose some generalizations.

The good report . . .

1. Arrives by the date it is due.
2. Arrives in good physical condition—properly labeled and packaged, and without postage due.
3. Makes a good impression when it is picked up, handled, and flipped through.
4. Has the necessary identifiers on the cover and title page, so the librarian can log it in.
5. Has the necessary preliminary or front matter to characterize the report and disclose its purpose and scope.
6. Has a substantial body to describe and document what was done.
7. Has a summary or set of conclusions to reveal the results obtained.
8. Has been so designed that it can be read selectively: by some users, only the abstract; by some users, only the summary; by other users, all the main sections.
9. Has a rational and readily discernible plan, such as may be revealed by the table of contents.
10. Reads coherently and cumulatively from beginning to end.
11. Answers readers' questions as these questions arise in their minds.
12. Conveys an overall impression of authority, thoroughness, soundness, and honest work.

Beyond all these basic characteristics, the good report is free from typographical errors, grammatical slips, and misspelled words. Little flaws distract attention from the writer's main points.

Reports Compared with Writing in General

Some people—especially those who have never tried their hand at it—believe that report writing is simply good writing. They believe that it has no qualities which distinguish it from other expository writings such as the essay, journal article, or news account. Some diehards take the opposite stand: they hold that the engineer and scientist must use a brand of writing so peculiar and specialized that it is essentially divorced from all other writing.

The truth, we believe, lies between these extremist attitudes: report writing—perhaps we should say technical writing—is most certainly a specialty within the field of writing as a whole. Beginning report writers have to serve an apprenticeship. They must gain a working knowledge of their new subject matter and its terminology. They must learn to develop a prose style that is clear, objective, and economical. They must learn report types, variations in format, standards for abbreviations, the rules that govern the writing of numbers, the kinds of people who read reports and their expectations. That is, they have to learn the whole special business of being report writers.

And yet a broad and sound foundation in other writing is a tremendous asset to report writers, for it gives them versatility both on and off the job. They can write a good letter, prepare a brochure, compose an essay. In the comprehensive sense, they are *writers*. It is no news to them that subordination and coordination, coherence and emphasis are important in *all* writing.

As for the diehards, we can only remind them that employees who write well are far more appreciated by their employers than employees who write badly. We gain nothing by quarreling with the necessities of our time, and one of these necessities is that the technician, engineer, and scientist make themselves and their work clearly understood.

EXERCISES

1. As your instructor directs, bring to class one or more magazine or newspaper clippings that you believe to be technical. In what respects is the writing technical? Subject matter? Purpose? Tone? Vocabulary? Other?

2. Rewrite a brief paragraph of technical prose (perhaps a clipping submitted in Exercise 1) to substantially lower its technical level. Explain what you have done and why.

3. Go to the periodical room of your local or school library. Examine one or more issues of the periodicals listed here and add any others of special interest to you.

Aviation Week	Chemical Engineering
Agricultural Engineering	Construction Digest
Journal of Forestry	Electrical Engineering
Architectural Forum	American Gas Journal
Industrial Hygiene	American Journal of Physics

In what ways and to what extent has your examination of selected periodicals confirmed or changed your first impressions of technical writing?

4. On a two-column page, list your present *assets* and *limitations* as a technical writer.

5. Turn to the job advertisements section of a large metropolitan newspaper such as the Sunday *New York Times*. What advertisements for technical writers do you find? What qualifications are demanded of them?

6. If you have the opportunity, talk with several engineers and scientists of your acquaintance. Ask them how much writing they do and what kinds. Ask them how much importance they attach to good writing in their profession.

CHAPTER **2**

Finding Your Way in the Library

No matter how well you write, no matter how good your style, organization, and mechanics, without adequate data you won't be convincing. Most of the papers you will write—proposals, process descriptions, feasibility reports—require library research. Even original physical research reports require library research before you can begin. After all, you have to establish what is already known in your area of experimentation. You'll want to determine which previous experimental techniques have worked, which have failed. The library is the primary source for such information.

Because you will use the library for almost every writing chore you do, we have placed this chapter early in the book. Chapter 3, "Gathering Information," further describes the research process by telling you how to narrow down your subject and how to take notes once you are in the library. It also explains nine ways to supplement library research. In addition, Appendix B, "Technical Reference Guides," provides an annotated list of the guides, reference works, and computer services particularly helpful to technical people.

This chapter explains how to work with the card catalog, reference works, periodicals, government publications, and computer services. Sometimes the task of library research can seem overwhelming, for good reason. Since Johannes Gutenberg first printed the Bible in 1456, over thirty million different *titles* have been printed. Roughly ten million books are contained in the main branch of the New York City Public Library. Widener Library at Harvard, the largest university library in the United States, holds over nine million volumes.

Even a small undergraduate library may hold as many as fifty thousand volumes. The figures are staggering, but the task is not hopeless. Given a system, the curiosity of a good reporter, and a willingness to expend some shoe leather, you, too, can master the library.

The Card Catalog

At the heart of every library is its card catalog, listing most of the library's holdings—its books, reference works, and periodicals. Any listing needs a classification scheme, and the one used in most card catalogs is subject, author, title. If you know any one of the three, you can find out if the work you want is in the library. In most libraries, cards for author, title, and subject are filed together in one alphabetical listing; some libraries may have a separate subject file. Check also to see if your library alphabetizes letter by letter or word by word. It can make a significant difference in finding material, as you can see in the following example:

Letter by letter	*Word by word*
elk	elk
Elkhorn	Elk Island National Park
Elkhorn City	Elk Mountains
elkhound	Elk Point
Elk Island National Park	Elk River
Elkland	Elkhorn
Elk Mountains	Elkhorn City
Elko	elkhound
Elk Point	Elkland
Elk River	Elko

One method or the other is also used in alphabetical listings in reference works such as encyclopedias and indexes. Always be alert for these techniques.

Finding a book by its author is the easiest and most obvious route if you are fortunate enough to have that information, so let's examine that entry to the card catalog first. Look carefully at Figure 2-1, and we'll explain the information that you can get from the card catalog without getting the book from the shelf.

In this case you knew the name of the author you were looking for, so you went directly to "Hitchcock," passed over "Hitchcock, Alfred" and "Hitchcock, Barton," and came to "Hitchcock, Christopher"

Figure 2-1   Author Catalog Card

> 581.192
> H631   **Hitchcock, Christopher.**
>   Plant lipid biochemistry: the biochemistry of fatty
>   acids and acyl lipids with particular reference to higher
>   plants and algae [by] C. Hitchcock and B. W. Nichols.
>   London, New York, Academic Press, 1971.
>
>   xiii, 387 p.   illus.   24 cm.   (Experimental botany: an international series of
>   monographs, v. 4)   £6.50
>
>   Bibliography: p. [306]–332.
>
>   1. Lipids.   2. Acids, Fatty.   3. Plants—Chemical analysis.   I. Nichols,
>   Brian William, joint author.   II. Title.   III. Series: Experimental botany, v. 4.
>
>   QK898.L56H5   581.1'9'247   77-170752
>   ISBN 0-12-349650-0   MARC
>
>   Library of Congress   72

printed at the top of the card. Directly under Hitchcock's name is the title of the book: *Plant Lipid Biochemistry*. A fuller explanation of the title tells you that the book is about "the biochemistry of fatty acids and acyl lipids with particular reference to higher plants and algae." You read on and discover that Hitchcock had a co-author, B. W. Nichols, and that the book was published in London and New York by Academic Press in 1971.

The publisher's name can be important to you. As you come to know your field better, you may discover that some publishers have better reputations in the field than do others. The date may also be important. In some fields, medieval history for example, a 1925 book may be as good as or even better than a 1980 book. But in the fast-moving sciences, often the later the date the more useful the book.

The next line tells you that the book has xiii (13) pages of prefatory matter, which includes such items as the table of contents, illustration list, foreword, preface, and so on. The main body of the book has 387 pages and contains illustrations. It is 24 centimeters high (about 9½ inches, useful information for the librarian who must shelve it) and is the fourth volume in a series called "Experimental Botany." The book costs £6.50, about fifteen dollars.

The next line gives the useful information that the book contains a bibliography on pages 306–332. Research is much like throwing a rock into a pond, producing an everwidening series of concentric circles. Each item researched usually leads to other items, in this case a 27-page bibliography.

The next listing on the card tells you the subject headings under which you can find this book in the card catalog: Lipids; Acids, Fatty; and Plants—Chemical analysis. But if you've already found the book, of what use are headings telling you where to find it? A good deal of use, actually. Under these subject headings you will find other books in the same field of knowledge—again, the widening concentric circles. The catalog also lists the book under the co-author, Nichols; the title of the book; and the series, "Experimental Botany"—another place where you might find related books.

The numbers and letters at the bottom of the card provide additional information for the librarian that you need not concern yourself about. But the numbers and letters at the upper lefthand corner—"581.192 H631"—are important to you. They represent the call number of the book. If you have access to your library's collection (open stacks), somewhere near the card catalog will be a map of the library showing you where the various numbered sections are. If your library has closed stacks, you will need to give the call number to a librarian who will get the book for you.

Figure 2-2 shows a title card. Notice that it is identical to the author card, except that the title of the book has been placed above the author's name. When looking for titles, ignore articles (a, an, the) and look for the entry under the first major word of the title.

Figure 2-3 illustrates a subject card. Again, it is the same as the author card, but in this case the subject has been placed above the

Figure 2-2   Title Catalog Card

> 581.192   PLANT lipid biochemistry.
> H631
>   **Hitchcock, Christopher.**
>   Plant lipid biochemistry: the biochemistry of fatty acids and acyl lipids with particular reference to higher plants and algae [by] C. Hitchcock and B. W. Nichols. London, New York, Academic Press, 1971.
>
>   xiii, 387 p.   illus.   24 cm.   (Experimental botany: an international series of monographs, v. 4)   £6.50
>
>   Bibliography: p. [306]–332.
>
>   1. Lipids.   2. Acids, Fatty.   3. Plants—Chemical analysis.   I. Nichols, Brian William, joint author.   II. Title.   III. Series: Experimental botany, v. 4.
>
> QK898.L56H5   581.1'9'247   77-170752
> ISBN 0-12-349650-0   MARC
>
>   Library of Congress   72

Figure 2-3   Subject Catalog Card

581.192   Acids, Fatty.
H631   **Hitchcock, Christopher.**
            Plant lipid biochemistry: the biochemistry of fatty
        acids and acyl lipids with particular reference to higher
        plants and algae [by] C. Hitchcock and B. W. Nichols.
        London, New York, Academic Press, 1971.

            xiii, 387 p.   illus.   24 cm.   (Experimental botany: an international series of
        monographs, v. 4)   £6.50

            Bibliography: p. [306]–332.

            1. Lipids.   2. Acids, Fatty.   3. Plants—Chemical analysis.   I. Nichols, Brian
        William, joint author.   II. Title.   III. Series: Experimental botany, v. 4.

        QK898.L56H5   581.1'9'247   77-170752
        ISBN 0-12-349650-0   MARC

        Library of Congress   72

author's name. You must be imaginative and alert when you are
searching under subject headings. For example, looking for information about federal funding of interstate highways we found nothing of
use under "Federal Funding" or "Funding, Federal." So we went at it
from the other end and tried "Interstate Highways." We did not find
any subject cards but did pick up two title cards: *Interstate Highway
Maintenance Requirements* and *The Interstate Highway System.* The
second card looked promising, with a cross-reference to "Road—
Economic Aspects." This avenue led to five other books which dealt in
part with interstate highway funding. On a catalog card for one of
these five books, we found a cross-reference to "Express highways—
economic aspects" that led to three more books. So it goes. Often the
problem is not in finding enough books but in knowing when to cut
the search off.

You can sometimes look up periodicals (see Figure 2-4) in the card
catalog. Some libraries list them there; some do not. Reference works
(see Figure 2-5) are found in the card catalog. Figure 2-5 illustrates a
few other points as well. Notice that you can look up reference works
(and periodicals also) under subject headings as well as titles. For the
Demographic Yearbook, "Population—Yearbook" is the subject heading. And notice that in this case, as in the case of many reference
books, there is no author as such to be listed. So the corporate author
(here, the United Nations) is often used for that purpose. Corporate
authors also include names of professional organizations ("The

Figure 2-4   Periodical Catalog Card

The National geographic magazine.   v. 1–
1888–
Washington, National Geographic Society.
  v. illus. (part col.) ports, maps (part fold., part col.) 26 cm.

 Frequency varies.
 Editors: Jan. 1896–Oct. 1901, J. Hyde.—Nov. 1901–Feb. 1903, H.
Gannett.—Mar. 1903–June 1954, G. H. Grosvenor.—July 1954–Jan. 1957,
J. D. LaGorce.—Feb. 1957–
INDEXES:
  Vols. 1–42, Jan. 1899–Dec. 1922. 1 v.
  Vols. 1–49, Jan. 1899–Dec. 1925. 1 v.
  Vols. 1–66, Jan. 1899–Dec. 1934. 1 v.
  Vols. 1–70, Jan. 1899–Dec. 1936. 1 v.
  Vols. 1–78, Jan. 1899–Dec. 1940. 1 v.
  Vols. 1–100, Jan. 1899–Dec. 1951. 2 v.

(Continued on next card)
         14—7038*

American Standards Association"), names of institutions, and names
of governmental agencies. Here, too, you can apply some imagination
to your card catalog search. If you are interested in world
demography—that is, vital statistics on births, deaths, and population change—it would be a good bet that the United Nations would
have published information on the subject, and as we see, that is the
case.

Figure 2-5   Reference Book Catalog Card

Population—Yearbook

Demographic yearbook. 1948–
Lake Success.

   v. 30 cm.
  Yearbooks for 1948– issued with the United Nations publication sales
no.: 1949.XIII.1:
  Prepared 1948– by the Statistical Office of the United Nations in
collaboration with the Dept. of Social affairs.
  In English and French.

  1. Population—Yearbooks. I. United Nations. Statistical Office.

HA17.D45   312.058   50—641

 Library of Congress   [69z³2]

The card catalog is your key to much that is in the library, and you should learn as much about its use as you can. If you need help, ask a librarian. Librarians are, almost without exception, friendly and eager to help you, particularly if they see some sign that you are trying to help yourself.

Reference Works

You are reading an article in the business and finance section of *Newsweek,* and the author tells you that $282 million worth of hair dryers were sold in the United States in 1976. An article in *Fortune* about the U.S. fishing industry reports that the value of the 1976 lobster catch in the United States was $60.2 million. An article in *Scientific American* states that in 1974, 94,700 psychologists were at work in the United States. Do you know how to locate such statistics for yourself when you need them? Or must you always rely on secondary sources such as magazines and newspapers? Do you know how to use reference books, books meant to be consulted for a specific piece of information? Beyond a doubt, you are familiar with a few of them, such as encyclopedias and dictionaries; but you may not be fully aware of the vast store of fundamental knowledge available in the reference section of even a small library. Or you may know the knowledge is there without knowing how to get at it, short of always going hat in hand to the reference librarian. Let us help you a bit.

The best introduction to the world of reference books is itself a reference book—Eugene P. Sheehy's *Guide to Reference Books.* Even a small library will have this book, and the librarian will know it by the familiar name of "Sheehy's." Ask your reference librarian for "Sheehy's" and spend a little time reading the introduction. You'll learn that Sheehy's covers about 7500 reference works. You'll receive some information on how reference books are arranged— alphabetical order, chronological order, tabular order, regional order, and so forth. You'll be advised always to begin by reading the introduction to any reference book to see how it is arranged, what abbreviations it uses, and what little tricks you should know to extract information from it. Looking through Sheehy's table of contents, you'll see that it provides a comprehensive listing of both general reference works and reference works in the humanities, social sciences, history, and the pure and applied sciences.

Check out some of the listings. The general reference works section is a good place to begin. There you'll find many books that will help you find other books, such as *Books in Print* and the *Cumulative*

Book Index (CBI). Both of these are found in every library and provide complete listings of books printed in America or, in the case of the CBI, printed in English anywhere in the world. Such bibliographical books supplement your library's card catalog. If a book you find in the CBI is not available in your library, don't be discouraged. Chances are good that you can get it through an interlibrary loan. Ask your librarian for help in this. But start early. Interlibrary loans do take time.

Now turn to the section of Sheehy's that covers your discipline. In the section on the pure and applied sciences, for example, there is a subsection on the biological sciences. Figure 2-6 is a reprint of one page of that section. It lists the major indexes and abstract journals that cover the field and lists biological dictionaries and periodicals. Other pages in the same section list guides to biological information, bibliographies, encyclopedias, handbooks, and style manuals. In short, Sheehy's lists all the essential reference works a student in the field would need. In every subject area it covers, Sheehy's lists whatever is available and appropriate.

The fact that Sheehy's *Guide to Reference Books* includes some 7500 works suggests the rich diversity that is available. Encyclopedias are often good starting places for information in a field in which you are not thoroughly familiar. The general encyclopedias such as *Americana*, *Britannica*, and *Colliers* provide a variety of articles written for nonexperts. In addition to the general encyclopedias, there are numerous specialized ones: *The Encyclopedia of the Biological Sciences*, the *Encyclopaedic Dictionary of Physics*, the *Universal Encyclopedia of Mathematics*, the *International Encyclopedia of the Social Sciences*, the *Encyclopedia of Textiles*, and so forth. Indeed, there is an encyclopedia for almost every subject you might be interested in.

For unadorned facts, the various almanacs, yearbooks, and handbooks are indispensable aids. You are probably familiar with general almanacs such as the *The World Almanac and Book of Facts*. It includes such diverse items as the altitudes of the world's cities; how to address an ambassador; the table of atomic weights; gazetteer information about every nation in the world, including population, area, and principal cities; the depth of the oceans; the casualties suffered in World War II (or the Boer War, for that matter); the public debt of the United States from 1870 ($2.5 billion) to the present (over $675 billion); and the odds against a royal flush in poker (649,739 to 1).

Yearbooks cover important events in any given year. There are general yearbooks, such as those encyclopedias produce annually to supplement their basic volumes. And there are specialized yearbooks that cover a political unit or a specialized area of knowledge. One of

Figure 2-6   Sheehy's Guide to Reference Books

Pure and Applied Sciences | Biological Sciences | General Works   EC17

and science information; a selected bibliography, 1957–1961
... prep. in cooperation with the American University Center
of Technology & Administration, School of Government &
Public Administration. Paul C. Janaske, ed. Wash., 1962.
unpaged.   EC4

Pt.2 of the report of the Seminar on Biological Science Communication, 1961. Pt.1 of the report was entitled *Information and communication in biological science.*

A selected, annotated list of 1,121 titles on data handling, information storage and retrieval, documentation, mechanical systems of information handling, etc. Arranged alphabetically by author with keyword title index.   Z699.2.A6

International catalogue of scientific literature: L, General
biology. 1st–14th annual issues, 1901–14. London, 1901–19.
14v.   EC5

For full description *see* EA20.

Murray, Margaret Ransome and **Kopech, Gertrude.** A bibliography of the research in tissue culture, 1884–1950; an index
to the literature of the living cell cultivated in vitro. N.Y.,
Academic Pr., 1953. 2v. (1741p.)   EC6

"Tissue culture has been defined as the maintenance of isolated
portions of multicellular organisms in artificial containers outside the
individual for considerable periods of time."—*Introd.* Represents
15,000 original articles from serials and books, expanded to 86,000
entries by cross-indexing, using a comprehensive subject classification scheme. Authors and subjects arranged in one alphabet. Entries
include references to abstracts in *Biological abstracts.*

———————— Supplementary author list, 1950. (Incomplete
and univerified, Oct. 1953) N.Y., Academic Pr., 1953. 11p.
                                                                                            Z6663.T5M7

Periodicals

U.S. Library of Congress. Science and Technology Division.
Biological sciences serial publications; a world list, 1950–
1954. Prep. under the sponsorship of the National Science
Foundation. Philadelphia, Biological Abstracts, 1955. 269p.
                                                                                            EC7

A classified listing with geographical index, comp. by John Henry
Richter and Charles P. Daly.   Z5321.U52

Abbreviations

Biological Council. Abbreviated titles of biological journals:
a list culled with permission from the world list of scientific
periodicals ... 3d ed., comp. by P. C. Williams. London, The
Council, 1968. 47p.   EC8

2d ed. (1954) had title *A list of abbreviations of the titles of biological journals.*
Based on standard abbreviations as approved by the International
Organization for Standardization. Listing is alphabetical by full titles. More than 1,400 entries. Reprinted with minor corrections in
1969.   Z6945.A2B5

Indexes

Biological & agricultural index, a cumulative subject index
to periodicals in the fields of biology, agriculture, and related
sciences, 1964– . N.Y., Wilson, 1964– . v.50– . Monthly
(except Aug.) with annual cumulation.   EC9

Continues the *Agricultural index* (EL31).
An alphabetic subject index to approximately 190 English-language periodicals in the agricultural and biological sciences. Title
inclusion is decided by subscriber vote. Publications of U.S. and state
government agencies and of university service and research facilities
are not included. The list of periodicals indexed gives subscription
information. No author entries.   Z5073.A46

Abstract journals

Biological abstracts from the world's biological research literature. v.1– , Dec. 1926– . Philadelphia, Biological Abstracts, 1926– . Semimonthly.   EC11

Subtitle varies. Frequency varies; now published semimonthly,
with semiannual (formerly annual) cumulative indexes.

An abstracting journal of theoretical and applied biology, covering
more than 5,000 periodicals published in 90 different countries.
Preceded by *Abstracts of bacteriology*, v.1–9, 1917–25 (Baltimore,
Williams & Wilkins), and *Botanical abstracts*, v.1–15, 1918–26 (Baltimore, Williams & Wilkins), which merged to form *Biological abstracts.*

Titles are given in the original language (except that the Oriental
and Russian titles are transliterated), with English translation. Abstracts are in English and usually are signed. Arrangement and indexing vary. Each issue contains abstracts arranged by section and
subsection, with subject (BASIC), author, systematic, generic, and
cross indexes.

BASIC (Biological Abstracts Subjects in Context) is an indexing
technique using computer methods. Each significant word is indexed
and alphabetically positioned to the center of a line which includes
several words preceding and following the keyword.

A list of new books and periodicals appears in each issue.

An annual "List of serials with word abbreviations" formerly appeared in *Biological abstracts.* This feature has been superseded by:
                                                                                            QH301.B37

BIOSIS ... List of serials with coden, title abbreviations,
new, changed and ceased titles. [Philadelphia, BioSciences
Information Service], 1971– . Annual.   EC12

International abstracts of biological sciences, v.1– , 1954– .
London, Pergamon, 1954– . Monthly.   EC13

Title varies: v.1–3, *British abstracts of medical sciences.*
For complete record *see* EK64.   QH301.I475

Dictionaries

Abercrombie, Michael, Hickman, C. J. and **Johnson, M. L.**
A dictionary of biology. 5th ed. Harmondsworth, Eng., Penguin, 1966. 284p. il.   EC14

1st ed. 1951.
For the student and layman.   QH13.A25

Cowan, Samuel Tertius. A dictionary of microbial taxonomic
usage. Edinburgh, Oliver & Boyd, 1968. 118p.   EC15

Intended for the microbiologist, but includes selected terms from
botanical and zoological systematics. Modeled on Fowler's *Dictionary of modern English usage*, so that comparisons of terms and
discussions of broader subjects generally accompany the definitions.
                                                                                            QR9.C65

Dumbleton, C. W. Russian-English biological dictionary.
Edinburgh, Oliver & Boyd, 1964. 512p.   EC16

Gives equivalent English terms for Russian terms in the biological
sciences, excluding pathology. Includes the scientific names of species. Assumes a knowledge of Russian grammar and syntax.
                                                                                            QH13.D78

Gray, Peter. The dictionary of the biological sciences. N.Y.,
Reinhold, [1967]. 602p. il.   EC17

A new work by the author of the *Encyclopedia of the biological
sciences* (EC26), made up largely of brief definitions for terms which
did not warrant entries in the *Encyclopedia.* Users should note that
all words from a given root are listed under that root, usually without
cross reference.   QH13.G68

Henderson, Isabella Ferguson and **Henderson, William Dawson.** A dictionary of biological terms: pronunciation, derivation, and definition of terms in biology, botany, zoology,
anatomy, cytology, genetics, embryology, physiology. 8th ed.

719

the most useful yearbooks is the *Statistical Abstract of the United States*, published by the U.S. Bureau of the Census. This book is the source for the information on hair dryers, lobsters, and psychologists which began our section on reference books. It presents tables and figures that abstract almost all the information gathered each year about the United States. It contains sections on population, education, geography and environment, energy, science, forests and forest products, and many more. Figure 2-7 presents a typical page of figures from the *Statistical Abstract*. In addition, the original sources of all statistics are supplied, so the reader can pursue the data further if needed. The United Nations publishes a similar work, the *Statistical Yearbook*, which offers data on population, finance, agriculture, trade, education, and much more for 150 countries.

Handbooks are books containing information needed in specialized fields. A very partial listing suggests the possibilities: *Handbook of Basic Economic Statistics*, *Handbook of Chemistry and Physics*, *Handbook of Latin American Studies*, *Handbook of Private Schools*, and *Handbook of American Natural History*. Students might be wise to look at the biennial *Occupational Outlook Handbook* published by the U.S. Bureau of Labor Statistics. It gives the latest information on the major occupations in the United States, including training required, earning potential, locations, and much more. You can find economic information in *Sources of Business Information* and the *Editor and Publisher Market Guide*. The last covers the United States. It is arranged by city and state and gives information for each city listed on location, transportation, population, industries, banks, retail outlets, newspapers—even the drinking water, as in this entry for Liberal, Kansas: "Alkaline, medium hard, not fluoridated."

Atlases are another rich source of information. To learn how to use them, begin with the general atlases that are a collection of maps and information about population, climate, resources, and so forth and read the introduction to any good atlas such as *Goode's World Atlas*. As with other types of reference works, there are many specialized atlases as well as general ones. *The Rand McNally Commercial Atlas and Marketing Guide* gives economic information of use to business people. Another example is the *Atlas of the Cereal-Growing Areas in Europe*, which cites areas in Europe where the soil and climate are suitable for grain growing and lists the grains grown in those areas.

We hope we have given you the idea. Many people in both the governmental and private sectors gather data about our world, its people, and their activities day by day, year by year. Most of these data find their way into a reference book. Such books are tools that, properly used, can take the guesswork out of many of the decisions that all of us have to make.

Fig. 8-2. FEDERAL BUDGET RECEIPTS AND OUTLAYS: 1960 TO 1975

(For years ending June 30. See table 370)

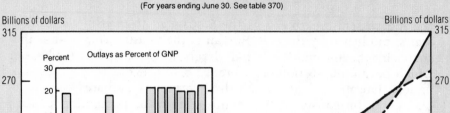

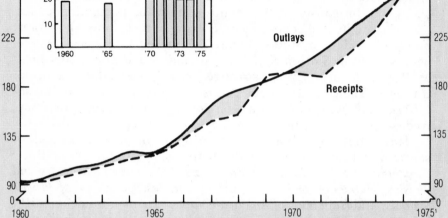

¹Estimate.

Source: Chart prepared by the U.S. Bureau of the Census. Data from U.S. Office of Management and Budget.

Fig. 8-3. THE ANNUAL FEDERAL BUDGET: 1970 TO 1975

(Average annual percent distribution, by function. For fiscal years ending June 30. See table 371)

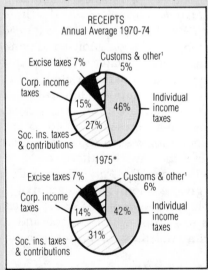

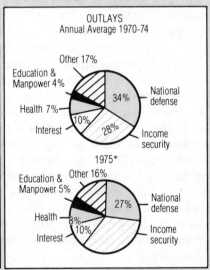

*Estimate.

¹Other includes estate and gift taxes and other receipts.

Source: Chart prepared by U.S. Bureau of the Census. Data from U.S. Office of Management and Budget.

From U.S., Bureau of the Census, *Statistical Abstract of the United States*, 96th ed. (Washington, D.C.: U.S. Government Printing Office, 1975), p. 224.

Figure 2-7   Statistical Abstract of the United States

Periodicals

For our purposes here, we'll define periodicals as commercial magazines, newspapers, and professional journals. Each is discussed below.

Commercial Magazines

The numerous commercial magazines—*Fortune, Scientific American, Sports Illustrated,* and so forth—provide a constant stream of
information on almost every subject imaginable. They often serve as
useful sources for initial research into a topic because they are written in easy-to-understand prose for the nonexpert. Also, they often
provide, within the text of an article, numerous names and titles that
can serve as sources for additional information if needed.

When you consider extracting information from any periodical,
you should refer to periodical indexes; that is, published guides to the
subjects, authors, and titles that you wish to find. One guide—the
Readers' Guide to Periodical Literature—covers over one hundred
American commercial periodicals of a broad, general nature in all
subjects. The *Readers' Guide,* published every two weeks, provides a
constant up-to-date source of information. The biweekly publications
(like many such guides and indexes) are gathered in cumulative volumes quarterly and yearly. (Some such cumulative system is standard for most guides and indexes of this sort.) In one alphabetical list,
the *Readers' Guide* provides a subject and author listing of all the
articles printed in the magazines it indexes during the time period
covered by that particular *Guide.* The subject index is carefully cross-
referenced. The entry for an article includes the title of the article, the
author's name, the magazine the article appeared in, the date and
volume number of the magazine, and the page number—in short, all
the information you need to find the article you are after in the library's periodical section.

Chances are that you are already familiar with the *Readers' Guide.*
If not, seek it out in the reference section of your library. Read its
introduction and then use it to find a few articles on some subject that
interests you. The *Readers' Guide* is a most useful index in its own
right, and it also will familiarize you with the correct procedure for
using most such indexes.

Newspapers

Finding information in newspapers is not nearly so easy as finding it in magazines. Fortunately, there is one reliable source to one
reliable newspaper—*The New York Times Index* for *The New York*

Times. The New York Times dates back to 1851. For most of that time, it has been indexed, and all back copies of the paper are currently available on microfilm. Most libraries have a copy of the *Index;* most university and large city libraries also have a set of the microfilms. *The New York Times Index* is organized like the *Readers' Guide* and provides some additional features that make it useful even if you don't have access to the microfilms. The *Index* summarizes major news stories and includes a chronological list of the major events in any story. The *Index* also contains the pictures, maps, and graphs that originally accompanied the stories. (See Figure 2-8.)

Once the *Index* has revealed the date of a story on the subject you are researching, you can turn to your local newspaper for national and international stories on the same topic or event. Most newspapers maintain an indexed file of their back copies in a library familiarly called the "morgue." The morgue is usually accessible to the public, often even by phone and mail. Many newspapers will provide photocopies of requested stories and articles for a small fee. Knowing the dates of a major story through your research in *The New York Times Index* will simplify your search for the same story in your local newspaper's morgue.

Professional Journals

As you advance in your field, the most useful periodicals for you will be the professional journals relating to that field. These journals report the latest research and frequently publish literature reviews that summarize the research over an extended period of time. Journals contain much information that may never get into book form or that may appear in a book too late to do you any good. Therefore, any serious member of a profession has to master research in its journals. This task may seem complex on the surface but is really not that difficult. Three reference books will simplify it considerably.

The *Applied Science and Technology Index* constitutes a subject index to over 200 journals in numerous scientific fields. It's a good starting point for more advanced research. Figure 2-9 shows you a typical page from this index. Notice that good cross-referencing is provided as under the entries for "Physical Geography" and "Physics." The entries provide a fair amount of information about each item. Look at the first entry under "History" in the right-hand column. After consulting the instructions and abbreviations in the front of the index, you would be able to translate this entry as follows: The title of the article is "Publications in the History of Physics during 1973." The author is S. G. Brush. The article contains a 5-page

Figure 2-8   The New York Times Index

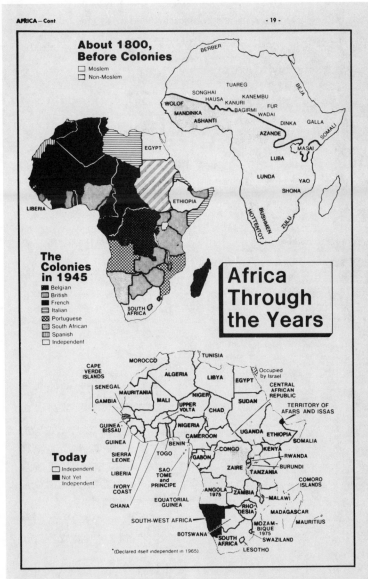

From *The New York Times Index 1976* (New York p. 19. Reprinted by permission.)

Figure 2-9   Applied Science and Technology Index

292   APPLIED SCIENCE & TECHNOLOGY INDEX

PHOTOMULTIPLIER tubes
  Double beam, single photomultiplier spectro-
    fluorimeter. W. H. Melhuish. diags Sci Instr
    8:815-16 O '75
  Inexpensive device for the determination of
    photomultiplier current gains. N. W. Bower
    and J. D. Ingle, jr. bibl diags Anal Chem
    47:2069-72 O '75
  Photomultiplier anode pulse integration circuit.
    G. S. Beddard and P. Williams. il diag Sci
    Instr 8:720-1 S '75
  Photomultiplier window materials under elec-
    tron irradiation; fluorescence and phosphores-
    cence. W. Viehmann and others. bibl App Op-
    tics 14:2104-15 S '75
PHOTOMULTIPLIERS. See Photomultiplier tubes
PHOTONS
  Simple approach to the multiphoton states of
    a thermal radiation field. S. A. Afsenius. bibl
    diag Am J Phys 43:882-4 O '75
  Spin and photon echoes. D. C. Hauelsen and
    T. D. Bonifield. bibl Am J Phys 43:824-9 S '75
  See also
  Mössbauer effect

                   Measurement
  Digital light scattering photometer. S. P. Lee
    and others. bibl diags R Sci Instr 46:1278-82 S
    '75
  High-sensitivity appearance potential spectro-
    meter for the detection of photons, electrons,
    and ions. A. Kluge. bibl diag R Sci Instr
    46:1179-81 S '75
  Simple digital clipped correlator for photon
    correlation spectroscopy. S. H. Chen and
    others. bibl il diags R Sci Instr 46:1356-67
    O '75
PHOTOOXIDATION. See Oxidation
PHOTOSENSITIVE glass. See Glass, Photosensi-
  tive
PHOTOSYNTHESIS
  Alternative sources for power generation. B.
    Francis. il diags Electronic Eng 47:37-9 O '75
  Optically detected zero-field magnetic resonance
    studies of the photoexcited triplet state of the
    photosynthetic bacterium rhodospirillum rub-
    rum. R. H. Clarke and others. bibl Am Chem
    Soc J 97:7178-9 N 26 '75
  Source of photosynthetic oxygen in bicarbonate-
    stimulated Hill reaction. A. Stemler and R.
    Radmer. bibl Science 190:457-8 O 31 '75
PHOTOTYPESETTING
  Putting words on paper. il Comp Air Mag 80:
    6-7 O '75
PHOTOVOLTAIC cell. See Photoelectric cells
PHOTOVOLTAIC effect. See Photoelectric effect
PHTHALATES
  Properties of materials; diallyl phthalates; tables.
    Materials Eng 82:140 Mid-S '75
PHTHALIC anhydride
  Control phthalic anhydride emissions. C. W.
    Moores. il diags Hydrocarbon Process 54:100-3
    O '75
PHTHALIMIDE
  Kinetics of intramolecular electron transfer
    between phthalimide groups linked by -(CH₂)ₙ-
    or -(CH₂CH₂O)ₙCH₂- chains; studies of
    chain flexibility. K. Shimada and others. bibl
    Am Chem Soc J 97:5834-41 O 1 '75
  Solvent effects in the kinetics of electron transfer
    between N-n-butyl-phthalimide radical ion and
    its parent molecule. Y. Shimozato and others.
    bibl Am Chem Soc J 97:5831-3 O 1 '75
PHTHALOYL group
  Azetidinone antibiotics; removal of phthaloyl
    protective group from acid and base sensitive
    compounds. S. Kukolja and S. R. Lammert.
    bibl Am Chem Soc J 97:5582-3 S 17 '75
PHYSICAL apparatus and instruments
  1975 AAPT apparatus competition. F. E. Chris-
    tensen. Am J Phys 43:935 N '75
  See also
  Astronomical instruments
  Electric instruments
  Geophysical instruments
  Goniometers
  Optical instruments
  Scientific apparatus and instruments
  Sound—Apparatus
PHYSICAL geography
  See also
  Climatology
  Drainage (physical geography)
  Lagoons
  Paleogeography
  Topography
  Valleys
PHYSICAL measurements
  See also
  Dimensional analysis
  Doppler effect
  Electric measurements
  Temperature—Measurement
  Vibration—Measurement
  Viscosity—Measurement
PHYSICAL metallurgy. See Metallography
PHYSICIANS
  Physician migration reexamined. R. Stevens and
    others. bibl Science 190:439-42 O 31 '75

PHYSICISTS
  Motivations of physicists. M. J. Moravcsik. Phys
    Today 28:9 O '75
PHYSICS
  See also
  Atoms
  Ballistics
  Compressibility
  Diffusion
  Dimensional analysis
  Dynamics
  Elasticity
  Electricity
  Electromagnetic theory
  Electrons
  Evaporation
  Expansion (heat)
  Fluids
  Fluorescence
  Force and energy
  Free energy
  Friction
  Geophysics
  Gravitation
  Gravity
  Heat transmission
  Inertia (mechanics)
  Ionization, Gaseous
  Ions
  Kinematics
  Light
  Liquids
  Magnetism
  Matter
  Mechanics
  Meteorology
  Molecules
  Music—Acoustics and physics
  Particles, Elementary
  Pauli exclusion principle
  Photoelasticity
  Plasma (Physics)
  Plasticity
  Porosity
  Pressure
  Radiation
  Relativity (physics)
  Relaxation time (physics)
  Reflection (optics)
  Sound
  Space and time
  Statics
  Steam
  Temperature
  Thermodynamics
  Turbulence
  Velocity
  Vibration
  Viscosity
  Vortex motion
  Wave mechanics
  Work function

                  Bibliography
  Publications in the history of physics during
    1973. S. G. Brush. Am J Phys 43:850-60 O '75

                  Examinations
  Repechage, a second chance at learning. M. S.
    Weinhous and W. R. Webber. Am J Phys 43:
    831-2 S '75

                  Experiments
  Best measuring time for a Millikan oil drop ex-
    periment. J. I. Kapusta. Am J Phys 43:799-800
    S '75
  Conservation of angular and linear momentum
    on an air table. R. R. Rockefeller. diags Am
    J Phys 43:981-3 N '75
  Kerr effect in nitrobenzene; a student experi-
    ment. A. W. Knudsen. bibl diags Am J Phys
    43:888-94 O '75
  Optical activity demonstration. G. Freier and
    B. G. Eaton. diag Am J Phys 43:939 N '75
  See also
  Nuclear physics—Experiments

                  History
  Publications in the history of physics during
    1973. S. G. Brush. Am J Phys 43:850-60 bibl
    (p856-60) O '75
  Role of music in Galileo's experiments. S. Drake.
    il Sci Am 232:98-104 Je '75; Discussion. E. Car-
    terette. 233:8-9+ N '75

                  Study and teaching
  Experience with a course in physics for visual
    arts majors. R. C. Bearse and D. W. McKay.
    Am J Phys 43:907-9 O '75
  Exploratorium and other ways of teaching
    physics. F. Oppenheimer. Phys Today 28:9+
    S '75
  Karate strikes. J. D. Walker. bibl il diags Am
    J Phys 43:845-9 O '75
PHYSICS, Mathematical. See Mathematical physics
PHYSICS laboratories
  Life science-related physics laboratory on geo-
    metrical optics. T. H. Edwards and others. il
    Am J Phys 43:764-5 S '75

bibliography. It appeared in the *American Journal of Physics*, volume 43, pages 850 to 860, in October 1975.

But most scholarly journals are not covered by the *Applied Science and Technology Index*, so you'll have to broaden your search. We've already mentioned Sheehy's *Guide to Reference Books*. Remember that this guide is arranged by disciplines. The indexing and abstracting services for a particular discipline are listed under that discipline. Look again at Figure 2-6, a reproduction of a page covering the biological sciences. Under the heading "Indexes," we find listed the *Biological and Agricultural Index*. If you read the entry, you'll see that this index covers 190 English-language periodicals in the field.

Below the indexes are listed the abstract journals. We might take a moment here to explain the term "abstract journal." An abstract journal provides the same service that an index does, plus one valuable additional feature. As the name implies, each article listed is also abstracted, that is, summarized. Sometimes the abstract itself will provide all the information you need. At the very least, you will know after reading it whether you want to read the original article. Abstracts are great time-savers, and whenever you can use them instead of indexes, it makes sense to do so. Under the heading "Abstract Journals," Sheehy's lists two such services. The first, *Biological Abstracts from the World's Biological Research Literature*, "covers more than 5000 periodicals published in 90 different countries."

The third major source in conducting a literature search in professional journals is *Ulrich's International Periodical Directory*. This reference, published every two years with supplements in the intervening years, is found in even the smallest libraries and is known to librarians simply as "Ulrich's" (pronounced ol-ricks). It covers approximately 55,000 in-print periodicals. Ulrich's classifies by subject and contains a title index as well. Its subject listing covers all fields of knowledge and is well divided and subdivided. For example, there are 12 subdivisions under biology ranging from "Biological Chemistry" to "Microscopy," "Ornithology," "Physiology," and "Zoology."

Figure 2-10 reproduces a page from Ulrich's. Look at the entry for the periodical *Economic Botany*. Again, after reading the instructions in the front of Ulrich's, we are able to translate this entry easily: *Economic Botany* is devoted to applied botany and plant utilization. It was first published in 1947 and is published quarterly. A subscription costs $22 a year. The editor's name and address are provided. Among its special features are abstracts, advertisements, bibliographies, book reviews, and illustrations. Its circulation is 1800.

The next bit of information given about *Economic Botany* is your key to easy access to the journal. Ulrich's tells which indexing and abstracting services (if any) cover the journal, in this case, *Biological Abstracts, Biological and Agricultural Index, Chemical Abstracts*, and

Figure 2-10   Ulrich's International Periodical Directory

206   BIOLOGY — BOTANY

CARIBBEAN RESEARCH INSTITUTE.
QUARTERLY REPORT. see *BIOLOGY*

581   US   ISSN 0008-7475
CASTANEA. 1936. q. membership. (Southern
Appalachian Botanical Club) West Virginia Univ.,
Morgantown, WV 26506. Ed. Jesse F. Clovis. adv.
bk. rev. bibl. index. circ. 800. Indexed: Biol.Abstr.

581
CATALOG OF FOSSIL SPORES AND POLLEN.
(Text in various languages) 1957. irreg.(3/yr) $13.50
to individuals; institutions $35. Pennsylvania State
Univ., Coal Research Section, Deike 517, College of
Earth & Mineral Sciences, Univ. Park, PA 16802.
Ed. A. Traverse. index. circ. 600. (looseleaf format;
also avail. in cards)

589.2   CS   ISSN 0009-0476
CESKA MYKOLOGIE. (Text and summaries in
various language) 1947. 4/yr. fl.44. (Ceskoslovenska
Akademie Ved, Czechoslovak Scientific Society of
Mycology) Academia, Publishing House of the
Czechoslovak Academy of Sciences, Vodickova 40,
112 29 Prague 1, Czechoslovakia (Distributor in
Western countries: John Benjamins B.V., Amsteldijk
44, Amsterdam (Z.), The Netherlands) Ed.Bd. bk.
rev. abstr. charts. index. circ. 850. Indexed:
Biol.Abstr. Chem.Abstr.

CHEEKWOOD MIRROR. see *ART*

CORNELL PLANTATIONS. see *GARDENING
AND HORTICULTURE*

CORRIERE FITOPATOLOGICO. see *BIOLOGY —
Entomology*

COTON ET FIBRES TROPICALES. see
AGRICULTURE

581   SP   ISSN 0011-2372
CUADERNOS DE BOTANICA CANARIA;
comunicaciones sobre flora y vegetacion del.
Archipielago Canario. (Text and summaries in
English and Spanish) 1967. 3/yr. 300 ptas.($5.)
(Excmo. Cabildo Insular de Gran Canaria, las
Palmas) Finca Llano de la Piedra, Santa Lucia de
Tirajana, Gran Canaria, Spain. Ed. Guenther
Kunkel. bk. rev. index. circ. 500. Indexed:
Excerp.Bot.

CURRENT ADVANCES IN PLANT SCIENCE. see
BIBLIOGRAPHIES

581   UK   ISSN 0011-4073
CURTIS'S BOTANICAL MAGAZINE. 1787. 2/yr.
$9.25 per no. Bentham- Moxon Trust, Royal
Botanic Gardens, Kew, Richmond, Surrey, TW9
3AB, Eng. Ed. D. Hunt. illus. index per vol. of 4
parts. circ. 600. Indexed: Biol.Abstr.

500.9   AG   ISSN 0011-6793
DARWINIANA. (Summaries in English, French,
German and Latin) 1922. s-a. Arg.$180.($23) ‡
(Academia Nacional de Ciencias Exactas, Fisicas y
Naturales de Buenos Aires) Instituto de Botanica
Darwinion, Lavarden y del Campo. San Isidro,
Buenos Aires, Argentina. Ed. Arturo Burkart. bk.
rev. bibl. illus. circ. 900. Indexed: Biol.Abstr.
Chem.Abstr. Excerpt.Bot.

581 635   CN   ISSN 0045-9739
DAVIDSONIA. 1970. q. $4. University of British
Columbia. Botanical Garden, Vancouver, B.C. V6T
1W5, Vancouver 8, B.C., Canada. Ed. Dr. C.J.
Marchant. bibl. illus. circ. 300. (tabloid format)
Indexed: Biol.Abstr.

DEUTSCHE BAUMSCHULE. see *GARDENING
AND HORTICULTURE*

581   GW   ISSN 0011-9970
DEUTSCHE BOTANISCHE GESELLSCHAFT.
BERICHTE. 1882. 12/yr. DM.156. Gustav Fischer
Verlag, Wollgrasweg 49, Postfach 720143, 7000
Stuttgart 72, W. Germany. Ed. Dr. A. Pirson. bibl.
charts. illus. index.

581   SX   ISSN 0012-3013
DINTERIA; contributions to the flora and vegetation
of south west africa. (Text and summaries in
Afrikaans, English & German) 1968. price varies.
South West Africa Scientific Society, Box 67,
Windhoek, South Africa. Ed.W.Giess. illus.

589.2   UK   ISSN 0012-396X
DISTRIBUTION MAPS OF PLANT DISEASES.
1942. m.(3 maps per mo.) £3($7.80) Commonwealth
Agricultural Bureaux, Commonwealth Mycological
Institute, Ferry Lane, Kew, Surrey, Eng. Ed. A.
Johnston. charts. maps.
Mycology

581   US   ISSN 0012-4982
DOKLADY BOTANICAL SCIENCES. English
translation of: Doklady Akademii Nauk S S S R.
vols.208-213,1973. s-a. $60. (Academy of Sciences
of the U. S. S. R. Botanical Sciences Section)
Consultants Bureau, 227 W. 17th St., New York,
NY 10011. Ed. A. I. Oparin. bibl. charts. illus.
index. circ. 292. (also avail. in microfilm) Indexed:
Biol. & Agr.Ind. Biol.Abstr. Chem.Abstr.

581   ISSN 0013-0001
ECONOMIC BOTANY; devoted to applied botany
and plant utilization. 1947. q. $22. New York
Botanical Garden, Bronx, NY 10458. Ed. Dr.
Richard E. Schultes. adv. bk. rev. abstr. bibl. illus.
circ. 1,800. Indexed: Biol. & Agri.Ind.
Chem.Abstr. Sci. Cit. Ind.

581
EGYPTIAN JOURNAL OF BOTANY. (Text in
English; summaries in English and Arabic) 1958. 3/
yr. $9.50. (Botanical Society, U.A.R) National
Information and Documentation Centre, (NIDOC),
Cairo, U.A.R. Ed. A. I. Naguib. charts. illus.
Indexed: Biol.Abstr. Chem.Abstr.
Formerly: Journal of Botany of the United Arab
Republic (ISSN 0021-9363)

EUPHYTICA; Netherlands journal of plant breeding.
see *AGRICULTURE*

581   GW   ISSN 0014-4037
EXCERPTA BOTANICA. SECTIO A:
TAXONOMICA ET CHOROLOGICA. (Text in
English, French and German) 1959. 1-2 vols per
year(7 nos. per vol.) DM.134. Gustav Fischer
Verlag, Postfach 720 143, Wollgrasweg 49, 7000
Stuttgart 72, W. Germany. Ed.Bd. adv. bk. rev.
cum.index.

581   GW   ISSN 0014-4045
EXCERPTA BOTANICA. SECTIO B:
SOCIOLOGICA. (Text in English, French and
German) 1959. 1 vol. per yr.(4 nos. per vol.)
DM.70. Gustav Fischer Verlag, Wollgrasweg 49,
Postfach 720143, 7000 Stuttgart 72, W. Germany.
Ed. Dr. R. Tuexen. adv. bibl. index.

632 581   UN   ISSN 0014-5637
F A O PLANT PROTECTION BULLETIN; a
publication of the world reporting service on plant
diseases, pests, and their control. (Editions in
English, French and Spanish) 1952. bi-m. $4. ‡
Food and Agricultural Organization, Plant
Production and Protection Division. Viale delle
Terme di Caracalla, Rome, Italy. Ed. Dr. A. V.
Adam. bibl. charts. illus. index. circ. 6,000.
Indexed: Biol.Abstr. Biol. & Agri Ind. Chem.Abstr.

581   GE   ISSN 0014-8962
FEDDES REPERTORIUM; Zeitschrift fuer
botanische Taxonomie und Geobotanik. (Text in
English, French and German) 1906. 10/yr. M.15
per no. Akademie-Verlag GmbH., Leipziger Str. 3-4,
108 Berlin, E. Germany. Ed.Bd. bibl. charts. illus.
index. Indexed: Biol.Abstr.
Formerly: Feddes Repertorium Specierum
Novarum Regni Vegetabilis.

581   US   ISSN 0015-0746
FIELDIANA: BOTANY. 1895. irreg. a change basis
or sale (price avail. on inquiry) Field Museum of
Natural History, Roosevelt Rd. & Lakeshore Drive,
Chicago, IL 60605. Ed. Patricia M. Williams. bibl.
charts. illus. circ. 500. Indexed: Biol.Abstr.
Chem.Abstr.

580   UR   ISSN 0015-3303
FIZIOLOGIIA RASTENII. (Summaries in English)
1954. bi-m. $36. Izdatel'stvo "Nauka", Podsosensky
21, Moscow, U.S.S.R. Ed. A. L. Kursanov. bk. rev.
charts. illus. index. circ. 2,800. Indexed:
Biol.Abstr. Chem.Abstr.

500.9   EC   ISSN 0015-380X
FLORA. (Text in English and Spanish) 1937.
exchange basis. Instituto Ecuatoriano de Ciencias
Naturales, Apartado 408, Quito. Ecuador. Ed. Dr.
M. Acosta-Solis. bk. rev. abstr. bibl. charts. illus.
Indexed: Biol.Abstr.

581   DK   ISSN 0015-3818
FLORA OG FAUNA. (Text in Danish; summaries in
English) 1894. q. Kr.40.($7.) Naturhistorisk
Forening for Jylland, Skjaersoevej 5, Risskov,
Denmark. Ed. Edwin Norgaard. bk. rev. illus.
index. circ. 1,000. Indexed: Biol.Abstr. Zoo.Rec.

581   SA   ISSN 0015-4504
FLOWERING PLANTS OF AFRICA. (Text in
Afrikaans and English) 1945. 2/yr. R 1.50 per no. ‡
Dept. of Agricultural Technical Services, Private
Bag X144, Pretoria, South Africa. Ed. Dr. D. J. B.
Killick. illus. cum.index. circ. 750. Indexed:
Biol.Abstr.

581   CS   ISSN 0015-5551
FOLIA GEOBOTANICA ET
PHYTOTAXONOMICA. (Text and summaries
inEnglish, French, German and Russian) 1966. q.
fl.75. (Ceskoslovenska Akademie Ved, Botanical
Institute) Academia, Publishing House of the
Czechoslovak Academy of Sciences, Vodickova 40,
112 29 Prague 1, Czechoslovakia (Distributor in
Western countries: Dr. W. Junk B.V., 13, van
Stolkweg, The Hague, Netherlands) Ed. S. Hejny.
bk. rev. charts. index. circ. 1,100. Indexed:
Biol.Abstr.
Formerly: Folia Geobotanica et Phytotaxonomica
Bohemoslovaca.

581   SA   ISSN 0015-847X
FORUM BOTANICUM. (Text in Afrikaans and
English) 1962. m. R.5 for non-members. South
African Association of Botanists, Private Bag X101,
Pretoria, 0001, South Africa. Ed. E.G.H. Oliver.
bibl. index. cum.index. circ. 350. (processed)
Formerly: South African Forum Botanicum.

581   US   ISSN 0092-1793
FREMONTIA. 1965. q. $8. California Native Plant
Society, 2490 Channing Way, Berkeley, CA 94704.
Ed. Margedant Hayakawa. adv. bk. rev. circ. 2,600.
(back issues avail)
Formerly: California Native Plant Society.
Newsletter.

589.2   DK   ISSN 0016-1403
FRIESIA; Nordisk mykologisk tidsskrift. (Text in
Danish, English, French, German or Swedish) 1932.
1-2 nos.per year. Kr.60($10) Svampekundskabens
Fremme - Danish Mycological Society,
Thorvaldsens Vej 40, 1871 Copenhagen V,
Denmark. Ed. N. F. Buchwald. bk. rev. charts.
illus. cum.index every 5 years. Indexed: Biol.Abstr.
Mycology

581   US   ISSN 0016-2167
FRONTIERS OF PLANT SCIENCE. 1948. s-a.
contr. free circ. ‡ Connecticut Agricultural
Experiment Station. New Haven, CT 06504. Ed.
Paul Gough. illus. circ. 7,500. Indexed: Biol.Abstr.
Chem.Abstr.

581   US   ISSN 0016-4585
GARDEN JOURNAL. 1951. bi-m. membership(non-
members $5.) New York Botanical Garden, Bronx,
NY 10458. Ed. Mary E. O'Brien. adv. bk. rev. illus.
index. circ. 6,500. Indexed: Biol.Abstr. Biol. &
Agri.Ind.   Sci. Cit. Ind.
Formerly: Garden Journal of the New York
Botanical Garden.

GARDENS ON PARADE. see *GARDENING AND
HORTICULTURE*

581   CL   ISSN 0016-5301
GAYANA: BOTANICA. 1961. s-a. Universidad de
Concepcion, Instituto Central de Biologia,
Concepcion, Chile (Subscr. To: Comision Editora,
Casilla 301, Concepcion, Chile) Ed. Bd. Indexed:
Biol.Abstr.

GENETICA POLONICA; Polish journal of theoretical
and applied genetics. see *BIOLOGY — Genetics*

581   IT   ISSN 0017-0070
GIORNALE BOTANICO ITALIANO. (Text in
English. French, German and Italian; summaries in
English and Italian) 1844. bi-m. L 12.000($20.)
Societa Botanica Italiana, Via Lamarmora N. 4,
50121 Florence, Italy. Ed. Prof. Pier Virgilio
Arrigoni. charts. illus. index. circ. 600. Indexed:
Biol.Abstr. Chem.Abstr. Excerp.Bot.

From *Ulrich's International Periodicals Directory*, 16th ed. (New York: R. R. Bowker
Company, 1975–76), p. 206. Reprinted by permission of the publisher.

Science Citation Index. You now know where to look up articles printed in *Economic Botany.* This indexing information is given for all the periodicals listed in Ulrich's when applicable; for a researcher, it is one of the most valuable features of the book.

Finding your way in the scholarly journals, then, is not that difficult. The *Applied Science and Technology Index* will locate many journal articles for you. Between them, Sheehy's and Ulrich's will help you locate any journals you're likely to need and tell you what indexing and abstracting services cover them. With that information, you're on your way.

Government Publications

Government publications are a motherlode of information waiting to be mined. The United States government publishes pamphlets, periodicals, research reports, and books. It addresses every audience, from the expert to the elementary school child. Most libraries have at least a limited collection of government publications. But several libraries in every state are designated depositories for government documents, and that is where you would find the most complete coverage. For the depository libraries in your state, see *Government Publications and Their Use* by Laurence F. Schmeckebier and Roy B. Eastin, or ask your local librarian.

Most libraries with a substantial collection of government documents catalog and shelve them separately from the rest of the collection. Before beginning any research in government publications, ask the librarian in charge of them to familiarize you with what is available and how to get at it. The brief coverage that follows should provide enough information about some of the basic tools to help you survive in what is sometimes a confusing research area.

Good starting points for any novice in the use of government publications are *Government Publications and Their Use* and Ellen Jackson's *Subject Guide to Major United States Government Publications.* Jackson's book is a guide to major and basic government documents on a wide variety of subjects ranging from "Aeronautics," "Aged," and "Agriculture" through "Korean War" to "Zoology." No such list can be comprehensive, but Jackson has done a good job hitting the highlights. Perhaps the most valuable feature of the Jackson book for the beginner is a section by W. A. Katz called "Guides, Catalogs, and Indexes," a directory of the guides that cover government publications. Browsing through both books is a good introduction to the world of government documents.

The best general guide to government documents is the *Monthly Catalog of United States Government Publications*, published by the U.S. Superintendent of Documents. Each monthly catalog lists the documents published that month. The list is organized by government agency—"Environmental Research Laboratories," "Federal Aviation Administration," "Federal Energy Office," and so forth. The agency-by-agency listing is followed by subject, author, and title indexes. The December catalog each year contains author, subject, and title indexes covering the entire year. The *Monthly Catalog* indexes most, although not all, of the documents printed by government agencies. It also tells the reader how to order government publications.

A supplement to the *Monthly Catalog* is *Selected U.S. Government Publications*, which lists publications of general public interest. Anyone can get on the mailing list for this useful publication by requesting it from the U.S. Government Printing Office, Public Documents Department, Washington, D.C. 20402. Figure 2-11, a reprint of a page from this guide, will give you some idea of the material available.

The Superintendent of Documents frequently issues separate, updated *Price Lists* covering the approximately 80 fields in which the government publishes, such as home economics, geology, maps, scientific tests, standards, and consumer information. These lists are particularly helpful in locating recent publications in a given subject area.

When looking for less recent government documents, the best guide is the fifteen-volume *Cumulative Subject Index to the Monthly Catalog of United States Government Publications 1900–1971*, compiled by William W. Buchanan and Edna M. Kanely. This collection indexes 72 years of government documents.

The federal government conducts and commissions research on an enormous number of subjects, such as "Severe Local Storms Research" (conducted by Purdue University) or a "Survey of Sonic Boom Phenomena for the Non-Specialist" (conducted by a private research company). For the report literature on this research, the best guide is *Government Reports Announcements and Index*, published every two weeks by the National Technical Information Service of the Department of Commerce. It indexes reports from 22 subject areas including agriculture, behavioral and social sciences, biology, chemistry, physics, and space technology. The introduction to each issue gives complete instructions on how to use the index. Figure 2-12 shows a typical page.

For keeping track of the work done by congressional committees, the *CIS/Annual*, published by the Congressional Information Service of Washington, D.C., is the best guide. It provides indexes and

Figure 2-11   Selected U.S. Government Publications

GOVERNMENT'S RESPONSE TO THE ELDERLY. Hearings before the House Select Committee on Aging.

71E Introduction to the Administration on Aging, September 24, 1975. Arthur S. Flemming, U.S. Commissioner on Aging, discusses the work of the Administration on Aging. 1975. 77 p. il.
Y 4.Ag 4/2:El 2/2
S/N 052–070–02950–1   95¢

72E National Clearinghouse on Aging, November 6, 1975. Donald F. Reilly, Deputy Commissioner on Aging, and Clark Tibbetts, Director of the National Clearinghouse on Aging, testify about the development and workings of the National Clearinghouse. 1975. 29 p. il.
Y 4.Ag 4/2:El 2/3
S/N 052–070–02953–6   45¢

73E HUD'S RESPONSE TO THE HOUSING NEEDS OF SENIOR CITIZENS. Hearings before the Subcommittee on Housing and Consumer Interests of the House Select Committee on Aging, September 25, 1975. Testimony on what the Department of Housing and Urban Development (HUD) is doing to assist the elderly with their housing problems is given by Sanford A. Witkowski, Director of the Office of Policy and Program Analysis and Development of the Office of Housing Production and Mortgage Credit. 1975. 40 p.
Y 4.Ag 4/2:H 81
S/N 052–070–02955–2   65¢

69E GROWTH OF GOVERNMENT SPENDING FOR INCOME ASSISTANCE, A Matter of Choice. The cost of income assistance programs is projected to the year 2000 in this report prepared for the Senate Budget Committee. 1975. 19 p. il.
Y 4.B 85/2:In 2
S/N 052–070–02945–5   45¢

129E THE SUPERVISOR'S GUIDE TO LABOR RELATIONS IN THE FEDERAL GOVERNMENT. Covers every aspect of labor relations, including working with union representatives, contract negotiations, handling grievances, and strike actions. 1975. 24 p. il.
CS 1.7/4:L 11/2
S/N 006–000–00907–7   80¢

130E MINORITY GROUP EMPLOYMENT IN THE FEDERAL GOVERNMENT. A report on the status of full-time minority group employment in the Federal Government. Provides a framework for studying general characteristics of the Federal work force. Among the factors reviewed are minority group designations, pay plans, grade or salary levels, and geographic locations. 1975. 566 p.
CS 1.48:SM 70–73B
S/N 006–000–00912–3 $6.20

18E NATIONAL BUREAU OF STANDARDS ANNUAL REPORT, FY 1975. 1975. 32 p. il.
C 13.1:975
S/N 003–003–01557–1 $1.05

102E VITAL STATISTICS OF THE UNITED STATES, 1973, VOLUME II, SECTION 5: LIFE TABLES. 1975. 20 p.
HE 20.6210/A:973/v.2/sec.5
S/N 017–022–00365–1   75¢

103E VITAL STATISTICS OF THE UNITED STATES, 1971, VOLUME III: MARRIAGE AND DIVORCE. 1975. 136 p.
HE 20.6210:971/v.3
S/N 017–022–00397–0   Cloth, $7.20

82E FEDERAL CATALOG SYSTEM POLICY MANUAL. The Federal Catalog System was designed in 1952 to simplify identification of items within the Department of Defense. This manual outlines the official policies and instructions for using the system. 1975. 184 p. il.
D 1.6/2:C 28
S/N 008–007–02690–9 $2.65

9E A STATEMENT OF NATIONAL TRANSPORTATION POLICY. A statement by Secretary of Transportation William Coleman, on the broad policy considerations that help dictate the Federal Government's response to U.S. transportation needs. Topics discussed include Government and the private sector; Federal expenditure programs; and international transportation. 1975. 60 p.
TD 1.2:P 75/2/975
S/N 050–000–00103–2 $1.15

26E FEDERAL AND STATE INDIAN RESERVATIONS AND INDIAN TRUST AREAS. A useful reference book of basic information about Indian tribes and Alaskan natives. Information is given about each reservation's population, land status, history, culture, government, tribal economy, climate, transportation and community facilities, and recreation. 1974. 604 p. il.
C 46.8:In 2/3/973
S/N 003–011–00076–3 $6.50

ENERGY & THE ENVIRONMENT

79E ENVIRONMENTAL WARFARE: Questions and Answers. Defines environmental warfare and environmental modification, and discusses the environmental outcome of the 1975 Geneva Conference of the Committee on Disarmament. 1976. 10 p.
AC 1.2:W 23
S/N 002–000–00053–9   40¢

75E THE ECONOMIC IMPACT OF ENVIRONMENTAL REGULATIONS. Hearings before the Joint Economic Committee, November, 1974, concerning the pros and cons of conservation and environmental regulations. Witnesses included representatives of the business community and of environmental groups. 1974. 228 p. il.
Y 4.Ec 7:En 8
S/N 052–070–03026–7 $2.15

61E FROM RAILS TO TRAILS. See page 10.
PR 37.8:En 8/T 68
S/N 040–000–00330–4 $1.50

59E CITIZEN ACTION GUIDE TO ENERGY CONSERVATION. See page 10.
PR 37.8:En 8/C49/3
S/N 040–000–00300–2 $1.75

141E REPORT TO THE PRESIDENT AND TO THE COUNCIL ON ENVIRONMENTAL QUALITY. See page 10.
PrEx 14.1/2:974
S/N 040–000–00333–9 $1.25

60E ENERGY IN SOLID WASTE. See page 10.
PR 37.8:En 8/En2/2
S/N 040–000–00319–3 $1.25

48E GEOTHERMAL ENERGY: ECONOMIC POTENTIAL OF THREE SITES IN ALASKA. 1975. 40 p. il.
I 28.27:8692
S/N 024–004–01802–0 $1.00

39E TECHNOLOGY ASSESSMENT OF RESIDENTIAL ENERGY CONSERVATION INNOVATIONS. This study examines the benefits of selected technical innovations intended to reduce residential energy consumption. Such innovations as storm doors, a furnace energy recovery device, and an open air cycle air conditioning system are analyzed. 1975. 248 p.
HH 12:En 2/11
S/N 023–000–00309–1 $3.10

7

From *Selected U.S. Government Publications* (Washington, D.C.: U.S. Government Printing Office, 1976), p. 7.

Figure 2-12   Government Reports Announcements and Index

Field 8—EARTH SCIENCES AND OCEANOGRAPHY

Group 8I—Mining Engineering

Prompt Copper Recovery from Mine Strip Waste.
Rept. of investigations,
B. W. Madsen, R. D. Groves, L. G. Evans, and G. M. Potter. Mar 75, 25p BuMines-RI-8012

Descriptors: *Copper ores, *Strip mining, *Leaching, Wastes, Materials recovery, Flotation, Leaching, Molybdenum, Silver, Gold, Copper, Size separation, Reclamation, Extractive metallurgy.
Identifiers: *Mine wastes.

In conventional dump leaching of sulfide waste, generally less than one-quarter of the copper is extracted. Gold, silver, and molybdenum that may be present in the dump are not recovered. Much of the sulfides in mine strip waste occurs along natural fracture planes. Some of these sulfides are liberated with enrichment of the fines fraction when the waste is broken during mining or in subsequent crushing. In research to take advantage of such enrichment, the treatment of three sulfide copper mine stripping waste samples was investigated by sizing or by crushing and sizing, followed by floatation of the enriched fines and leaching the remaining coarse rock. In the treatment of a chalcocite-bearing monozite mine waste sample, as much as 35 percent of the copper, 38 percent of the molybdenum, and 19 percent of the silver originally in the waste were recovered by floatation of the fines. An additional 33 percent of the copper was recovered by leaching the coarse fraction for 500 days.

PB-241 218/7GA   PC$3.25/MF$2.25
Bureau of Mines, Albany, Oregon. Albany Metallurgy Research Center.
Recovery of Nickel and Cobalt from Low-Grade Domestic Laterites.
Rept. of investigations,
R. E. Siemens, P. C. Good, and W. A. Stickney. Mar 75, 19p BuMines-RI-8027

Descriptors: *Cobalt, *Nickel, *Laterites, Reduction(Chemical), Carbon monoxide, Leaching, Extraction, Electrowinning, Ammonium hydroxide, Ammonium sulfate, Extractive metallurgy, Magnesium.
Identifiers: Low grade deposits.

A process is being developed by the Bureau of Mines to selectively recover nickel and cobalt from low-grade domestic laterites. In laboratory evaluation of the process, the oxides in the laterite were selectively reduced with carbon monoxide at temperatures from 350 to 600 C. For material containing more than about 5 percent magnesia, pyrite additions or post reduction heat treatments were necessary to achieve satisfactory nickel and cobalt extraction for this range of reduction temperatures. Multistage leaching of the reduced material at ambient temperature and pressure in the presence of oxygen, ammonium sulfate, and ammonium hydroxide extracted up to 92 and 87 percent of the contained nickel and cobalt, respectively. The nickel was selectively recovered from the leach solution by solvent extraction and was then stripped from the loaded organic with dilute sulfuric acid to provide a nickel-rich electrolyte. Treatment of the raffinate with hydrogen sulfide, resulted in the recovery of cobalt as a sulfide. The only contaminant in solution was magnesium which was removed by ion exchange or precipitation.

PB-241 343/3GA   PC$4.75/MF$2.25
Missouri Univ., Rolla. Dept. of Geological Engineering.
Effect of Mining Operations on Ground Water Levels in the New Lead Belt, Missouri.
Completion rept.,
Don L. Warner. Dec 74, 92p W75-06986, OWRT-A-060-MO(2)
Contract DI-14-31-0001-3825

Descriptors: *Ground water, *Mining, Water supply, Aquifers, Lead ore deposits, Mine waters, Dewatering, Pumping, Missouri.
Identifiers: *Water levels.

In order to work the underground mines of the New Lead Belt, it is necessary to dewater the Bonneterre Formation of southeast Missouri. This requires pumpage of several hundred to several thousand gpm, depending on the mine location. The effect of this pumping on groundwater levels was evaluated. The study showed that it is probable that mine pumping does not effect groundwater levels in the deep aquifer beyond a distance of about five miles from any of the mines. Major areas of influence are still more restricted. As mining continues, the area influenced will extend further from north to south, but will probably not expand much eastwest.

PB-241 504/0GA   PC$3.75/MF$2.25
Continental Oil Co., Ponca City, Okla. Research and Development Dept.
Seismic Mine Monitor System. Phase II.
Research rept. Aug 73-Jun 74,
James C. Fowler. Oct 74, 46p BuMines-OFR-24-75
Contract H0133112

Descriptors: *Seismic detection, *Monitors, *Search and rescue, Mining, Safety, Mines(Excavations).
Identifiers: Mine safety.

The research studied the possibility of using a seismic listening system to detect and locate miners trapped in areas where other communications were not available. This report covers the installation and testing of a permanent seismic monitor system at the Loveridge Mine in West Virginia. The testing of the system showed the following: Getting the seismic data back to the central computer in a usable form is possible; Detecting small explosions is possible; Using this system as an auxiliary system to detect hammer blows from miners trapped in the area of one of the geophones is possible.

PB-241 629/5GA   PC$4.75/MF$2.25
Geological Survey, Denver, Colo. Geologic Div.
Selected Bibliography Pertaining to Uranium Occurrence in Eastern New Mexico and West Texas and Nearby Parts of Colorado, Oklahoma, and Kansas.
Interim rept.,
Warren I. Finch, James C. Wright, and Michael W. Sullivan. 1975, 100p USGS/GD-75/003

Descriptors: *Uranium ores, *Bibliographies, Geology, Groundwater, Stratigraphy, Geophysical prospecting, New Mexico, Texas, Colorado, Kansas, Oklahoma.

Nearly 500 selected references to uranium and to stratigraphy, structure, and groundwater geology related to uranium-bearing formations in eastern New Mexico and West Texas and nearby parts of Colorado, Kansas, and Oklahoma are indexed topically and geographically. The list is nearly complete through 1972 and contains some references with later dates.

8J. Physical Oceanography

AD-A009 442/5GA   PC$3.25/MF$2.25
Woods Hole Oceanographic Institution Mass
Measurements of Vertical Fine Structure in the Sargasso Sea.
Technical rept.,
S. P. Hayes, T. M. Joyce, and R. C. Millard, Jr. 15 Jul 74, 10p Rept nos. WHOI-75-24, WHOI-Contrib-3453
Contract N00014-66-C-0241
Office of Naval Research, Arlington, Va.
Availability: Pub. in Jnl. of Geophysical Research, v80 n3 p314-319, 20 Jan 75.

Descriptors: *Sargasso Sea, *Internal waves, Microstructure, Temperature, Depth, Sea water, Bathythermograph data, Electrical conductivity, Salinity, Spectrum analysis, Thermoclines, North Atlantic Ocean, Reprints.

The vertical fine structure statistics in the northwest Atlantic (midocean experiment area) have been studied by using the WHOI/Brown CTD (conductivity, temperature, and depth). Five depth intervals including water masses representative of the entire water column have been subjected to spectral analysis of the temperature fine structure. All intervals show a similar power law dependence upon vertical wave number of -2.5, although spectral levels vary by more than a factor of 1000, whereas differences of only a factor of 2 exist on estimated vertical displacement spectra. Results are not inconsistent with the supposition that internal waves are responsible for much of the variability. On scales smaller than 10 m in the main thermocline, features with the characteristic signature of sheets and layers begin to appear.

AD-A009 444/1GA   PC$3.25/MF$2.25
Woods Hole Oceanographic Institution Mass
The Temperature and Salinity Fine Structure of the Mediterranean Water in the Western Atlantic.
Technical rept.,
S. P. Hayes. 14 Mar 74, 14p Rept nos. WHOI-75-27, WHOI-Contrib-3273
Contract N00014-66-C-0241
Availability: Pub in Deep-Sea Research, v22 p1-11 1975.

Descriptors: *Sea water, *Sargasso Sea, North Atlantic Ocean, Mediterranean Sea, Microstructure, Mixing, Temperature, Salinity, Advection, Density, Depth, Gradients, Temperature inversion, Reprints.
Identifiers: Pycnoclines, West Atlantic Ocean.

The temperature and salinity fine structure of the Mediterranean Water in the Sargasso Sea (28 deg N, 70 deg W) have been studied with the Woods Hole Oceanographic Institution/Brown CTD microprofiler. The features observed are shown to be due to the advection of water with different temperature and salinity characteristics along isopycnal surfaces. The vertical density gradient of the water column is unaffected by the presence of the temperature inversion. (Author)

AD-A009 526/5GA   PC$3.25/MF$2.25
Scripps Institution of Oceanography La Jolla Calif
NORPAX Highlights. Volume 3, Number 3, April 1975. Large-Scale Thermal Structure during POLE. Pacific Sea Level Stations,
T. P. Barnett, S. D. Rearwin, and Klaus Wyrtki. Apr 75, 19p
Contract N00014-69-A-0200-6043
See also report dated Feb 73, AD-755 755.

Descriptors: *Ocean tides, *Bathythermograph data, *Pacific Ocean, Temperature, Bathythermographs, Heat, Flowmeters, North Pacific Ocean, Ocean currents, Sea level.

Contents:
Large-scale thermal structure during POLE; Pacific Sea level stations.

AD-A009 602/4GA   PC$3.25/MF$2.25
Oregon State Univ Corvallis School of Oceanography
Temporal Variability of Suspended Matter in Astoria Canyon,
William S. Plank, J. Ronald V. Zaneveld, and Hasong Pak. 3 Apr 74, 7p Rept no. Ref-74-21
Contract N00014-67-A-369-0007
Availability: Pub. in Jnl. of Geophysical Research, v79 n30 p4536-4541, 20 Oct 74.

From U.S., Department of Commerce, National Technical Information Service, *Government Reports Announcements and Index* 75, no. 14 (1975), p. 70.

abstracts to House and Senate hearings and reports. It's a valuable guide to the many documents that grow out of congressional committee work. The *CIS/Annual* also provides both legislative histories and index references for public laws under consideration.

A good many states publish yearbooks, annuals, data books, statistical abstracts, and so forth that contain a wealth of information on various subjects. The table of contents from *The State of Washington Pocket Data Book 1976* reprinted in Figure 2-13 illustrates the range of material available. States also publish reports, brochures, monographs, and similar items in a manner similar to the federal government. For a listing of the major publications of the 50 states, see the *Monthly Checklist of State Publications* published by the Library of Congress.

Many libraries also carry some of the documents published by the United Nations. A good many of these documents pertain to science and technology. A sensible procedure here is to read through *The Yearbook of the United Nations* to survey the activities of the UN's major scientific agencies during that year—the International Atomic

Figure 2-13 The State of Washington Pocket Data Book, 1976

TABLE OF CONTENTS

	Page
I. STATEWIDE INFORMATION	1-162
Population	1-10
POPULATION TREND	3
POPULATION DISTRIBUTION	5
BIRTH RATE	7
AGE GROUPS	8
The Economy	11-28
PERSONAL FINANCE	12
UNEMPLOYMENT	16
PUBLIC EMPLOYMENT	19
EMPLOYMENT	20
FOREIGN TRADE	22
BUSINESS ACTIVITY	25
BANKING	26
General Government	29-40
GENERAL GOVERNMENT BUDGET	30
STATE REVENUE	33
STATE EXPENDITURES	39
LIQUOR REVENUE	43
PROPERTY TAXES	45
COURTS AND CRIME	46
MILITARY ORGANIZATIONS	49
Human Resources	51-80
HUMAN RESOURCES BUDGET	52
PUBLIC ASSISTANCE	54
ADULT CORRECTIONS	64
JUVENILE REHABILITATION	66
DEVELOPMENTAL DISABILITIES	69
INSTITUTIONS	71
PUBLIC HEALTH	73
EMPLOYMENT PROTECTION	78
Education	81-112
EDUCATION BUDGET	82
COMMON SCHOOLS	85
HIGHER EDUCATION	93
COMMUNITY COLLEGES	105
VOCATIONAL EDUCATION	109
Natural Resources	113-142
NATURAL RESOURCES BUDGET	114
LAND USE	117
FORESTRY	119
AGRICULTURE	124
FISHERIES	126
OUTDOOR RECREATION	131
HUNTING	134
FISHING	136
ECOLOGY	139
Transportation	143-162
TRANSPORTATION BUDGET	144
HIGHWAYS	147
ACCIDENTS	153
MOTOR VEHICLES AND DRIVERS	158
AVIATION	160
FERRIES	162
II. COUNTY INFORMATION	163-251
INDIVIDUAL COUNTY SUMMARIES	166-243
TABLES: PUBLIC ASSISTANCE	244
EMPLOYMENT	247
PROPERTY TAXES	248
EDUCATION	250
TRANSPORTATION	251
III. CITIES AND TOWNS	253-262
POPULATION	254
PUBLIC EMPLOYMENT	260
STATE REVENUES DISTRIBUTED	262
IV. SOURCES	263-272
V. INDEX	273-282

From Washington, *State of Washington Pocket Data Book, 1976* (Olympia, Wash.)

Energy Agency (IAEA), Food and Agriculture Organization (FAO), United Nations Educational, Scientific, and Cultural Organization (UNESCO), World Health Organization (WHO), and World Meterological Organization (WMO).

Armed with information gathered from *The Yearbook*, you can find your way around in the *United Nations Documents Index*, which keeps track of UN publications. Also check your library's card catalog under "United Nations" or the names of the agencies listed above. Check also to see if your library carries any of the more specialized indexing and abstracting services published by the UN, such as *FAO Documentation* or *World Fisheries Abstracts*. Both of these are published by the FAO and exemplify the services that the UN provides.

Computerized Information Retrieval

Many libraries now offer access to computerized sources of reference material. That is, instead of your doing all the work normally involved in tracing useful sources of information, you can hire a computer to do much of it for you. Computerized information retrieval systems are now available in such areas as agriculture, business, chemistry, environmental studies, and education. For a complete listing, see Appendix B.

The United States government, through its National Technical Information Service (NTIS), now offers computer searches covering over 600,000 documents published by the federal government or under its sponsorship. The way in which NTIS operates is typical of many such services.

Figure 2-14   NTIS Catalog Entry

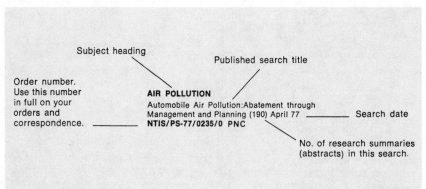

From U.S., Department of Commerce, National Technical Information Service, *NTISearch* (Springfield, Va., 1978), p. 3.

NTIS offers both published searches and custom searches. For a published search, you look in the NTIS catalog, *NTISearch*, for published searches in your field. For example, under the subject heading *Air Pollution* you'll find the catalog entry concerning automobile air pollution (reproduced in Figure 2-14). If you order this search, using the appropriate number listed, you'll receive 190 bibliographical citations and abstracts for research reports concerning automobile air pollution. Figure 2-15 shows one such citation and abstract. The cost for such a search is presently $28.00.

NTIS, and most other computerized retrieval systems, can go beyond merely giving you citations and abstracts. Most will offer you, in either paper or microfiche, copies of the cited reports. Microfiche is a form of microfilm that saves both money and storage space. Look at the bottom line of Figure 2-15. If the cited report appears useful to

Figure 2-15   NTIS Citation and Abstract

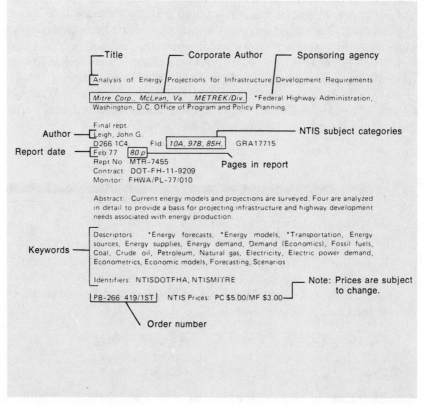

From U.S., Department of Commerce, National Technical Information Service, *NTISearch* (Springfield, Va., 1978), p. 4.

you, a copy of it can be obtained from NTIS by using the appropriate order number. In paper it will cost you $5.00; in microfiche, $3.00. Obviously, for a price, you can save yourself a lot of legwork.

NTIS also offers customized searches. Using this system, you (usually after consultation with an information retrieval specialist) provide NTIS with a list of key words called descriptors. (See Figure 2-15 for a list of typical descriptors.) NTIS then searches its entire data base for any articles stored under the descriptors you have furnished. Such searches can cost $100.00 and up.

Computerized information retrieval systems have multiplied significantly during the '70s. Their success indicates that such services will play an increasing research role in the '80s and beyond.

Conclusion

Our aim in this chapter has been to help you find your way in any library, regardless of its size. Only when you are armed with the facts can you really be a good writer. A knowledge of the facts makes you a more valuable member of your profession and, indeed, a better citizen. Knowledge is power, and libraries contain most of our knowledge. If you want to go beyond the basic information provided here, check out from your library some of the books listed under *Library Research* in Appendix D, "A Selected Bibliography." Also look through Appendix B, "Technical Reference Guides."

EXERCISES

1. See if your library prints a guide to its holdings and operation. If it does, obtain a copy for future use.

2. Go to your library's card catalog. Locate a book of interest to yourself. Find the book in the stacks, take it to the circulation desk, and check it out. Write out full bibliographical information for the book:

author(s) or editor(s)	publisher
title	city of publication
edition	date of publication

3. Go to your library's reference division. Obtain the following information:

    a. From Sheehy's, list the indexes and abstract journals in the discipline closest to your intended major or in a discipline in which you think you may be interested.

b. In Ulrich's, find a journal in your field of major interest. Give the following information for the journal:

editor's name
circulation
full names (not abbreviations) of indexes and/or abstracts that cover the journal

c. From the *Oxford English Dictionary*, copy the first three lines of a definition of a word that begins with the same first two letters as your last name (example, Smith—smack).

d. In each of the following categories, furnish the names and call numbers of at least one reference work that you found in the reference division:

encyclopedias	bibliographies
dictionaries	yearbooks
atlases	indexes
biographical works	abstracts

4. See if your library has *The New York Times* on microfilm. If so, find the following information:

a. Two major headlines from the front page of *The New York Times* for the day that you were born.

b. The name and price of an article of clothing listed in an advertisement in *The New York Times* on the day that you were born. Name also the firm that placed the ad.

5. Go to your library's periodical division and find the following information:

a. The cover story of a news magazine such as *Time* or *Newsweek* for the month and year in which you were born.

b. The name of the first periodical shelved in the reference division whose title begins with the same letter as your last initial. In the event that no periodical begins with the same letter, go forward or backward in the alphabet until you find one.

6. See if your library has any holdings in government publications. If so, find the following information:

a. From *Government Reports Announcements and Index* find the descriptors from an article in the discipline closest to your major.

b. Find the name and price of the first entry in a section that begins with your last initial in the last *Selected U.S. Government Publications.*

7. See if your library has access to any of the computerized information retrieval services listed in Appendix B. If so, list the services available.

CHAPTER **3**

Gathering Information

As you begin this chapter, your mind may be bursting with some subject matter that you earnestly want to communicate to others. If you were in Florida or the Bahamas this past summer, you may have had an encounter with a trio of Mako sharks. Did they behave as sharks are supposed to? Are there devices and methods of deterring sharks from attacking a swimmer or diver?

Or you may be interested in sports cars. Three cars may be for sale at your neighborhood used-car lot—a Porsche, an Austin-Healey, and a Fiat. Which of the three would be the best for you to retool for rally competition?

Again, your instructor in Chemistry 12 last term may have outlined a theoretical procedure for producing acetic acid by the methanol carbonylation process. By digging into the details and practical aspects, could you develop this topic for your research project?

If you are already equipped with a topic for research and later writing, consider yourself fortunate. Propose it to your instructor if you are in a report writing class. Both of you may be very happy with it, for the spontaneous topic ensures the writer's interest.

On the other hand—and this is the more common situation—your mind may be utterly blank as you face the requirement to locate, research, and write up a topic for submission several weeks hence. If this is your predicament, we have several additional suggestions to make.

Choosing a Topic

You can talk over possibilities with students and instructors in your other courses. You can go to the library and examine the periodicals in your major field. You can read the daily newspapers to find what is worrying the world. Or perhaps your imagination will be triggered by scanning this list of topics other students have used in recent months:

Conversion of methanol to gasoline

Two-cycle engines for automobiles

Basic woodworking tools for the hobbyist

Versatility of vinegar

Importance of determining SWR for CB antennas

Determining the value of diamonds for use in jewelry

Home insulation for summer and winter

Thermostats for temperature control

Synoptic weather forecasting

Forecasting tides and currents

Diseases of parrots and other pet birds

The most over-prescribed drugs

How doctors calm fearful patients before surgery

Hypothermia: risk for the aged

Potential of solar and wind power sources

Causes of failure in automobile mufflers

Future of the family farm

Preparing your house for sale

Most common mistakes in amateur photography

Glazing and firing pottery

Clock escapement movements

Pros and cons of studded tires

Emergency electric power generators

Tuning your car engine at home

What to look for in a home sewing machine

Of mice and men: how carcinogens are identified

Collecting antiques for future resale

How to estimate the safety of a cruise ship

Leasing versus buying a car

Importance of humidity level in office and home

Heel heights versus posture and health

Keeping your house plants healthy

Selecting a ski resort

Cooking methods that waste vitamins

Psychology of defensive driving

Preparing for a backpacking trip

Casinos and lotteries for state revenue

Imports and the American car market

Btu values of competitive fuels

Best investments for hedging against inflation

How wineries start and cultivate vineyards

Economics of the home freezer

Remember that we have listed these topics for their suggestive value only. They are topics that other students have used with success in the past. But do not simply grab hold of one of them out of desperation. As worded, not a single one of them may be right for you.

Here is another technique you might try. First select a large "field" and give it a covering label, say Current Problems. Next, within the large field, identify more specific "cells" or problems: (1) health, (2) sanitation, (3) national, (4) state, (5) community, (6) racial, (7) financial, etc. Your graphic thinking aid might now look like the drawing below; give particular consideration to cell 5, community problems.

Figure 3-1   Topic Cells

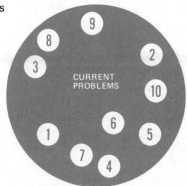

For the rest of this chapter let us assume that you have decided to deal with some problem in your own community (a very good bet, by the way). If you are normally observant, you have reacted to many features and peculiarities of your neighborhood. Without attempting to be exhaustive or logical at this time, simply list whatever problems have bothered you most. Your list might read like this:

Inadequate street lighting—intersections only
Shade trees in grass strips—few, mixed, unkempt
Cinders and other debris along curbs
Many streets not named at intersections
Community-owned plots of land serving no useful purpose
Sewage disposal plant sometimes smelly
Police force in aged and assorted uniforms
No community-owned street decorations at Christmas
Once-a-week-only garbage and trash collection

Your own list of community problems may take you only half an hour to construct. Obviously, all items in the list above are not equally costly or significant or urgent, but any of them might serve you well for research and writing. However, since time flies and research eats up time, you would probably fasten upon some one of the topics and let the others go for now.

As you review the community problems you have listed, your overall preference may gradually swing to the second on the list: trees. (Perhaps this topic is appealing because you spent your early years on the family farm, enjoy the outdoors, and have thoughts of transferring from liberal arts to forestry.)

Before spending research time on trees, however, you should check your topic for usability as described under the next heading.

Checking Your Proposed Topic for Usability

You would be unwise to commit yourself to research a topic without first observing several safeguards, including these five:

1. You must have the necessary technical background to handle the subject matter. Otherwise, you have to acquire specialized information—in chemistry, banking, highway engineering, acoustics—even before you begin work on the project you have tentatively selected. To obtain this supporting information would place an unnecessary burden upon you, one that does not ordinarily trouble the working specialist.

2. You must have access to the necessary means and facilities. These may include physical equipment, laboratory setups, time and money for travel, pilot models, sites for inspection, and so on. Again, from the re-porting view, you must restrict yourself to reasonable requirements.

3. You must be able to complete the project within the hours and calendar time available to you. If you have only ten or fifteen weeks (the usual length of a college term or semester), do not obligate yourself to grow a crop of corn, test some material under conditions of the four seasons, or record weather data for the coming ten years. To do so would be to guarantee failure before you get started.

4. Insofar as your particular circumstances allow, choose a general subject and a specific problem of interest to you. Subjects that at the beginning are of only mild interest may become absorbing; but subjects that are genuinely distasteful tend to become intolerably so with the passage of time.

5. Because, ordinarily, investigations depend to some degree on informa-tion already provided by other persons, go to your local libraries to run a check on the sources of information available in the form of periodical literature, standard reference works, previous reports, and pamphlets. If you can locate three or four potentially good sources within an hour, you can reasonably conclude that enough additional sources to make the project feasible will come to light if you spend several more hours of effort.

If the trees topic passes the usability criteria, you are ready for the next step: further limitation of your topic. Because you have spent some time touring the streets of your town, your main interest may finally settle on the topic of planting shade trees in grass strips. This limitation to shade trees will give you a topic of workable size, neither sprawling nor too confined.

Of course, you will need a stated purpose or set of objectives to guide your research and define what you are to accomplish. Without an agreed-upon destination you would be simply wandering around in a forest of facts. For the shade-tree problem, here are some possible objectives:

Should your town undertake a shade-tree planting program?

Should such a program, if adopted, be carried out by local means or should commercial experts be hired?

What species of trees should be selected for planting?

Should the program, if accepted, be accomplished in one planting season or should it be spread over several years?

Community problems tend to develop a history, leading from the slightest first suggestion to full-scale completion. For the shade-tree problem you will cut into this chain of developments at some particular point. Therefore you will need to determine where your community stands at the moment. If sentiment clearly favors a planting program, take it from this point. Do not waste effort needlessly by rearguing old and settled questions, but on the other hand do not attempt to leap into the distant future. It could well be that letters to the editor of your town newspaper revolve at the moment around this single question: Which species of trees should be selected? If so, *that* is the primary question your report will have to answer. Tailor your material and thesis to the timely question. Unless unanticipated difficulties arise, therefore, the specific question to be answered by your research and the ensuing report will be:

What species of shade trees should be planted along the grass strips of your town?

Gathering Information

Once you have selected your research topic and have clarified your intentions in dealing with it, you are ready to make your next big decision: *What means should I use to gather information bearing on my problem, so that I can carry out my intentions?* You will usually have some choice. However, you should not lose sight of the total possibilities, which include but are not limited to the following:

Calling Upon Your Memory
Searching the Literature
Generalizing from Particulars and Particularizing from Generalities
Inspecting Local Sites and Facilities
Administering a Questionnaire
Checking Customer Attitudes and Requirements
Interviews
Letters of Inquiry
Performing Calculations and Analyses
Reviewing the Information Already Gathered

Calling Upon Your Memory

As a researcher you will already have some facts and ideas about the new subject you are beginning to explore. At the start of the investigation it will help to jot down whatever information is lodged in your memory. This is a period of speculation and appraisal, a time of making the first rough acquaintanceship with the new project. Among other things, you should endeavor to recall when and where you learned what you already know, for you may thereby obtain promising leads to more information from the same and related sources.

What do you know about trees? Here are some assorted facts you may already know:

Some trees will survive and thrive only within a limited climate range.
Some trees, like the chestnut and elm, are especially subject to disease and/or insect attack.
Different trees vary greatly in their rate of growth, longevity, and proportions.
Some varieties are notorious for the damage they cause to sewers, sidewalks, etc.
The weeping willow and the mulberry are said to be "dirty trees."
Nursery whips cost less than branched saplings but often have a higher rate of mortality.

If trees have been your hobby for several years, you may be able to extend this list to several dozens of items. It is amazing, in fact, the variety of information to be found in the average person's head!

Searching the Literature

Now let's turn back to the usability criteria on page 44. Do your local libraries have enough reading matter to support your research on shade trees? Very probably they do, but it is better to check now than to be stymied later by a poverty of "hard" information usually obtainable from published literature.

If you are a college student, you may have a choice of three or more libraries: your town library, the main library on campus, the library maintained by each school of your college and, sometimes, by a department within a school.

Because of its nearness to your dormitory, or some equally practical consideration, you decide to try Old Main Library on central campus. After an hour of searching through the subject card files, you are pleased to have located these five items as a start on your "working bibliography":

Working Bibliography

Little, E. L., Jr. 1953. Checklist of the native and naturalized trees of the United States. Washington, D.C.: U.S. Department of Agriculture.

Fenska, R. R. 1956. The complete modern tree expert's manual. New York: Dodd, Mead and Company.

U.S. Department of Agriculture, Forest Service. 1965. Silvics of forest trees of the United States. Agricultural Handbook no. 271. Washington, D.C.: U.S. Government Printing Office.

Petrides, G. A. 1958. A field guide to trees and shrubs. Boston, Mass.: Houghton Mifflin.

Maino, E., and Howard, F. 1955. Ornamental trees. Berkeley, Calif.: University of Calif. Press.

We hope that you would eventually be able to extend this initial working bibliography to several times its present length. But, at the start, you have evidence that a sizable quantity of reading matter will be available to you.

Now you have an important decision to make—which information source to read first? In general, it is best to select a published source that is large, recent, authoritative. On the basis of these criteria, the third item probably should be your first choice. Make a bibliographic identification card for it:

> Silvics: USDA '65
>
> U.S. Department of Agriculture, Forest
> Service. 1965. Silvics of forest trees of
> the United States. Agricultural Handbook
> no. 271. Washington, D.C.: U.S. Government
> Printing Office.

Notice the brief identifier placed in the upper-right corner. Such "code" identifiers will save you much time and writing later, as you will shortly see.

Now you start to open the book. But hold on a moment! How are you going to extract and use any pertinent information it contains? Having a method of doing this will prove critically important to you later as you turn to compilation of the report.

Many researchers and report writers find that converting recorded data into prose reports is unbearably torturous and time-consuming. It is true, of course, that there are no magical shortcuts or miracle methods that will produce "instant reports." However, there is no sense whatsoever in plunging blindly into the information-recording and compiling process or in continuing with methods that have proved cumbersome and inefficient in the past. Do you have a method that works well for you? Do you, in fact, have a method? If not, try the method outlined in the paragraphs that follow.

Has it been your practice to copy research information onto the sheets of a notebook? If so, you know the battle that ensues when your individual notes have to be sorted and somehow placed in usable order for compilation into a prose report. You have to leaf back and forth through the notebook to identify and locate notes bearing upon the topic of present interest. Some notes may escape your attention: others may accidentally be used twice. Still worse, the resulting text is likely to be disjointed and badly organized.

Abandon the notebook practice for taking notes. Instead, use cards of four by six inch size, or even larger. Limit the information on each card to a single narrow topic—one that you may label with an identifying word or two such as *Installation Cost, Service Procedures, Maintenance.* Place the topic identifier prominently on each card, perhaps at the upper-left corner. You have thus made it possible to sort through the entire collection of cards, and can place in subpacks all cards bearing the same topic identifier. Now you can copy onto a sheet of paper all of the topic identifiers and thus be able to inspect the extent of your information in topical array.

At this stage you may be able to draft one or more tentative outlines to hold and arrange the information for the body of your report—and at the same time be able to detect omissions, duplications, overlaps, and irrelevancies. (For further information on organizing your material, see Chapter 7.)

Once you have created the subpacks, you can make a still finer sorting of cards in each of the subpacks until you have them arranged in order for actual composition. It is a relatively straightforward process to pick up Card #1 of Subpack #1 and consume its information in order to write the first paragraph of your first body chapter. Prior sorting, permitted by the card notetaking process and the use of topic identifiers, will have freed your mind of the necessity to hunt and sort while composition is in progress. Hence you can give your full attention to the digestion of information and to its clear and coherent expression in prose.

If you do not have a notetaking and note-using method that works for you, try the method we have outlined. The method works for thousands of researchers and report writers, both student and professional.

At this point, you have selected your first book of readings, you have made a bibliographic identification card for it, you have devised a short label for it (*Silvics:* USDA '65), and you have adopted a systematic procedure for taking notes on cards. Let your reading and notetaking begin.

The third paragraph on the first page of *Silvics* catches your particular attention. It is a likely prospect for inclusion in your final report. Because the language of this paragraph is economical and informative, you decide to copy it verbatim onto a card as shown in Figure 3-2.

The label in the upper-right corner of the card in Figure 3-2 identifies the source as that fully described by the bibliographic identification card you made out earlier. The label in the upper-left corner (*Environment*) is the topic identifier. The quotation marks indicate that this is a verbatim extract. The label *p. 1* in the lower-right corner indicates the page from which the extract was taken. Be sure to set down all these labels and identifiers. You will need them later.

You read on. The first full paragraph in the right column on page 2 catches your attention. Because the sense of this paragraph can be condensed, you decide to paraphrase it (Figure 3-3).

Figure 3-2   A Verbatim Extract

the environment is the same for all trees. In the following reports of the silvics of each tree species, known responses of the species to a detailed or specific environmental condition will be given. In this section, we present background information of the general responses of trees to factors of the environment.

THE TOTAL ENVIRONMENT

The total environment of the tree is a complex integration of numerous interrelated physical and biological factors. Physical factors include those of the climate, such as various measures of radiation, of precipitation, and of movement and composition of the air. They include also factors of the soil such as texture, structure and depth, moisture capacity and drainage, nutrient content, and topographic position. Biological factors are the plant associates; the larger animals that use the forest as a source of food and shelter; the many small animals, insects, and insectlike animals; the fungi to which the trees are hosts; and the myriads of micro-organisms in the soil, the functions of many of which are beneficial to the tree.

Because of the complexity of the total environment and the difficulty with which some factors are measured, complete and exact

> ENVIRONMENT   SILVICS: USDA '65
>
> "The total environment of the tree is a complex integration of numerous interrelated physical and biological factors. Physical factors include those of the climate, such as various measures of radiation, of precipitation, and of movement and composition of the air. They include also factors of the soil such as texture, structure and depth, moisture capacity and drainage, nutrient content, and topographic position. Biological factors are
> (over)

> plant associates; the larger animals that use the forest as a source of food and shelter; the many small animals, insects, and insectlike animals; the fungi to which the trees are hosts; and the myriads of micro-organisms in the soil, the functions of many of which are beneficial to the tree." p. 1

Figure 3-3   A Paraphrased Extract

temperature is about 40°, the maximum 105°, and the optimum 77° to 86°. For tropical plants the minimum is 50°, the maximum 122°, and the optimum 86° to 95°.

In a dormant or resting state plants can endure extremes much greater than the minimum and maximum temperatures for growth. Evergreen trees endure winter temperatures of -60° to -70°F., but temperatures of 25° to 30° kill twigs during the growing season. During summer in the temperate zones, growth is often completed before maximum temperatures occur. These reach 115° or more in some forested areas of the United States.

Resistance to freezing temperatures, or frost hardiness, may result from a change in the protoplasm. The osmotic concentration of the cell sap increases with the hydrolysis of insoluble carbohydrates to soluble sugars. Dehydration of the protoplasm leads to an increase in the apparent bound water content

> TEMPERATURES   SILVICS: USDA '65
>
> Dormant temp. tolerance: down to -70° F.
> Growing min. temp. tolerance: 30° F (twigs).
> Growth attained before max. summer temp.
> of 115° F. p. 2

Your reading and notetaking continue. After finishing *Silvics of Forest Trees of the United States,* you read the other books in your working bibliography. You return to the subject card catalog to add to your original list of readings. You refer to the *Readers' Guide to Periodical Literature.* You read, take notes, make plans.

Generalizing from Particulars and Particularizing from Generalities

Now that you have done some reading in the literature, you may want to lean back in your chair and take a long, hard look at what you have so far learned. In older communities, shade trees along the streets were selected and planted at the whim of property owners. Some of their results have been excellent: broad avenues with a canopy of sycamores. At other times, the results have been distressing: low-branching trees that obstruct traffic, trees such as the weeping willow that clutter the streets with debris, trees so unsuited to the climate and conditions of growth that they chronically look half-dead.

As you recall the information so far acquired, you are convinced that a knowledgeable group of persons (shade-tree commission?) will be needed for your town if a program of shade-tree planting is to succeed. Members of the Department of Horticulture of State University should be consulted. Magnolias are marvelous in Savannah, but would they thrive in Central Pennsylvania? Members of the Department of Horticulture would know. What is the average rainfall in your locality? The Department of Meteorology would know. Must pruning and other surgery be done by trained specialists? Someone in the Department of Grounds would surely have the answer.

In other words, by now you have a fair supply of information, consisting of both generalities and particulars. Much of this information was gained from localities other than your own; some of it is old information that may be outdated. Other parts of it represent "rule-of-thumb" generalities that may or may not apply to your own geographical area.

The kinds of reasoning we have just touched upon have their formal names in logic. Reasoning from one or more particular facts to reach a covering generalization is called *induction.* Suppose you learn by observation that the pin oaks planted several years ago along Fairlane Road are in terribly poor health. So you telephone your local arborist and ask for information on good growing conditions for this species of oak. You are rewarded with a host of facts, one being that pin oaks prefer a mildly acid soil. The next day you go back to Fairlane and find limestone outcroppings. You take a soil sample and learn it has a pH of 8.6. You therefore conclude that the high alkalinity is a probable cause of the poor growth. As the next step, you decide that before any additional trees are planted in town, acidity tests should always be made. This generalization, obtained by inductive reasoning, will be a useful guide for the future.

The other kind of reasoning leads from the general to the particular, and is called *deduction*. To see this process at work, let us continue our illustration.

Suppose the question arises as to whether pin oaks could properly be planted on Nimitz Avenue. An on-site check shows that this area has a highly alkaline soil. You know from your previous investigation that pin oaks require an acid soil (this is your generalization). Therefore you immediately rule out the planting of this species on Nimitz. In this process you have applied your earlier generalization to derive (deduce) a particular answer.

Your experience will also suggest another generalization: to continue your research, you must delve into an additional source of information, namely, the local scene. A grass-roots study is in order.

Inspecting Local Sites and Facilities

On your tour of inspection, take along a notebook to jot down whatever you observe. At this time do not try to be particularly logical: your on-the-spot notes can always be organized later.

After an hour or so of tramping the streets of your town, both downtown and in outlying residential areas, your jottings might read as follows:

College Dr. and Main St.—former grass strips mostly covered with asphalt. Plots might be dug for trees and then loosely bricked over or dressed with pebbles.

Young trees would require fencing guards to ward off injury.

Downtown, because of restricted passages, trees could not be allowed to branch below the eight-foot level, for head clearance.

Heavy foliage on the maples along Garner St. suggests that the town will have to get leaf-vacuuming trucks.

Some grass strips (e.g., Fenwick Ave.) are badly eroded. Can this problem be solved?

Large elms on private property may interfere with trees planted in grass strips.

Some older trees have had their tops looped off to clear power and telephone lines. Importance of tree placement and eventual height. Can storm sewers handle the load of water-borne leaves?

Depending upon your powers of observation and your knowledge of the facts of life insofar as trees are concerned, you might fill several notebook pages with such comments in an hour or so of walking. In any case, you will have familiarized yourself with the local scene, so do not be discouraged if you end up with more questions than answers. That is entirely normal, and wholesome, in the early stages of research.

To continue your research, several courses are now open to you. In this state of open-mindedness, it occurs to you that many other residents of your community may also have thoughts about shade trees. Obviously, a questionnaire administered to the neighboring residents could yield some interesting results. Let's try it.

Administering a Questionnaire

It is not easy to construct and administer questionnaires and tabulate their replies without years of study and experience, yet even the "experts" do not pretend to have solved all the problems. We do not intend to make an instant expert of you, but we can offer some basic advice that should enable you to handle questionnaires of limited complexity and scope.[1]

It is always an imposition to ask someone to fill out a questionnaire. You should keep this fact in mind when the time comes to design and administer a questionnaire yourself. Fortunately, you can usually approach your subjects in such a way that the questionnaire does not seem an imposition.

First, be sure that a questionnaire is really necessary. Questionnaires should be used only when the desired information cannot be obtained from another source—interviews, personal observation, searches through daily newspapers, courthouse and library files, and so on. Time, money, and effort can be saved by avoiding the questionnaire process whenever possible.

Whether or not you decide a questionnaire is indicated, the research you do before making that decision is not wasted. If you find no questionnaire is needed, that means your research yielded a lot of the answers you thought might require a questionnaire, and you're farther along than you had thought. If you *do* need to use the questionnaire approach, the information you've obtained about your topic will be invaluable in designing that questionnaire.

---

[1] For detailed treatment we suggest you read *Questionnaires: Design and Use,* by Douglas R. Berdie and John F. Anderson (Metuchen, N.J.: The Scarecrow Press, Inc., 1974).

In a sense, this preliminary exploration is comparable to the advance work a lawyer does in preparing to face a jury. You, too, have a "case" to make. You must convince your respondents that you are well-informed on your subject, so the time they take to assist you will be well spent. And, like trial attorneys with their briefs, you must be sure that all material in the questionnaire gives a favorable impression of you as the informed author.

The extent of your preliminary research will vary, of course, depending on the amount of time involved. If the questionnaire is part of a project for a one-semester or one-term course, you will have to tailor your topic to suit the available time and plan your research accordingly. With a relatively narrow, straightforward topic (like our shade-tree example), a week or so of study should suffice to establish your credentials as an authority and enable you to write an authoritative questionnaire. But if the questionnaire is part of a graduate-level thesis, your topic will undoubtedly be much broader—and you may need to spend a year or more simply acquiring the information that will serve as a basis for the questionnaire.

Drafting the questions. When you've thoroughly studied your topic and determined that a questionnaire is definitely the best way to obtain at least some of the data you need, it's time to make up a list of questions. This part of the project is much more difficult than it sounds, for the reasons outlined below.

First, your questions have to take into account the factor of *reliability*. That is, they should be worded so the respondent's state of mind when the questionnaire is administered will affect the answers as little as possible. (The exception would be a questionnaire specifically aimed at judging public opinion on a certain matter at a certain time.) If the shade-tree questionnaire asked, "Do you think the town council has ignored the need for a major tree-planting effort along our streets?" most of the answers wouldn't reflect the respondents' actual feelings about the need for more trees. Rather, they might reflect political bias (support of or opposition to the town council members); or fear of the tax increase suggested by "major planting effort"; or enthusiasm for trees in general among conservation-minded citizens; or all manner of similar emotional responses. The answer of any one respondent could vary from one day to the next, depending on who was elected to the council, whether taxes or environmental concerns had precedence at the moment, and so on. Thus, none of the responses would be *reliable*, and the questionnaire would be practically useless. As we will see in the next section, our shade-tree researcher did manage to word the questionnaire so that fairly reliable responses could be anticipated.

Second, your questions must be worded to ensure a *valid* interpretation by respondents—that is, an interpretation that matches your own. This requirement overlaps the first to some extent, but also goes beyond it. In trying to write questions that express your meaning so that respondents cannot possibly misinterpret it, you must try to read each question through their eyes. Remember that although you have studied the topic in depth, the questionnaire may be their first introduction to it. Some people will know more than others, of course, but the idea is to formulate questions that will be clear to *everyone*. Returning to the shade-tree problem, you would not learn much from the answers to the question, "Would you favor expanding the park maintenance staff to permit increased landscaping efforts?" That question asks the respondent to make several related decisions on the basis of very little information before arriving at an answer. What is meant by "landscaping efforts"? Are such efforts necessary? Why or why not? Assuming they are, is "expanding the maintenance staff" the best way to proceed? How much would it cost? And so on. Questions that are open to interpretation, like those that encourage emotional responses, practically guarantee unreliable answers.

Third, you have several different *types* of questions to choose among in drafting your questionnaire. Once you have selected the type that best suits your purpose, stick to it throughout the questionnaire. Changing from one type of question to another in midstream will only confuse respondents. The five types of questions most commonly used on questionnaires are as follows:

1. *Dichotomous.* A *dichotomy* is a division into two separate parts, and a dichotomous question offers the respondent a choice between two answers—*yes* or *no, before* or *after, manual* or *electric,* and so on. Thus, this type of question provides a relatively limited span of responses, and is useful only when the questionnaire topic can be analyzed in black-and-white terms.

2. *Multiple-choice.* In this case, the respondent may choose among several possible answers for each question:
   In preparing copy for a printer, if you must attach printed pages to a backing sheet, which of these methods do you prefer?
   stapling ☐   taping ☐   gluing ☐   using paper clips ☐
   If you think there may be other possible answers, you can add a blanket category, followed by a line for the respondent's answer: other

   This approach is probably the most common for questionnaires; it permits a fairly wide range of responses, but the choices are sufficiently controlled by the administrator to permit easy interpretation of results.

3. *Ranking.* Here, the respondent is asked to rank several possibilities in order of personal preference:

   If your company required you to transfer to one of the following cities, what would be your order of preference?
   Denver ☐   Miami ☐   Chicago ☐   Boston ☐
   The respondent would write the appropriate number in each box.

4. *Fill-in-the-blanks.* This short-answer approach may be used to elicit either factual answers or opinions.
   How many children do you have? (If none, write "none.") _____
   How long have you held your present job? _____
   What is your gross annual income (to the nearest thousand)? _____
   Do you approve of the mayor's fiscal policies? Why or why not? _____

   Obviously, this approach gives respondents more latitude in answering questions than any of the methods listed above.

5. *Essay.* This type of questionnaire allows respondents maximum freedom in composing their answers. Questions are relatively general and followed by an inch or more of space for the answers. Despite the obvious advantage of complete and detailed answers, this method is the least efficient of the five described here, and therefore the least suitable for short-term projects. (By the same token, it is probably the best for a long-term project where thorough individual responses are important.) Essay questionnaires are time-consuming for both administrator and respondent, handwriting may be illegible and difficult to decipher, and replies may be very difficult to tabulate.

Preparing the questionnaire. Once the questions are drafted— with no chance of reader misinterpretation or emotional response, and a consistent format used throughout—you have nothing more to worry about. Nothing, that is, except: How many dozens or thousands of questionnaires should be administered? Should the administration procedure involve mailing, telephoning, or personal interviews? Will the local telephone directory supply enough respondents or should another source be used? Is there any guarantee that the questionnaire respondents will be representative of the population you hope to study?

There is no set formula for solving these problems. The answers will vary with each project, depending on the issue being investigated, the time period involved, any special circumstances that must be dealt with, and so on. For the sake of this discussion, however, let us assume that our shade-tree researcher has not only drafted a satisfactory group of questions but also decided to approach a small sampling of respondents by means of personal interviews. The final questionnaire might resemble the following illustration.

Even the most intelligently devised questionnaires sometimes miss their mark, and the responses prove unrewarding or hard to

TYPICAL OPENING

Good afternoon! My name is _____ .   I live
at _____.   I am not selling anything, but I would
like about three minutes of your time to try to make our town a
better place to live in.   Would you be willing to answer a
very few questions?   I assure you that your replies will be
kept anonymous.   I need this information for a paper that I
want to submit to Town Council.

QUESTIONNAIRE

1. How long have you lived in this community?

Less than 1 year	1 to 2 years	2 to 5 years	Longer? No. of years
☐	☐	☐	☐

2. Would you say that your street and your general area are attractive?

Very	Fairly	No
☐	☐	☐

3. Have you ever done any gardening or tree planting?

Much	Some	None
☐	☐	☐

4. Have you noticed that some grass strips of our town have no trees at all or else a mixture of trees in poor condition?

Yes	No	Not sure
☐	☐	☐

5. Would you be in favor of having the town council establish a regular and controlled program of planting trees in the grass strips?

Yes	No	Doubtful
☐	☐	☐

6. Judging by the street frontage of your lot, I would say that it would accept either one or two trees. How much would you be willing to pay to have each tree bought, planted, and cared for by the town?

Nothing	No more than $5	No more than $10	No more than $15
☐	☐	☐	☐

7. What variety or varieties of tree would you prefer?
   Indicate first and second choices.

Norway Maple	_	Scarlet Oak	_
Red Maple	_	Pin Oak	_
Sugar Maple	_	American Elm	_
Ginkgo	_	Other(s)	
Honey Locust	_	_____	
White Ash	_	None _	

TYPICAL CLOSING

That is all, then. I thank you for giving me this chance to talk with you about our town. Your responses have been very (helpful, interesting, informative). Of course they will be kept in strict confidence. Goodbye.

tabulate. Therefore we suggest that you administer the first version of your questionnaire on a trial basis—possibly to members of your own family or your immediate neighbors. Then make whatever revisions seem to be indicated.

Do not expect your questionnaire to produce miracles. If you mail it to members of your community (stamped return-address envelope included, of course) do not be shocked if only thirty-five percent respond. (Of course, if you plan to mail out questionnaires instead of administering them in person, you would revise the opening and closing paragraphs of the sample questionnaire above accordingly.) House-to-house canvassing may or may not produce a higher percentage of responses, depending upon the time of day and the mood of the community, among other things. Administering your questionnaire by telephone is a tedious procedure, but permits you to cover a large geographical territory without pounding the streets. A combination of these means of administration, including the polling of membership of clubs to which you belong, may prove to be the best overall answer.

Having licked the questionnaire problem, you should ask what other avenues of exploration are open to you (see page 45). Because you cannot do everything at once, you have to settle upon one avenue at a time. You therefore think of local individuals who should be informed and interested and probably would consent to interviews.

Checking Customer Attitudes and Requirements

From time to time in your research, especially while talking with residents, some inklings of preference and prejudice have slipped through. You decide to sum up these attitudes as well as your notes and memory allow.

1. Some residents, frankly, place neatness above all else. A dandelion for them is a call to arms. They like their grounds clear and clean of plantings. Probably they would be vexed by having to mow around trees in the grass strip.

2. Some residents feel they have absolute dominion over their property to the curb, including the grass strip. Probably they would resent having the town select a species of tree, plant it in the grass strip, and charge the property owner for it. Isn't that an invasion of privacy?

3. Some people resist change per se. Nothing old must ever be taken away, and nothing new must ever be added. Wouldn't trees in the grass strip be a pretty radical innovation?

4. For some people every silver cloud contains many dark linings. Trees might be nice, in a way, but wouldn't they heave sidewalks that would have to be repaired and replaced at the owners' expense? Shade trees kill lawns. Shade trees make sidewalks dusky by early evening and thus invite hoodlums to mug. Shade trees mean hours of hard raking in the fall. And so on.

These negative attitudes, though expressed only by a minority, suggest that you should compile information on the assets stemming from a well-designed shade-tree program. These benefits come immediately to mind: summer cooling, wind-breaking effects, contrast and attractiveness of greenery, sun screening, and possible increase in resale value of properties. You decide to investigate these and other possible beneficial aspects further.

Interviews

To move ahead on your problem you decide to list several persons who may be able to provide both attitudes and information relating to your community:

Dr. Strickland, Associate Professor of Horticulture
Ms. Strand, local representative of Empire Tree Nurseries
Mayor Peabody
Mr. Clarke, Director of Parks and Recreation
Mr. Ambler, senior member of the Business Association

Your instincts suggest that Mayor Peabody, who has held office for twenty consecutive years, may very well have the broadest view of community attitudes, interests, and problems. A brief telephone conversation with his secretary elicits a half-hour interview with Mayor Peabody on Thursday, three days hence.

If you have watched interviews on TV you probably have noticed that most experienced interviewers keep a tablet or clipboard in front of them, and from time to time glance at their prepared notations. This tactic is helpful for the expert—it will be essential for you.

Prepare a list of questions in advance. In general, stay clear of "dead end" questions that can be answered with a simple *Yes* or *No*. Instead, phrase your questions in a way that invites the person being interviewed to expand upon his or her first answers. Pass the ball to the interviewee and let that person do practically all of the carrying. Phrase your questions clearly and economically: avoid ten-line "essay" questions that will eat up valuable time when your interviewee should be doing the talking. On the other hand, do not insist upon asking every question on your prepared list. If the person catches fire and discourses productively, let the discourse continue until the interviewee runs out of things to say. Also, do not insist upon posing the questions in their prepared order; instead, use the cues provided by what has just been said to guide you in selecting a reasonably relevant next question.

When you arrive for the interview, introduce or reintroduce yourself, giving your name, business or professional connections, subject matter for discussion, reasons for the interview, and the use to be made of the information gained. You should both sense the mood of the occasion and try to color the mood. For a half-hour interview, the opening five minutes may be given to generalities and personalities, apparently far afield from the real reasons for your interview. Once rapport has been established, take some key remark to lead into one of your prepared questions. What follows may read, in part, like this:

Interview

Q. Mayor Peabody, you have been mayor of this town for twenty years, I understand. Over these two decades, what would you say the major changes have been, for better and for worse?

A. Well, certainly we are a lot bigger than we were twenty years ago, about double in population and more than double in area. You could say that we have changed from a village to a town.

Q. Yes, my memory of this town goes back to 1966, when my parents drove us through on summer vacation. Now I hardly recognize it—new buildings, a much larger business district. Mayor, what would you like to have back again from twenty or so years ago?

A. Some things are gone forever, and should be. But I do believe that we have lost our sense of leisure. Now we are all pepped up like people in a big city. A sense of graciousness, a touch of the rural, has gone.

Q. When this town was laid out, the original planners wisely provided for wide streets and generous sidewalks. But some visitors, I have heard, find our streets bare and unattractive. My next question, then, and it is my main reason for requesting this interview, is this: Do you believe that planting shade trees along our streets would be a worthwhile investment?

A. That very thought has occurred to me, but I have never taken it up in town council. Yes, when I think of the pleasantest towns I have visited, I have a vision of their tree-lined streets.

Q. Very good, I agree. What I have in mind is to put together a report setting forth the pros and cons of shade trees bordering the streets of our town. Suppose I do just that and let you have several copies at the spring council meeting.

A. I'd be very pleased, and grateful to you. And if I can be of any help, let me know.

During an interview such as this, you are looking for cues and clues. Do not expect blueprints and detailed estimates. Save these for later. If, as we anticipate, your interview with Mayor Peabody has gone well, you may want to make at least a tentative arrangement for a later interview. In Mayor Peabody, you have a good public-minded citizen on your side. Keep him informed of developments.

Letters of Inquiry

Familiarity, it has been said, breeds contempt. Be that as it may, familiarity most certainly breeds blindness—in that you may become less sensitive to your surroundings. You begin to wonder, then, what former residents of your town think about it in retrospect. They could have a clearer view than present residents. So you make a list of former residents whom you have known.

Tom Showers—nursery operator, moved to Minnesota two years ago.

Harry Foreman—former police chief, retired to Miami three years ago.

Dr. Mary Paulding—superintendent of schools until 1978, new position on West Coast.

To gain a detached view, separated by geography and time, you decide to write to Tom Showers:

```
Dear Mr. Showers:

I am a long-time resident of State College, and take an active
interest in most community affairs.  At present I am doing some
voluntary research on the selection of trees for planting in
our grass strips.
```

I am writing to you because you, as a former resident of State
College, have personal knowledge of our town and operated The
Empire Tree Nursery for nearly two years.

Would you please send me any brochures or other literature you
have available on this subject. Would you let me know what tree
species are currently preferred for community planting and,
also, any species that we should avoid.

In my written report to the town council I will give you full
credit for use of any information you supply, and will send you
a copy of the report upon request.

Any help you can give will be most gratefully appreciated.

Sincerely,

While writing a letter of inquiry, keep in mind that you are im-
posing upon the other person, asking for a little time and attention.
Therefore, brevity without curtness is in order. Make use of existing
bonds to place that person in a receptive mood. Sometimes a list of
specific questions is appropriate—plus a general solicitation for advice and information. If you ask for too much, you may get nothing at
all. If there is any favor you can offer in return, introduce this thought
toward the end of the letter. An expression of gratitude is the very
least you can offer.

For additional information on letters of inquiry and reply, see
Chapter 13, pages 280–286.

Performing Calculations and Analyses

Two or three weeks ago, when you began your research on shade
trees, you may have felt that you had a tidy little problem. With the
passing days, however, your original problem has gained in depth
and has reached out in unexpected directions. It now occurs to you to
consider the financial aspects of your project. Would the cost amount
to hundreds of dollars? Thousands of dollars? Tens of thousands? Let
us make some rough calculations.

First, you get your Empire Nurseries catalog from the shelf by the
fireplace. You learn that sunburst locusts and sugar maples, for
example, average height twelve to fourteen feet, cost about $18.00
each, but with a twenty percent reduction on orders of 100 and multiples thereof. At first approximation, then, trees suitable for planting
in grass strips will cost about $14.50 each.

Trees have to be planted, of course. So you telephone Heavy Equipment Rentals in Centre Hall. From the manager you learn that the "back hoe" would be ideal for scooping out holes and refilling afterwards. There will be a $20 delivery fee, via flat-bed truck, for each period of use. The service charge will be $26 per hour, including a trained operator. In addition, two laborers, at $6 per hour, will have to be in constant attendance.

What else will be needed? Every tree will need three guy wires, for a total of six yards, at 18¢ per yard, or $1.08 per tree. A planting-soil mixture, at $1.25 per bushel, will cost $2.50 per tree.

Let's say that each tree will require thirty minutes to place in position—a conservative estimate. Rough computation suggests these amounts:

Cost per tree	$14.50
Back hoe, half an hour	13.00
Laborers, two, for half an hour	6.00
Guy wires	1.08
Planting soil	2.50
Total	$37.08

This total may surprise you, for when you prepared your questionnaire you did not anticipate a cost of more than $15 per tree. But your new estimate seems to be more realistic. It follows that every property owner could be assessed at least $37 per tree if the contemplated planting program should be adopted.

What will be the dollar amount for the whole operation? Looking at a map, you learn that your town spreads about one and one-half miles east and west and about one mile north and south. Fifteen streets traverse the town in the north-south direction. Ten streets traverse east-west. You are astonished to discover, as the result of multiplication, that your town has about thirty miles of streets. However, about one-third of this mileage has no curbs or sidewalks, and eventually will be regraded. That means some twenty miles of streets are in condition for tree planting at present.

A statute mile contains 5280 feet. The average lot width is seventy-five feet. That would mean about seventy trees per side per mile, or a total of 2800 trees for twenty miles. At $37 per tree, the grand total would be about $103,600, not including delivery fees for the back hoes! Now you realize that you no longer have "a tidy little problem."

Reviewing the Information Already Gathered

At many stages in your research, perhaps weekly, you should take stock of your situation: what you have learned, where you are now, what step to take next. Without an occasional review you will most assuredly waste much of your time and go off on tangents. Therefore, with the opinions and information you have so far gained, you have established these points on the shade-tree problem:

1. The need for a shade-tree program in your town is not earthshaking, but researching it is within your means, and its accomplishment will be well worth the effort.
2. Most, if not all, of the information you will require is locally available, from individuals and libraries.
3. Somewhat to your surprise, you already possess substantial information on shade trees, all of which must be checked out and verified.
4. You have located library readings and have begun the note-taking process.
5. You have "furrowed your brow" to move back and forth between generalities and particulars. In the process, you obtained many leads pointing to additional information.
6. Your on-site inspection of the town uncovered facts you have never heretofore appreciated.
7. Your questionnaire promises to give you a cross-sectional impression of the views of the neighboring residents on the shade-tree question.
8. You learned a lot from your interview with Mayor Peabody. For one thing, you learned that a meritorious cause may never catch on because no one takes the initiative to push for it. Can local enthusiasm be generated for the shade-tree problem?
9. Letters of inquiry (you are hoping for a reply from Tom Showers next week) are a useful means of reaching out to persons who are not conveniently available by phoning or visiting. Will Tom Showers' response be perfunctory or rewarding?

The stocktaking represented by the nine items listed above may seem to you to be distressingly short on "hard" information to fill out the body of your final report. We agree. It is the merest beginning. Understand, however, that all research tends to have a snowballing effect: information begets information, and the more you already have learned, the more adroit you will be in devising new means of adding to your knowledge.

Understand that the material under each heading of this chapter is simply illustrative. We have shown you one questionnaire; you may need to prepare and administer several of them, at different stages of your research. You may need to arrange for a dozen interviews, though we have illustrated only one. In other words, you will probably have to cycle back and forth through the information-gathering methods listed on page 46.

Within a few weeks, if your shade-tree research progresses as it should, you will have built up a special vocabulary, including such terms as *silvical, dioecious, bipinnately compound.* Within a few weeks you will have built up a lengthy working bibliography of perhaps twenty library readings. And within another week or so you will have constructed a writing outline for the final report, and have loads of information to give body and substance to that outline.

EXERCISES ▮▮▮▮▮▮▮▮▮▮▮▮▮▮▮▮▮▮▮▮▮▮▮▮

1. All research must focus upon some subject matter ranging from *a* (for example, astronomy) to *z* (zoology). Somewhere between *a* and *z* you will find a subject that will prove to be compatible and engrossing.

    Sit down, close your eyes, and take thought. Ask whether you have already felt an interest in automobiles or airplanes, ballooning or blacksmithing, childhood diseases or cartography, and so on through the alphabet.

    As a result of this introspection, list five or more possibilities in rough order of preference. As your instructor directs, bring this list to class. You will probably be asked some probing questions, and the resulting dialogue will clarify your own thinking.

2. As the next step we suggest you do this: take a large card or half-sheet of paper, and for the most promising subject from Exercise 1 set down entries paralleling the following example, leading from the general to the specific.

```
General field: Aviation

Limited field: Airports

Specific topic: Runway lighting

Purpose of study: To determine whether a small local
                  airport would be justified in installing
                  a runway lighting system

Possible customer: University Airport Administration
```

3. Go to the library and, using the reference tools available to you, construct a working bibliography. We suggest that you write your bibliography on cards. (See page 48.)

4. Consult one of the sources listed in the bibliography you constructed for Exercise 3. On a note card (see pages 48–50), write a direct quotation from the source. On another card, paraphrase the quote on the first card.

5. Using the subject that interests you, complete one of the following assignments.
   a. Interview some expert on the subject. Write up the interview in a report of about 500 words.
   b. Construct a questionnaire that you could send to people who can provide you information in the subject area.

6. Refer to the list of research techniques on page 46 of this chapter. Can you predict which four or five of the ten listed should be most productive?

7. The questionnaire in this chapter (pages 57–58) was administered by a door-to-door canvass of the town. What considerations (including "facts of life") led the student researcher to use this method rather than the U.S. mails?

8. Under other circumstances the researcher might have mailed the questionnaire (perhaps 200 copies). On this assumption, rewrite the heading to the questionnaire and prepare a cover letter to accompany it. What else would have to be included with the mailed envelope?

CHAPTER 4

Analyzing Your Audience

You have been told many times, we suspect, that you can't write clearly and well until you have defined the purpose of your writing. We agree and have something to say about purpose in writing in many places in this book. But perhaps you have never been told to think about the *purpose of your audience* in reading what you write. *Why* do readers want to read what you write—what do they hope to learn from you, and what do they intend to do with that knowledge?

You must understand not only the purpose but also the background of your audience. You must know who your readers are, what they already know, and what they don't know. You must know what your readers will understand without explanation and without definitions. You must know what information you must elaborate, perhaps with simple analogies. You must know when you can use a specialized word and when you cannot. You must know when to define a specialized word that you can't avoid using. All this requires a lot of thought, but the successful writer always remembers that readers bring their experience *and their experience only* to their reading.

To be a good writer, then, you must know your audience—its purpose and knowledge. Perhaps in no other kind of writing is this business of matching a particular piece of writing to a particular audience as important as it is in technical writing.

Engineers of the nineteenth century were aware of the problems facing audiences of different levels. In 1887, the American engineer Arthur M. Wellington had different parts of his book, *The Economic*

Theory in the Location of Railways, set in three different type sizes: large type for the lay reader, medium-sized type for the reader who could understand some technical data, and small type for the readers who needed the most detailed scientific data. Readers, knowing their own limitations and interests, had clear signals as to what to read and what not to read.[1]

In this century, Philip W. Swain, an engineering professor at Yale and later chief editor of *Power* magazine, expressed the distinctions that exist between audiences this way:

> Note the many languages within our language. The college freshman learns that "the moment of force about any specified axis is the product of the force and perpendicular distance from the axis to the line of action of the force." Viewing the same physical principle, the engineer says: "To lift a heavy weight with a lever, a man should apply his strength to the end of a long lever arm and work the weight on a short lever arm." Out on the factory floor the foreman shouts, "Shove that brick up snug under the crowbar and get a good purchase; the crate is heavy." The salesman says: "Why let your men kill themselves heaving those boxes all day long? The job's easy with this new long-handled pinch bar. With today's high wages you'll save the cost the first afternoon."[2]

Mr. Swain here defines essentially four audiences: the scientist, the engineer, the technician or the operator, and the executive. Each audience has different interests and understands the same problem through a different language and from a different perspective.

In this chapter we break audiences down a bit differently. We tell you how to deal with laymen, executives, experts, and technicians. Also we tell you how to put together a combined report for an audience in which these groups are mixed. Before we discuss these four audiences, let us caution you that no audience is uniform, falling readily into a neat category. We speak of a lay audience, or an executive audience, but these are by no means totally homogeneous units. An audience might be compared to an aggregate of rocks of all shapes and sizes, as opposed to a mass of smooth marble.

This chapter, therefore, is not audience analysis made simple, foolproof, and mechanical. Rather, it contains a series of generalizations that should help you understand the process of audience analysis. But only you can analyze your own audience.

---

[1] Walter James Miller, "What Can the Technical Writer of the Past Teach the Technical Writer of Today?" *IRE Transactions on Engineering Writing and Speech* 4, no. 3 (1961): 69–76.

[2] Philip W. Swain, "Giving Power to Words," *American Journal of Physics* 13, no. 5 (1945): 318–320.

Laymen

Who are laymen? They are fourth graders reading a simplified explanation of atomic fission in terms of mousetraps and Ping-Pong balls. They are the bank clerk reading a Sunday newspaper story about desalination and the biologist with a Ph.D. reading an article in *Scientific American* entitled "The Nature of Metals." In short, we are all laymen once we are outside our own particular fields of specialization.

We can make only a few generalizations about laymen. They read for interest. They read to tune in more accurately on the universe. They are not very expert in the field or they would not be reading an article written for a lay audience. In these days of environmental concern and consumerism, they may be reading as a prelude to action. Certainly, their main reason for reading is practical. They are much more concerned with what things do than how they work. Their interest is personal. What impact will this new development have on them? They are more interested in the fact that widespread computer networks may invade their privacy than the fact that computers work on a binary number system. They are more interested in the efficiency and cost of automobile antipollution devices than the theory of such devices.

Many laymen are more attuned to fiction and television than to scientific exposition. They like drama. For this reason, narrative is often an effective device when writing for them. Anecdotes and incidents can be a useful way to illustrate what something is and what it does.

Beyond these generalizations, laymen present a bewildering complexity of interests, skills, educational levels, and prejudices. How then can we define exactly how to write for them? The truth is that we cannot—completely. But we can make some broad statements about their needs, interests, likes, and dislikes that—to paraphrase Lincoln—apply to all of the laymen some of the time, some of the laymen all of the time, but *not* all of the laymen all of the time.

To simplify matters a little, let's get a picture in our minds of typical laymen. (For our present purposes, we are excluding specialists with advanced degrees, like the biologist mentioned above, from the "typical laymen" category—though, as noted, specialists may be defined as laymen when outside their particular fields.) They are fairly bright and interested in science and technology. They have at least a high school education. They read well and have a smattering of mathematics and science but are a little vague about both subjects. How do we treat them? What approaches are best when we write for them?

Background

To begin with, laymen need to be given background material in the subject. We must assume they know little or nothing about the specialty. An Atomic Energy Commission booklet entitled *Atomic Energy in Use* is a simplified explanation for a lay audience of how a nuclear reactor works. Chapter 1 gives a history of uranium, and Chapter 2 explains nuclear radiation. At the start of Chapter 2, the writer does a good job of presenting background information for nonspecialists:

> Light is radiation that we can see. Heat is radiation that we can feel. Radio and television waves and X-rays are electromagnetic waves of radiation that we can neither see nor feel, but with whose usefulness we are well acquainted.
>
> Now we are hearing more and more about another kind of radiation as a result of man's continuing scientific and engineering achievements.
>
> This is nuclear radiation.
>
> Nuclear radiation consists of a stream of fast-flying particles or waves originating in and coming from the nucleus, or heart, of an atom. It is a form of energy we have come to call atomic, or nuclear, energy.[3]

The author begins with the familiar—light radiation and heat radiation. There is no assumption that the readers know all about light and heat radiation; that they could, for example, construct mathematical models of such phenomena in the way a physicist could. Nor does the writer attempt to give the readers such complicated theory. The writer knows that the average reader dimly understands that light, heat, and electromagnetic waves are forms of energy that somehow travel from point A to point B and can therefore be *used* in various practical ways. Nuclear radiation is a similar form of radiation and that's the only point the writer wants to convey here.

To make this point the writer has relied upon analogy—comparing the unfamiliar to the familiar. There is no better device to help the lay reader. In the everyday world around us there are countless things—such as light bulbs, radios, garden hoses, faucets, windows, mirrors, trees, tennis rackets, baseballs, hot-air registers, clay, loam, granite, ocean waves—known and somewhat understood by everyone that the writer can use to explain about every law of science. It's a question of using one's imagination and knowing how to talk in lay terms without being condescending.

---

[3] U.S., Atomic Energy Commission, *Atomic Energy in Use* (Washington, D.C., 1967), p. 11.

Often, background information is used to stress the importance of the subject at hand. A *Harvard Medical School Health Letter* on alcoholism begins this way:

> Consider that half of all deaths in automobile accidents, half of all homicides, and a fourth of all suicides are related to alcohol abuse; that persons with a "drinking problem" are seven times more likely to be separated or divorced than those in the general population; that the total cost of alcohol abuse in this country may exceed 44 *billion* dollars; that an alcoholic's life span is shortened (on average) by 10–12 years; and that at least ten million persons in this country abuse alcohol. No wonder some have labeled alcoholism the most devastating socio-medical problem faced by human society short of war and malnutrition.[4]

Give your readers, then, a grounding in your subject. When possible use familiar things as points of comparison. If your background also wakens your readers' concerns and interest, so much the better.

Definitions

Nonspecialists need specialized words and terms defined. There are at least two reasons why you shouldn't force readers to go to the dictionary while reading your work.

First, you are the host. You have invited your readers to come to you. You owe them every courtesy, and defining difficult terms is a courtesy. If you force them to the dictionary every fourth line, their interest will soon flag.

Second, if you give the definition, you can limit or expand the term in the way that is most useful to you. In the second half of Chapter 2 of *Atomic Energy in Use* the author stresses radiation control and radiation safety. His definition of an alpha particle reflects that emphasis: "Alpha particles are comparatively heavy particles given off by the nuclei of heavy radioactive materials such as uranium, thorium, and radium. They can travel about an inch in air and can readily be stopped by the skin or by a thin sheet of paper."[5]

If the writer had left his readers to look up the word themselves, the desired emphasis would have been lost. Compare this definition from *Webster's Seventh New Collegiate Dictionary* and you'll see the difference: "Alpha particle: a positively charged nuclear particle identical with the nucleus of a helium atom that consists of 2 protons

---

[4]*The Harvard Medical School Health Letter*, July 1978, p. 1. Reprinted by permission of the publisher.
[5]Atomic Energy Commission, *Atomic Energy in Use*, pp. 11–12.

and 2 neutrons and is ejected at high speeds in certain radioactive transformations." Nothing is said about the relative ease with which alpha particles can be blocked, a fact important to the writer's later emphasis on radiation control.

Depending upon the needs of both writer and audience, terms can be defined either briefly, usually by the substitution of a more familiar term, or at length. The use of brief definitions is seen in this segment of *The Harvard Medical School Health Letter* on alcoholism:

> Alcoholics have a much higher incidence of *peptic ulcers* and *pancreatis* (inflammation of the pancreas) than non-alcoholics. In addition, many suffer from repeated episodes of nausea, vomiting, and abdominal distress—most often due to superficial *gastritis* (inflammation of the lining of the stomach).[6]

Here, the lay definition substitutes simpler, more familiar language for the physician's technical terms. In the same article is this more extended definition:

> If a group of experts attempts a definition of "alcoholism," as many definitions as experts usually emerge, which, of course, points out the complexity of this particular "-ism." Some argue the relative merits of biological (it is a "disease") versus social-psychological (it is a "behavior disorder") theories, but most settle for descriptive definitions—such as the one from the Rutgers University Center of Alcohol Studies: "An alcoholic is one who is unable consistently to choose whether he shall drink or not, and who, if he drinks, is unable consistently to choose whether he shall stop or not." Ultimately, all will agree to descriptive definitions which point out that for the person with a serious drinking problem, tremendous disruption occurs in terms of health, interpersonal relationships, and the basic activities of life—eating, sleeping, and working.[7]

Notice how this definition is used to stress one of the major points of the article—that alcoholism, however caused, disrupts the lives of alcoholics and the lives of those around them.

Be careful not to distort the true meaning of terms when simplifying for a lay audience. One researcher, for example, felt his work was distorted by this lead in a newspaper story:

> A research group reported Friday that marijuana causes chimpanzees to overestimate the passage of time, and a single dose can keep them befuddled for up to three days.

[6]*The Harvard Medical School Health Letter*, pp. 1–2.
[7]Ibid., p. 1.

The researcher commented:

> The term "befuddle" was not employed in our scientific report, and the statement in the news article "and a single dose can keep them befuddled for up to three days" is erroneous and misleading. Three days were required to recover normal baseline performance following administration of high doses.[8]

Scientists choose words very precisely and for good reason. While their findings must be interpreted for lay readers, to distort or to sensationalize their work is a disservice both to them and to the reader. Define, then, for clarity and understanding and to aid your own exposition, but do it with care. Refer also to the section on definition on pages 99–102.

Simplicity

There are several ways to keep an article simple for laymen. Two of them we have already discussed: give needed background and define those specialized terms that you must use. Most often, avoid specialized words for which you can find simple substitutes. Experts can read certain meanings into the word *homeostatis* and you should use it for them, but for a lay audience *stable state* or *equilibrium* will serve as well. But another caution here: most people like to enrich their vocabularies. So, don't avoid technical terms altogether. Just don't put them together in incomprehensible strings with a "reader-be-damned" attitude.

Some scientific specialties are loaded with mathematics. Others, such as biochemistry, are full of formulas, complicated charts, and diagrams incomprehensible to laymen. Mathematics, formulas, and diagrams are useful shorthand expressions for experts. Through them, experts find a precision impossible to obtain in any other way. But what experts sometimes forget are the years spent learning how to handle such precise tools. The average person, lacking those years of training, rarely *can* handle them. When you write for laymen, you must force yourself to express your ideas in plain language. It can be done. Here is an example taken from an article written primarily for experts, but which the author had reason to believe might be read by technicians and even laymen:

> Now let's examine a representative nuclear fission reaction:
> Uranium-235 plus a neutron,

---

[8] Michael Ryan and James W. Tankard, Jr., "Problem Areas in Science News Writing," *Journal of Technical Writing and Communication* 4, no. 3 (1974): 225–235.

reacts to give:
radioactive fission products,
plus two or three neutrons,
plus ENERGY
$$_{92}U^{235} + _0n^1 \rightarrow _{42}Mo^{95} + _{57}La^{139} + _2 _0n^1 + E$$
in which molybdenum and lanthanum are the final stable products of the reaction.

In shorthand language, this equation tells us that a special type of uranium atom called uranium-235 (U^{235}), when struck by a neutrally charged particle called a neutron, will undergo a reaction called nuclear fission. This nuclear fission reaction is the basic process by which nuclear energy is generated. The heavy U^{235} fuel atom splits into lighter atoms of entirely different chemical elements, which are usually highly radioactive, and are designated fission products.

The fissioning of the atom also releases from within its nucleus two or three subatomic neutron particles which, as mentioned, are necessary to maintain the reaction by interaction with another U^{235} atom. In addition, the reaction releases tremendous amounts of energy.[9]

This excerpt is a first-rate example of how you can express an idea in a manner that will be understood by several different audiences. Most nonspecialists would be satisfied with the first plain-language explanation given. The expert needs and wants only the mathematical expression. More curious nonspecialists would be pleased with the further plain-language explanation that follows the mathematics. As this example shows, it usually takes more work to be simple, but why takes shortcuts that defeat your purpose?

A final way to achieve simplicity is in the way you handle language. By this we mean keep your sentences and paragraphs short and rely on the basic subject-verb-object order in your sentences. If you examine the quotation from *Atomic Energy in Use* found on page 71, you will find:

1. The sentences average 15 words each.
2. The paragraphs average 26 words each.
3. Eighty-six percent of the sentences start with the subject.

The rest of the chapter pretty much follows the same style. Admittedly, this example is rather extreme in its simplicity. But it does illustrate that complicated subjects can be discussed with simple syntax. We say more about this matter of simple style in Chapter 8.

---

[9]David F. Cope, "Nuclear Power: A Basic Briefing," *Mechanical Engineering* 89, no. 6 (1967): 48–53.

Human Interest

Most of us, at any educational level, have an interest in other human beings and in human personalities. Most writers for lay audiences recognize this interest and use it to gain acceptability for their subject matter. For example, an article in *Time* about inflation will give us statistical information about the rise in the cost of living. But the writer of the article knows that many of us do not relate very well to bare, abstract statistics, even when they affect us directly, as do the facts about inflation. Therefore, the *Time* writer will also usually introduce people into the story. Perhaps Bill and Mary Gould, a "typical" couple living in Houston, Texas, will be cited. We'll learn what effect inflation has had on their food bills, clothing bills, recreation, and so forth. We are interested in learning about what happens to real people, and by extension can then better understand the impact of inflation on ourselves.

Even fairly complex scientific material for a high-level audience can benefit by the introduction of human interest. In a *Saturday Review* article on new fuel sources, the author tells about some complex technology, but when possible introduces the human beings behind the technology, as in this excerpt:

> Among the newer systems are "hydrocarbon miscible flooding," "carbon dioxide flooding," and "miscellar polymer flooding." But while a number of systems have succeeded in varying degrees of getting at the oil, none has been able to do so with a sufficient net-energy gain or at an economic price.
>
> Enter two Texans, Frank DeFalco and Charles McCoy, and a Hungarian-born scientist, George Merkl. In 1974 DeFalco and McCoy created the Molecular Energy Research Company (MERCO) for the purpose of finding a technologically efficient and economically advantageous method of capture.
>
> Merkl, 48, has credentials in theoretical and solid-state physics, nuclear chemistry, petroleum geology, and electronics. The contribution he is making to the solution of America's energy problem is reminiscent of the role of other foreign-born scientists—Einstein, Szilard, Fermi, and Wigner among them—in helping the United States to use science as a decisive factor in the war against Nazi Germany. Merkl came to the United States in 1957 after advanced studies in Vienna. He worked in several American research laboratories and immersed himself in research on catalysts, agents that speed up chemical reactions.
>
> Merkl teamed up with DeFalco and McCoy in 1976. Like many others, he was obsessed with the possibility that a way could be devised to get at the estimated 350 billion barrels of petroleum locked into America's natural underground vaults. Almost intuitively, he turned to inorganic polymers—large-moleculed hard substances, like silicon or

graphite, with a very high melting point. The process involved a polarized hydrogen bond. The bonding releases atomic hydrogen, which cracks the light end of the oil, creating natural gas. This gas forces the trapped oil out of the rock and causes it to form a pool or reservoir, making it accessible to conventional collection methods.[10]

The language is direct, written predominantly in subject-verb order. *Saturday Review* is written for a high-level lay audience. Therefore, the writer can use a wide-ranging vocabulary. But really technical terms such as *catalysts* and *inorganic polymers* are defined. Most important for our purposes here, the writer whets our interest by telling us something about George Merkl. We learn about his education and experience and that he is "obsessed with the possibility" of getting at America's trapped petroleum. We are more interested in the technology because we see it as a human activity.

Illustrations

You can use illustrations in many ways in a report designed for a lay audience. You can use bar charts in place of equations to explain mathematical concepts or in place of formulas for chemical concepts. Or you can combine tables and pictographs that establish facts, formulas, or definitions quickly, clearly, and in a way that interests readers. The two pictographs from *Atomic Energy in Use* (Figure 4-1) illustrate this function. You can supplement such simple pictographs with text, or simply refer to them and let them carry the weight of the explanation.

Figure 4-2 was used by the Jet Propulsion Laboratory to illustrate the paths of two unmanned space missions to Jupiter and then around the poles of the sun. It graphically illustrates what is difficult to express in words. The illustration is both clear and a bit dramatic—note the presentation of flares erupting from the surface of the sun.

We have a good deal more to say about graphics in Chapter 11, "Graphical Elements."

The Hardest Audience

Laymen present the hardest audience to write for because their needs and capabilities are so difficult to pin down and define. We can say, in general, that in writing for them we should provide ample

---

[10]Norman Cousins, "Growing Your Own Fuel," *Saturday Review*, 30 September 1978, pp. 23–28.

Figure 4-1   Sample Pictographs

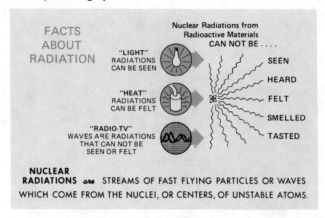

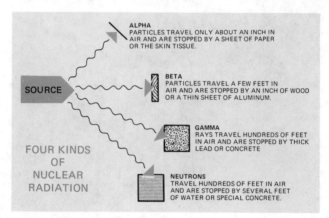

From U.S., Atomic Energy Commission, *Atomic Energy in Use* (Washington, D.C., 1967).

background (without mathematics or complicated formulas), definitions, photographs, and simple charts. Remember, they are reading mainly for interest and their interest is mainly practical: how will this scientific development affect our lives? So your cardinal rule in writing for laymen should be to keep things uncomplicated, interesting, human, practical, and personal; and, if possible, to provide a touch of drama.

Examples of good writing for laymen are easy to find. Magazines such as *Time, Newsweek,* and *National Geographic* regularly print articles on scientific subjects for their readers. Reading and analyzing such articles is the best school we know for learning how to write for a lay audience. And good technical books that can be understood by

Figure 4-2 Paths of Two Unmanned Space Missions Around Jupiter and the Sun

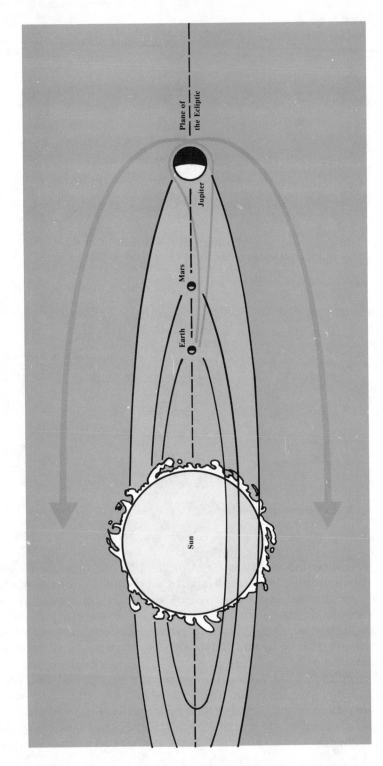

From U.S., National Aeronautics and Space Administration, *Solar Polar* (Washington, D.C., 1978).

laymen, such as Joseph Weizenbaum's *Computer Power and Human Reason* or Kenneth Cooper's *Aerobics*, are worth reading as much for their style as for their content.

Executives

Much of what we have said about laymen also applies to executives. You cannot assume that executives possess very much knowledge in the field you are writing about. While most executives have college degrees and many have technical experience, they represent many disciplines—not necessarily including the one you are writing about. Some may have training in management, accounting, a social science, or the humanities, but little or no technical background.

Even more than laymen, executives' chief concerns are with practical matters—what things do, as opposed to how they work. They want to know how technological development will affect the development of their companies, and they probably could use more technical background than laymen. James W. Souther, who has made an extensive study of executives and their needs, suggests that writing for them approximate the level found in *Scientific American*.[11] Important technical terms should be used and defined, but shop jargon avoided entirely. Executives are busy people; don't force them to use a dictionary any more than you would laymen.

When writing for executives, write in plain language, using sentences averaging about 20 words. Avoid mathematics. Use simple illustrations of the types suitable for laymen: bar graphs, pie charts, and pictographs.

But while executives resemble lay readers in many ways, there is a significant difference. Most of the time laymen are reading primarily for interest. While what they read obviously influences their lives and their decisions, they seldom have to act directly upon it. Executives, however, must often make decisions based upon what they read. People and profits figure largely in executive decisions. How much money will a new technological development cost? What new markets will it open? How much profit will a new technological development provide? Does the development call for restaffing a plant? Will new people have to be hired or old people retrained?

Executives must also consider the social, economic, and environmental effects of their decisions upon the community at large.

[11] James W. Souther, "What Management Wants in the Technical Report," *Journal of Engineering Education* 52, no. 8 (1962): 498–503.

Aesthetics, public health and safety, and conservation are key decision-affecting factors today, and few executives would consider a report complete that did not deal with them.

Executive Needs

What are executives interested in? What questions do they want you to answer in a report written for them?

They want to know how a new process or piece of equipment can be used. What new markets will it open up? What will it cost, and why is the cost justified? What are the alternatives?

Why did you choose the new equipment over the other alternatives? Give some information about the also-rans. Convince the executive that you have explored the problem thoroughly. For all the alternatives include comments on cost, size of the project, time to completion, future costs in upkeep and replacement, and the effects on productivity, efficiency, and profits. Consider such aspects as new staffing, competition, experimental results, problems likely to arise. What are the risks involved? What environmental impact will this new development have? Figure 4-3, "What Managers Want to Know," taken from "What to Report" printed in the *Westinghouse Engineer*,[12] illustrates what information Westinghouse executives feel they need in a report.

Experts, in particular, often find writing for executives a difficult task. Experts are often most interested in methodology and theory. Executives are more interested in function. Excerpts from a salmon study done for the Alaskan Fish and Game Department illustrate the frame of mind the expert researcher should have while writing for the executive. In the introduction, the researcher poses the questions that will be answered in the report. They are the questions an executive would ask:

> Why have they gone? Can the runs be restored to any significant degree? Is it reasonable to base a large industry on the harvest cycle of a wild resource? What should be done? What should be done now?[13]

---

[12] Richard W. Dodge, "What to Report," *Westinghouse Engineer* 22, no. 4–5 (1962): 108–111. The information in "What to Report" is based upon a study made at Westinghouse in 1959–60 by Professor James W. Souther of the University of Washington. All material reprinted from "What to Report" is done so through the courtesy of the Westinghouse Electric Corporation.

[13] Reported by Mary B. Coney, "The Use of the Reader in Technical Writing," *Journal of Technical Writing and Communication* 8, no. 2 (1978): 97–106.

Figure 4-3   What Managers Want to Know

Problems
What is it?
Why undertaken?
Magnitude and importance?
What is being done? By whom?
Approaches used?
Thorough and complete?
Suggested solution? Best? Consider others?
What now?
Who does it?
Time factors?

New Projects and Products
Potential?
Risks?
Scope of application?
Commercial implications?
Competition?
Importance to Company?
More work to be done? Any problems?
Required manpower, facilities and equipment?
Relative importance to other projects or products?
Life of project or product line?
Effect on Westinghouse technical position?
Priorities required?
Proposed schedule?
Target date?

Tests and Experiments
What tested or investigated?
Why? How?
What did it show?
Better ways?
Conclusions? Recommendations?
Implications to Company?

Materials and Processes
Properties, characteristics, capabilities? Limitations?
Use requirements and environment?
Areas and scope of application?
Cost factors?
Availablity and sources?
What else will do it?
Problems in using?
Significance of application to Company?

Field Troubles and Special Design Problems
Specific equipment involved?
What trouble developed? Any trouble history?
How much involved?
Responsibility? Others? Westinghouse?
What is needed?
Special requirements and environment?
Who does it? Time factors?
Most practical solution? Recommended action?
Suggested product design changes?

The stated purpose of the report further reassures the executive that the researcher is on the right track:

> Our approach has been first to gather and understand as much relevant information as could reasonably be found; and then to organize, interpret, and project toward the goal of defining a conceptual framework for successful actions by the State of Alaska through its Department of Fish and Game.[14]

Here it is obvious that scientific findings are going to be wedded to executive needs. Function—"successful action"—lies at the heart of the report.

Be honest. Remember that if your ideas are bought, they are *your* ideas. Your reputation will stand or fall on their success. Therefore, don't overstate your case. Qualify your statements where necessary.

Give your conclusions and recommendations clearly. In writing any report for the executive, remember that you must interpret your material and present its implications, not merely give the facts. Souther points out that "the manager seldom uses the detail, though he often wants it available. It is the *professional judgment* of the writer that the manager wants to tap."[15] The researcher who amasses huge

---

[14] Ibid.
[15] Souther, "What Management Wants."

amounts of detail but neglects to state the implications, conclusions, and recommendations that follow from the facts has failed to do the complete job. Many technical people who have aspired to executive rank have not made it because they failed to grasp this simple fact.

Organize your report around executive reading habits. The author of "What to Report," Richard W. Dodge, writes that "Every [Westinghouse] manager interviewed said he read the *summary* or abstract; a bare majority said they read the *introduction* and *background* sections as well as the *conclusions* and *recommendations;* only a few managers read the *body* of the report or the *appendix* material." Figure 4-4 from "What to Report" illustrates how managers read reports.

This research done at Westinghouse seems to indicate that if you want executives to read your conclusions and recommendations you should repeat them in the abstract and again in a separate section at the *front* of the report. And in modern practice that is where they are usually found.

If you feel you must report a large amount of technical data, put it in an appendix where executives can read it if they want to (or assign experts to read it).

By way of a conclusion, here is a checklist from "What to Report" that describes an effective approach to writing reports for executives:

Figure 4-4   How Managers Read Reports

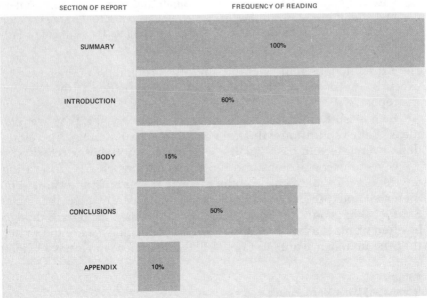

- State the purpose of this exercise.
- Give some background to set the stage.
- Explain the alternatives considered.
- Isolate and support the alternatives selected.
- Describe the next action to be taken or being recommended.

Good articles specifically written for executives are easy to find and worth examining; two excellent sources are *Fortune* and *Dun's Review and Modern Industry*, magazines aimed at executives in top and middle management. And our chapters on Proposals, Progress Reports, and Feasibility Reports deal with specialized executive reports.

Experts

To many laymen, writing meant for experts seems incredibly dull. But to the experts themselves, nothing could be more exciting. The well-written report in their specialty offers them solid facts to chew on—and experts are fascinated by facts. The report also draws inferences from those facts that experts can hail as new and valid or dispute as examples of faulty reasoning. An expert is akin to the symphony conductor who, while reading a musical score, can *hear it* and judge its potential. Laymen, like the musical audience, must wait for the interpretation by the conductor and the full orchestra before they can appreciate the same score.

Who are experts? For our purposes we will define them as senior scientists or engineers with either an M.S. or a Ph.D. in their fields or a B.S. and years of experience. They may be college professors, industrial researchers, or engineers who design and build. They know their fields intimately. When they read in their own fields, they seldom look for background information. You may offer background information that you feel is particularly pertinent to the narrow subject at hand—such as a review of the experiments leading up to the one you have conducted. But this background will not be presented in simple terms, as it would be in an article for nonexperts. In some cases, rather than give background, you can refer experts to other sources—books or articles—where they can pursue background if they need to.

Experts are very concerned with how and why things work. They want to see the theoretical calculations and the results of basic research. They want your observations, your facts—what you have seen, what you have measured. In reporting such things be as complete as time, space, and human patience allow. Facts that do not seem immediately important, that may even seem trivial, may assume great importance at a later time. Experts will suspect any conclusions drawn if they feel the facts are not complete. The technological explosion that boggles the mind with its enormity is built upon a foundation of many people cooperatively working upon an accretion of facts, most of which seem trivial standing alone.

When writing for experts you may use any shorthand methods such as abbreviations, mathematical equations, chemical formulas, and scientific terms that you are sure your audience can comprehend. Complicated formulas and equations needed to support the conclusions, but not essential for understanding them, are often not placed in the body of the report. Modern practice usually places them in an appendix.

Another shorthand device found in expert reports is the use of tables and graphs. Tables provide an excellent way to lift classifications and groups of closely related facts out of the text and display them clearly. The graphs used most often are line graphs. They best portray the relationship between variables. Scientific concepts can be expressed in drawings such as the one in Figure 4-5. Maps and photographs of unusual equipment also aid the expert reader significantly.

You do not normally have to define terms unless you have used them in some new or unusual way. However, a word of caution here. Abbreviations and symbols used should be the standard ones for a given field, as defined by the authorities in that field. When they are not they should be defined as in this example:

From Einstein's mass-energy equation, one can write the relations:

E (Btu) = Δm (lb) × 3.9 × 10

Δm being the loss of mass.[16]

The writer does not define E, *Btu*, and *lb* because these are standard symbols and abbreviations for energy, British thermal units, and pounds. He does define Δm because it is not standard. Writers who do not define nonstandard symbols cannot expect readers to know what they are. In fact, in a few years, returning to their own reports, *they* may not know what the symbols mean. Symbols and abbreviations may be defined as they are used, or, if there are many of them, they can be defined in a glossary at the front of the report.

The fact that most professional and scientific fields have grown increasingly specialized leads to another caution. Experts in the same

---

[16] Cope, "Nuclear Power: A Basic Briefing."

Figure 4-5   Imaging Radar Geometry

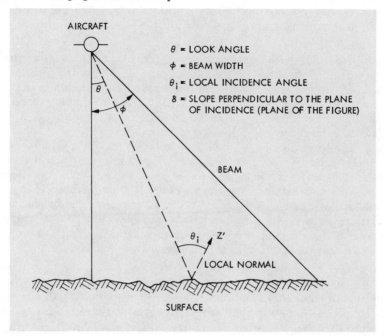

From M. Daily et al., *Application of Multispectral Radar and LANDSAT Imagery to Geologic Mapping in Death Valley* (Pasadena, Calif.: Jet Propulsion Laboratory, California Institute of Technology, 1978).

field but who have specialized in different aspects of the field may find they have difficulty in understanding one another. Not too long ago, we sat in on a meeting of agronomists being addressed by a fellow agronomist—one who specializes in plant genetics. The question-and-answer period that followed the talk made it painfully obvious that few of the agronomists had understood the speaker. He had not taken into account how far removed they were from his specialization. In this and similar cases, some of the same techniques recommended for the lay audience, such as giving background and defining terms, would be appropriate.

Expert reports are, of course, not merely calculations and facts. Experts as well as executives want your inferences and conclusions. When you do draw inferences from your facts and observations be sure to make no unwarranted leaps. Stay within the bounds of the scientific method. In presenting your conclusions be careful that your language shows where you are certain and where you are in doubt. Because of this need for scientific honesty and caution, most expert discussions and conclusions do contain qualifications as well as positive statements. The following excerpt is an example of the balance between fairly positive statements and qualified statements that is typical of scientific discussion. (We have italicized the qualifiers.)

> Use of a LANDSAT intensity image provided additional information on varnish distribution on the alluvial fans, thus permitting discrimination of two fan gravel units that were not distinguishable in the radar data. The use of all four bands of LANDSAT *should* add the capability to distinguish compositions. We thus have seven dimensions of imagery available for computer processing techniques.
>
> Surely, particulate alluvial and colluvial regoliths, evaporites, and eolian deposits represent widespread geologic units on terrestrial planetary surfaces. On cloud covered planets (e.g., Venus), multiconfiguration radar data *appears* capable of discriminating many of the surficial geologic units likely to exist. On other planets, additional use of optical sensors will permit a more thorough compositional mapping. LANDSAT/radar combinations offer considerable *promise* for making high-quality geologic maps of lithologic-textural units.[17]

Newswriters, when writing stories about scientific achievements, sometimes forget this need for caution—to the dismay of the scientists involved. A newspaper story concerning research on the skin ailment psoriasis carried this headline: "Psoriasis Cure Breakthrough Seen." The lead of the story announced, "Scientists Wednesday announced a breakthrough in treating psoriasis, the skin disease which causes misery for about 6 million Americans." The scientist involved criticized the story, saying that nowhere in the report presented by the scientists was the word "breakthrough" used. He concluded by saying, "The last sentence of our writeup said cure of psoriasis is probably 50 years away. Yet the title of this article you sent says 'Psoriasis Cure Breakthrough Seen.' All I can say is, [censored]!"[18]

Experts know that certainty in science is a hard-won achievement. They are content with probabilities until thorough experimentation and observation remove all reasonable doubt. Their style reflects their basic caution and honesty.

Our Chapter 17, "Physical Research Reports" describes the most predominant type of expert report. You can find good samples of expert reports in the journals that cover your particular discipline. For examples of expert reports that aim at a somewhat more

---

[17] M. Daily et al., *Application of Multispectral Radar and LANDSAT Imagery to Geologic Mapping in Death Valley* (Pasadena, Calif.: Jet Propulsion Laboratory, California Institute of Technology, 1978), p. 44.
[18] Ryan and Tankard, "Problem Areas in Science News Writing."

generalized but still expert scientific audience, we recommend to you *Science,* the journal of The American Association for the Advancement of Science.

Technicians

Technicians are at the heart of any operation. They are the people who finally bring the scientist's imaginative research and the engineer's calculations and drawings to life. They build equipment, and after it is built they maintain and use it. They are intensely practical people, perhaps with years of experience in a special area. They are the people who can say, "You know, if we used a wing nut here, instead of a hexagonal, the operator would have a much easier job getting that plate on and off," and they'll be right. Technicians are well worth listening to, and certainly you should write well for them.

Technicians' educational levels vary. Most typically they will range anywhere from a high-school graduate to a junior engineer with a B.S. degree. They may have been trained in one of the many vocational schools. The high-school graduate will probably have a great deal of on-the-job training and experience. The junior engineer may be better trained in theory but have less practical experience. Technicians have limitations. They may not be able to follow complicated mathematics, and they'll grow restive with too much theory. Also, you will want to keep your sentence length down when writing for technicians. Unless you are sure that your technicians are college-trained, give them sentences that they can readily handle— about 17 words or less on the average.

You can assume a good deal of knowledge on the part of this audience, but not as much as with the expert audience. Usually, with this audience, you will need to supply some background information and some definitions. An article from *Bell Laboratories Record,* a magazine for electrical technicians and junior engineers, begins this way:

> Waveguide systems are often loosely called "plumbing." The name implies a network of empty pipes where electrical energy flows unimpeded. Actually a waveguide is a precisely-designed, electrically-tuned structure for propagating electromagnetic waves. It transmits certain determinable frequencies well, does not transmit some frequencies at all, and transmits others only with large losses.[19]

---

[19] J. J. Degan, "Microwave Resonance Isolators," *Bell Laboratories Record,* April 1966, p. 123. Copyright © 1966 by Bell Telephone Laboratories, Incorporated. Reprinted by permission.

The author of this article has assumed general knowledge in the field on the part of his audience. For example, he does not define "electrically-tuned structure" or "electromagnetic waves." His audience will recognize these terms. But he begins his article with background information on the subject: waveguide systems. He uses analogy—"plumbing"—and later describes in great detail what waveguides look like. Generally speaking, the technical audience cares more about the practical application of a theory than about the theory itself.

The author of the *Bell Laboratories Record* article on waveguides uses no equations whatsoever. Rather, he presents mathematical information in visual form or in simple tables. We took the two examples in Figure 4-6 from his article.

But the technician audience is educated and will want some theory, particularly in the background sections. Figure 4-7 shows two pages of a John Deere manual concerning the operation of the equipment used in hay and forage harvesting. Most of the manual concerns practical advice about such things as mower preparation—for example, setting and adjusting springs and blades. But, as the reproduced pages show, the writers took the time to explain some theory also. The technician reading these pages now knows, for example, why a smooth cutting section on a mower cutterbar is superior in some applications to a serrated one and vice versa. Note also the excellent use of graphics.

Keep theory explanations simple and fairly nonmathematical, however. While a junior engineer normally can handle simple calculus, the skilled technician may not be able to. For both groups, the wisest choice is to put most of the math needed in graph form. If you feel equations must be given the technical audience, definitely put them in an appendix. The equations should supplement the graphical presentations. They should provide further depth for the reader who can use it. They should not be necessary for understanding the basic text.

Depend upon analogy when writing for the technical audience. Analogy bridges the gap between a reader's general information and the particular object or theory you are trying to explain. Once again, we turn to an excerpt from the *Bell Laboratories Record* article on waveguides:

Every electron orbiting about an atomic nucleus gives rise to magnetic fields. In some materials the field comes largely from the motion of the electron around the nucleus, but in ferromagnetic materials it depends more on the spin associated with the electron itself (see the drawing above). The spin creates a small magnetic movement that is precisely

Figure 4-6   Sample Table and Diagram

FREQ (Gc)	TYPE	TYPICAL VALUES				
		ISO (db)	INS LOSS (db)	RETURN LOSS (db)	BAND-WIDTH (%)	FIGURE OF MERIT
4	H-Plane	32	0.15	33	13	210
5	H-Plane	25	0.14	30	10	180
6	E-Plane	60	0.6	32	8	100
6	E-Plane	30	0.3	32	8	100
6	E-Plane	60	0.6	32	8	100
11	E-Plane	70	0.6	27	10	120
11	E-Plane	30	0.3	32	10	100
11	E-Plane	70	0.6	27	10	120
12	H-Plane	15	0.3	25	10	50

Resonance isolators developed at Bell Laboratories for Bell System use include designs for the 4-, 6-, and 11-gigacycle common carrier bands and other designs for special applications.

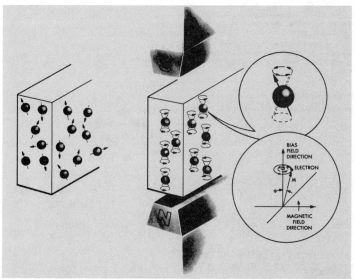

Operation of a resonance insulator hinges on the absorption of energy by electrons in a magnetic field. The left-hand drawing portrays spinning electrons in a piece of magnetic material. The arrows represent the spin axes and the magnetic moments that coincide with them, randomly oriented in the absence of any magnetizing influence. On the right, under the influence of an external magnetic field, the moments (spin axes) tend to align with the field. In lining up they precess about the magnetic lines of the external field as indicated in the upper detail. An rf field polarized in the same direction as the electrons precess can keep them precessing (lower detail).

From Degan, "Microwave Resonance Isolaters." Reproduced by permission.

Figure 4-7   Theory Section for Technicians

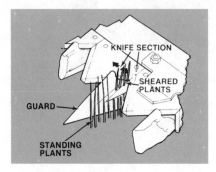

Fig. 8—Shearing Action of Mower Cutterbar

Fig. 9—Cutterbar Mower Cutting Hay

CUTTERBARS

The most commonly used hay-cutting device is a cutterbar with reciprocating knife that shears plant stems (Fig. 9). Cutterbars are used not only on mowers, but also on windrowers, forage harvesters, and grain combines. All cutterbars have the same basic components, but design variations are available to deal with specific problems. They are also called sicklebars, or sicklebar mowers.

Mower size is commonly designated by specifying the cutterbar length. Lengths commonly available are 5, 6, 7, 8, and 9 feet; however, most current mower cutterbars are 7 or 9 feet long.

CUTTERBAR COMPONENTS

The cutterbar is the heart of a mower and a complete understanding of its components (Fig. 10) and functions is essential. Basic cutterbar components are:

- Knife assembly
- Guards and ledger plates
- The bar
- Inner shoe
- Outer shoe
- Knife clips
- Wear plates
- Grass board and stick
- Yoke (cutterbar hinge)

Fig. 10—Basic Cutterbar Components

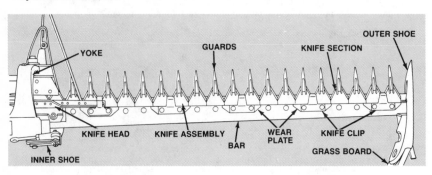

From *Fundamentals of Machine Operation—Hay and Forage Harvesting* (Moline, Ill.: Deere and Company, 1976). Copyright © 1976 Deere and Company, Moline, Illinois. Reprinted by permission.

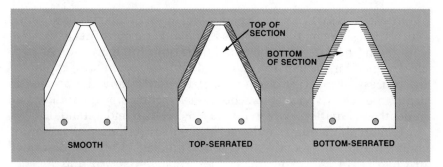

Fig. 11—Common Types of Knife Sections

KNIFE ASSEMBLY

Cutting is done between the knife assembly and the guards, so these components are the most important part of the cutterbar and will be discussed first. Proper knife and guard adjustment, operation, and maintenance are necessary for efficient mower operation. The knife is also known as a sickle.

The *knife head,* at the driven end of the knife, connects the knife and knife drive. The knife head is riveted securely to the knife back to permit easy replacement if needed.

The *knife back* is a flat steel bar to which all knife sections are securely riveted. Knife sections may be replaced if broken or badly worn.

A *knife section* is an individual cutting unit which must be matched to specific crop conditions (Fig. 11). Four types of sections are:

- Smooth
- Top-serrated
- Bottom-serrated
- Armored or hard-surfaced

Smooth sections are used in fine-stemmed crops, particularly where juices released during cutting tend to produce a buildup of residue. Chromed sections resist buildup of plant residue better than plain sections.

Top-serrated sections are used in coarse crops such as alfalfa, clover, timothy, straw, and other stiff-stemmed crops. The serrations have a crop-holding tendency which helps prevent stems from being pushed forward by the section movement instead of being cut. These sections retain their cutting ability without sharpening. Chrome surfacing provides a slick face; it resists buildup of plant-juice residue.

Bottom-serrated sections are used in somewhat the same conditions as are top-serrated sections. Be-

cause serrations are on the bottom, sections may be sharpened when edges become dull.

Armored sections (also called hard-surfaced) are coated on the underside with tungsten carbide: These specially-hardened sections may be used in all conditions suitable for serrated sections, and in most conditions suitable for smooth sections. Hard-surfaced sections are more susceptible to rock breakage than normal sections.

GUARDS

Mower guards are usually spaced on 3-inch centers. The most common knife stroke is also approximately 3 inches. However, the stroke varies on different mowers from 2½ to 3¾ inches—also with 3-inch guard spacing.

Guards have three functions. They protect the knife from solid objects (Fig. 12), hold the stationary *ledger plates* for efficient shearing of the crop, and divide the plants and guide them into the knife for easy cutting.

Fig. 12—Guards Protect Knife Sections

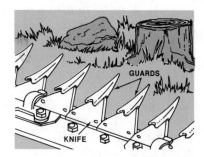

aligned with the axis of spin (this can be visualized as similar to the alignment of the earth's magnetic field between the north and south poles.)[20]

In addition to the above explanation, the article contains a simple illustration showing the spin axes and magnetic moments associated with them (the second of the two illustrations in Figure 4-6). An article designed for the expert audience doubtless would have explained the process in a series of equations. To understand the process in depth, the scientist needs the equations. Technicians do not need an in-depth explanation, nor would they be likely to understand it.

Technical manuals and operating instructions are the most common applications of writing for technicians. Both are described in Chapter 14. Also, the magazines *Popular Science* and *Popular Mechanics* offer fine examples of writing for the technician.

The Combined Audience

Certainly the most difficult audience for a technical writer is the combined audience—perhaps composed of executives, experts, and technicians. Yet in industry such an audience is a common one. How do you avoid bombarding the manager with detail and at the same time satisfy the needs of the experts?

The best approach to the problem is to consider how your readers will use the report. It is probable that the executive will use the report to make a decision. The experts, perhaps, will use it to guide the executive in the decision-making process. Both experts and technicians may use information in the report to implement the decision.

Suppose, for example, you were writing a report for your company to a television executive in another company proposing the adoption of a new microwave transmission system. Your report must satisfy the executive and his staff of experts and technicians. Obviously, you could not begin by talking about the transmission system in terms incomprehensible to the manager. Just as obviously you do not want to bore or irritate the experts with needless information. Also, there is a difference in point of view between the executive and the experts. The executive's major concern is whether the new system is a practical one. Will it represent enough of an improvement over previous systems to justify its cost? The executive may need a cost analysis and a cost comparison to previous systems. The experts need enough detail to understand the new design. The experts will be look-

---

[20]Ibid., p. 125.

ing for holes in your theory. You must present it completely. The technicians will be worrying about construction and maintenance problems.

How do you satisfy all these varied needs? The answer is that you must compartmentalize your report. Many technical reports are simply not designed to be read straight through by one reader as is a novel or a short story. By use of the captioning techniques described in Chapter 10 of this text, you divide your report into different segments designed to do different jobs.

You might begin with a general appreciation of the new system designed for the executive. Follow this with a section on cost analysis. Conclude with a summary and a recommendation in favor of the new system. These sections may provide all the information the executive needs or wants.

Following the material designed for the executive will be sections designed for the experts and technicians. Here you should use the techniques described in this chapter for these audiences. The experts and technicians will probably skim the parts designed for the executive and concentrate on the parts meant for them.

Executives will base their decisions on the sections you wrote for them and on consultations with the experts and technicians.

Just as the decision will be a team decision, so the report may very well be a team effort. A technical editor might organize the report into the segments we have suggested. The team leader might write the general appreciation for the manager. Cost analysis would be farmed out to an accountant. Experts and technicians would write their appropriate sections. The technical editor would edit the sections making sure that they can be read by their intended audiences. Such team techniques are common in industry.

No matter how you approach the problem—as an individual or with a team—realizing that you can segment a report to suit the needs of a combined audience will simplify your task.

You may not be ready yet to think in terms of a major project, like the proposal from one firm to another we have just described. However, you may need to reach less complicated combined audiences in your college work. Quite frequently these days, colleges and universities require technical reports to be graded by both an English teacher and an expert in the field. Try a segmented report on this audience. Start with a general appreciation or summary that your English teacher can understand. Follow this with a scientific or technical section designed for the expert. Be sure to make clear which section is for the nonspecialist, which for the expert. Both teachers will be happier with you, because you have considered their needs.

No matter what your audience, put its needs before your own. Be thinking of your intended audience at every stage of the writing process from research to final draft. Perhaps the advice that new staffers on *Life* magazine used to hear best sums up the attitude you should maintain toward all the audiences you may want to reach: "Never underestimate the intelligence of your readers and never overestimate their knowledge."

EXERCISES

1. Compare and contrast an article or report written for one kind of audience with another written for a different audience—perhaps an expert article with an executive one or an executive article with a lay one. Try to get articles on similar subjects and in your field. Ask and answer some of these questions about the articles.

    a. What stylistic similarities do you see between the two articles? What dissimilarities?

        (1) Which article has the greater sophistication?

        (2) What indicates this sophistication?

        (3) What is the average sentence length in each article? Paragraph length?

        (4) Which article defines the most terms?

    b. What is the author's major concern in each article? How do these concerns contrast?

    c. What similarities and dissimilarities of format and organization do you see?

    d. Are there differences in the kinds of graphic aids used in the two articles?

    e. Which article presents the most detail? Why? What kind of detail is presented in each article?

    f. How much background information are you given in each article? Does either article refer you to other books or articles?

2. Take an expert article from a journal in your field and rewrite it into a memorandum for an executive. Use the memorandum format shown on page 304.

3. Perhaps you are already planning a long paper for your writing course. Write an analysis of the anticipated audience.

CHAPTER 5

Technical Exposition

In this chapter we discuss exposition—in the next, narration, description, and argumentation. These are the basic modes of discourse from which all technical reports stem. The modes are strategies designed to persuade an audience to react in the way the writer wants it to. At that statement, you may grow uneasy. Do honest prose and persuasive strategies go together? The answer is *yes*. All writing, all purposeful communication, is persuasive. Even a simple set of instructions has to persuade the reader that the process described is a reasonable way to go about the task. The literature review of a well-written physical research report does far more than narrate the history of past experiments. It must persuade the reader that the present experiment is one more step in a historical process, a step taken by a careful experimenter with a regard for the past and an eye on the future. No tricks are involved. If the experimenter merely piles up historical facts to fill up space, no rhetorical skill will gloss over the emptiness of the endeavor. But if the writer's historical research does relate to the present experimentation, a rhetorically sound literature review will demonstrate the relationship in a persuasive way.

The modes of discourse are useful human acts that have evolved over time and are honest strategies of persuasion. They are not plug-in devices to be used thoughtlessly. Rather, they are methods to help you display your thought and your research in a persuasive manner. If you realize from the very beginning of any task that will end in a piece of writing that persuasion is your ultimate goal, your research and your thinking may be better than it otherwise would be.[1]

In exposition you persuade through explanation. In this expository paragraph from *Scientific American*, the writer explains the difficulty in understanding how language is learned:

> In order to understand how language is learned it is necessary to understand what language is. The issue is confused by two factors. First, language is learned in early childhood, and adults have few memories of the intense effort that went into the learning process, just as they do not remember the process of learning to walk. Second, adults do have conscious memories of being taught the few grammatical rules that are prescribed as "correct" usage, or the norms of "standard" language. It is difficult for adults to dissociate their memories of school lessons from those of true language learning, but the rules learned in school are only the conventions of an educated society. They are arbitrary finishing touches of embroidery on a thick fabric of language that each child weaves for herself before arriving in the English teacher's classroom. The fabric is grammar: the set of rules that describe how to structure language.[2]

Most of the writing in this book, like that you see every day in newspapers and magazines, is expository. Most of the writing you have done in high school and college has been expository. You can use various rhetorical devices in your exposition: topical arrangement, exemplification, definition, classification and division, comparison and contrast, and causal analysis.

You can use any of these devices as an overall organizing principle for an entire paper. For example, an entire paper could be written as a causal analysis or a definition. In order to justify some choice, you might write a report based upon a comparison-and-contrast organizational plan.

But you also use these devices as subordinate methods of development within a larger framework. For example, within a paper organized topically you might have small sections based upon exemplification, definition, causal analysis, and so forth.

This interlocking nature of rhetorical devices should become clear to you as we explain each device and give you examples. But we point it out to you now so you'll understand that the two uses of the rhetorical devices are mutually supportive and not in conflict with one another.

---

[1] For a series of interesting articles on this point of view, see *Technical Communication*, 4th Quarter, 1978.

[2] Breyne Arlene Moskowitz, "The Acquisition of Language," *Scientific American*, November 1978, pp. 92–108.

Topical Arrangement

As we pointed out on pages 42–44, writing projects often begin with a topic. Your topic might be "a choice between electric and oil heat." Or it might be something like "the production and marketing of Christmas trees." As you'll see when you get further into this chapter (pages 107–110), the heating-choice topic probably lends itself to a definite organizational plan based upon comparison and contrast. The Christmas-tree topic, however, does not immediately suggest a definite plan. When such is the case, a topical plan is often the answer. That is, you look for subtopics under the major topic. These subtopics should help you reach the objectives of your paper. They should serve as umbrella statements beneath which you can gather yet smaller sub-subtopics and related facts. In the case of the Christmas-tree topic, the subtopics might very well be "production" and "marketing." "Production" could be broken down further into "planting," "maintaining," and "harvesting." "Marketing" could be broken down into "retail," "wholesale," and "cut-your-own." With some thought, you can break most topics down into umbrella-sized subtopics.

While you are settling upon the topic and subtopics for your paper, you should also be aware of the need for topic limitation. Students, in particular, often hesitate to limit their topic sufficiently. They fear, perhaps, that if they limit their material too severely they will experience difficulty in writing essays of a sufficient length. The truth of the matter is really the reverse. You will find it easier to write a coherent, full essay of any length if you *limit the scope of your topic.* With your scope limited and your purpose clearly defined, you can fill your paper with concrete, specific facts and examples. When your scope remains broad and your purpose vague, you must deal in abstract generalizations.

Suppose you want to deal with the topic of lasers in about a thousand words. If you keep the purpose simply as "explaining lasers," what can you say in a thousand words? Probably just a few simpleminded generalizations about laser theory, history, and applications. Suppose, however, you limit the subject by asking yourself what smaller topics you can break it into. We have already indicated a few: theory, history, and applications. Any one of these is probably limited enough for a fairly decent thousand-word paper.

For the best paper, you would probably limit even more. Let's take "laser applications." What topics can this be broken into? To name just a few, communication, industrial uses, weaponry, medicine, and radar. You could go still further. Medicine could be broken down into eye surgery and cancer treatment. You could choose one of these. With a few minutes' thought, *before writing or even organizing*, you can limit your topic to manageable size.

Keep in mind that once your organizational plan is set, you're not restricted from using the other rhetorical devices in carrying out your plan. You can support a topical arrangement with exemplification, definition, classification and division, comparison and contrast, and causal analysis.

Exemplification

Expository writing sometimes consists largely of a series of generalizations supported by examples. Many writers deal too often in generalizations, failing to support them with the concrete examples that would make them understandable and believable. We have just given you a generalization about writers. We now have two options. We can trust that you will believe us and move on, or we can provide you with an example. One example will not prove our case, but it will make it more understandable and persuasive. We feel that, normally, a writer faced with this choice should stop to give an example. Look at the following faulty piece of student writing. The student is giving some background information for a paper on "Lasers in the Field of Communications."

> Since its recent development, the laser has been experimented with by hundreds of different groups. Many different materials have been used to make lasers, and efforts are continuing to find more new substances that can be used to make lasers. Besides lasers made of different materials, different types of lasers have been built.

The student continues in this vein, generalization upon generalization, without ever giving any solid examples of the point in question. With nothing to hold on to, the reader is left asking: "What is recent?" "What different groups?" "What materials?" "What types of lasers?" In the following example, the student gives background information for a paper on "Gyroscopes in Modern Aircraft Instruments."

> Gyroscopes have been in practical use since the 18th century. In 1774 the gyroscope was used as an artificial horizon so that navigators could take accurate sextant readings on hazy days. In 1896 Obry installed the gyro in the guidance system of a self-propelled torpedo. Not long after, Elmer Sperry developed the gyrocompass, an instrument which aligns itself at a right angle to the earth's axis of rotation. During World War II the Germans used an inertial guidance system with a platform stabilized by

gyros and gimbals in their V-2 rockets. The United States acquired some of the German scientists, and as a result our early guidance systems were much like those of the Germans. The gyro-stabilized platform is currently used in radar antennas, attitude control of planes and ships, and gunlaying equipment.

The second example is greatly superior to the first. It contains only one generalization, "Gyroscopes have been in practical use since the 18th century." The writer then follows the generalization with a host of examples, providing a good historical introduction to the uses of gyroscopes.

There are two common ways to use examples. One way is to give one or more extended, well-developed examples. We have used this method in showing you two large samples of student writing. The other way is to give a series of short examples that you do not develop in detail. The writer on gyroscopes has used this approach, carrying it to an acceptable extreme in the last sentence of the paragraph.

Examples give your writing life and believability. They give your audience something concrete to hold on to. Readers will remember and understand your generalizations when you have amply illustrated them.

Definition

As a writer, you will constantly be defining terms. Of course, it is true that you can sometimes substitute simpler, everyday words for specialized vocabulary. But, often, to do so would cause the reader more trouble than to use the specialized term and define it. Mark Twain illustrated the problem nicely by describing, without a single technical term, how to hitch up a team of horses:

> It may interest the reader to know how they "put horses to" on the continent. The man stands up the horses on each side of the thing that projects from the front end of the wagon, and then throws the tangled mess of gear on top of the horses, and passes the thing that goes forward, through a ring, and hauls it aft, and passes the other thing through the other ring and hauls it aft on the other side of the other horse opposite to the first one, after crossing them and bringing the loose end back, and then buckles the other thing underneath the horse, and takes the other thing and wraps it around the thing I spoke of before, and puts another thing over each horse's head, with broad flappers to keep the dust out of his eyes, and puts the iron thing in his mouth for him to grit his teeth on, up hill, and brings the ends of these things aft over his back, after buckling another one around under his neck to hold his head up, and hitching

another thing that goes over his shoulders to keep his head up when he is climbing a hill, and then takes the slack of the thing which I mentioned a while ago, and fetches it aft and makes it fast to the thing that pulls the wagon, and hands the other things up to the driver to steer with. I never have buckled up a horse myself, but I do not think we do it that way.[3]

You should define any term that you feel is not in your reader's normal vocabulary. The less expert your audience, the more you will need to define. Definitions range in length from a single word to long essays or even books. Sometimes, but not usually, a synonym inserted into your sentence will do. You might write, "The oil sump, that is, the oil reservoir, is located in the lower portion of the engine crankcase." Synonym definition serves only when a common interchangeable word exists for some bit of technical vocabulary you wish to use. Most often you will want to use at least a one-sentence definition containing the elements of a logical definition: *term = genus or class + differentia*. Do not let the technical terms frighten you. You have been giving and hearing definitions cast in the logical pattern most of your life. In the logical definition you state that something is a member of some genus or class and then specify the differences that distinguish this thing from other members of the class:

Term	=	*Genus or Class*	+	*Differentia*
An ohmmeter	is	an indicating instrument		that directly measures the resistance of an electrical circuit.
A legume	is	a fruit		formed from a single carpel, splitting along the dorsal and the ventral sutures, and usually containing a row of seeds borne on the inner side of the ventral suture.

The second of these two definitions, particularly, points out a pitfall you must avoid. This definition of a legume would satisfy only someone who was already fairly expert in botany. Real laymen would be no further ahead than before, because such terms as "carpel," "ventral sutures," and so forth are not familiar to them. When writing for nonexperts you may wish to settle for a definition less precise, but more understandable, such as, "A legume is a fruit formed of an easily split pod that contains a row of seeds, such as a pea pod." Here you have stayed with plain language and given an easily recognized example. Both of these definitions of a legume are good. The one you would choose depends on your audience.

To make sure you are understood, you will often want to extend a definition beyond a single sentence. The most common devices for

---

[3] Samuel Clemens [Mark Twain], *A Tramp Abroad* (Hartford, Conn., 1880), pp. 329–330.

extending a definition are description, example, and comparison. The following definition taken from *Chamber's Technical Dictionary* goes beyond the logical definition to give a description:

> *Anemometer.* An instrument for measuring the velocity of the wind. A common type consists of four hemispherical cups carried at the ends of four radial arms pivoted so as to be capable of rotation in a horizontal plane, the speed of rotation being indicated on a dial calibrated to read wind velocity directly.

In this example a drawing might also prove useful.

In the laymen's definition of legume given above, an example was given: "such as a pea pod." Often comparison is valuable: "A voltmeter is an instrument for measuring electrical potential. It may be compared to a pressure gauge used in a pipe to measure water pressure." Extend your definition as far as is needed to ensure reader understanding.

Sometimes you may wish to begin a definition by telling what something is *not,* as in the following definition, again from *Chamber's:*

> *Metaplasm.* Any substance within the body of a cell which is not protoplasm; especially food material, as yolk or fat, within an ovum.

You should, of course, avoid circular definition. "A botanist is a student of botany," will not take the reader very far. However, sometimes you may appropriately repeat on both sides of a definition common words you are sure your reader understands. In the following *Chamber's* definition it would be pointless to drag in some synonym for a word as readily understood as "test."

> *Gmelin's Test.* A test for the presence of bile pigments; based upon the formation of various colored oxidation products on treatment with concentrated nitric acid.

You have several options for placement of definitions within your papers: (1) within the text itself, (2) in footnotes, (3) in a glossary at the beginning or end of the paper, and (4) in an appendix. Which method you use depends upon the audience and the length of the definition.

Within the text. If the definition is short—a sentence or two—or if you feel most of your audience needs the definition, place it in the text with the word defined. Most often, the definition is placed after the word defined as in this example:

> The word Bantu is an exclusively linguistic label and has no other primary implications, either of race or of culture. The family of dialects was named by a 19th-century German philologist, Wilhelm H. I. Bleek.

The words *ki-ntu* and *bi-ntu* in these African dialects respectively mean "a thing" and "things"; the words *mu-ntu* and *ba-ntu*, "a man" and "men." Thus *bantu* can be construed as "people" or "the people," and this seemed to Bleek a name eminently suitable for a language held in common by so many.[4]

Sometimes, the definition is slipped in smoothly before the word is used—a technique that helps break down the reader's resistance to the unfamiliar term.[5] The following definition of *supernova* is a good example of the technique:

> The vast majority of stars in a typical galaxy such as our own are extremely stable, emitting a remarkably steady output of radiation for millions of years. Occasionally, however, a star in an advanced stage of evolution will spontaneously explode, and for a few months it will be several hundred million times intrinsically more luminous than the sun. Such a star is a supernova, and at the time of its greatest brilliance it may emit as much energy as all the other stars in its galaxy combined.[6]

When you are using key terms that must be understood before the reader can grasp your subject, define them in your introduction.

In footnotes. If your definition is longer than a sentence or two, and your audience is a mixed one—part expert and part layman—you may want to put your definition in a footnote. A lengthy definition placed in the text could disturb the expert who does not need it. In a footnote it is easily accessible to the layman and out of the expert's way.

In a glossary. If you have many short definitions to give and if you have reason to believe that all members of your audience will not read your report straight through, place your definitions in a glossary. (See pages 211–215.) Glossaries do have a disadvantage. Your readers will be disturbed by the need to flip around in your paper to find the definition they need. When you use a glossary be sure to draw your readers' attention to it, both in the table of contents and early in the discussion.

In an appendix. If you need one or more extended definitions (say, over 200 words each) for some but not all members of your

---

[4] D. W. Phillipson, "The Spread of the Bantu Language," *Scientific American*, April 1977, pp. 106–114.

[5] For this idea we are indebted to W. Earl Britton, and we suggest that you read his "What to Do About Hard Words," *STWP Review*, October 1964, pp. 13–16.

[6] F. Richard Stephenson and David H. Clark, "Historical Supernovas," *Scientific American*, June 1976, pp. 100–107.

audience, place them in an appendix. At the point in your text where readers may need them, be sure to tell them where they are.

Classification and Division

Classification and division are useful devices for bringing order to any complex body of material. You may understand classification and division more readily if we explain them in terms of the *abstraction ladder*. We borrow this device from the general semanticists— people who make a scientific study of words. We will construct our ladder by beginning with a very abstract word on top and working down the ladder to end with a concrete term:

Factor: Almost anything can be a factor. You could be a factor in someone's decision. So could wealth, climate, and geography, to name but a few significant concepts. Without more concrete references, we cannot determine what is specifically meant at this rung of the ladder.

Wealth: Now we have moved down a rung on the ladder. We have added concrete information. Wealth could be money. It also could be stocks and bonds, land, furniture, or any of the other numerous material objects that people value highly.

Furniture: We now have become much more concrete. We can mean beds, tables, chairs, desks, lamps, etc.

Table: Now we are zeroing in on the object. However, *table* still refers to a huge class of objects: coffee tables, kitchen tables, dining room tables, library tables, end tables, etc.

Kitchen table: Now we know a good deal more. We know the function of the object. People eat at kitchen tables. Cooks mix cakes on them. We can even generalize somewhat about their appearance. Kitchen tables are usually plain objects, compared to end tables, for example. Many of them are made of wood and have four legs.

John Smith's kitchen table: Now we know precisely what we are talking about. We can describe the size, weight, shape, and color of this table. We know the material of which it is made. We know that John Smith's family eats breakfast at this table, but not dinner. In the evenings John Smith's daughter does her homework there.

We must keep one important distinction in mind: even "John Smith's kitchen table" is not the table itself. As soon as we have used a word for an object, the abstraction process has begun. Beneath the word is the table *we see* and beneath that is the table *itself*, consisting of paint, wood, and hardware that consist of molecules that consist of atoms and space.

In classification you move *up* the abstraction ladder seeking higher abstractions under which to group many separate items. In division you move *down* the abstraction ladder breaking higher abstractions down into the separate items contained within them. We will illustrate classification first.

Suppose for the moment that you are a dietician. You are given a long list of foods found in a typical American home and asked to comment on the value of each. You are to give such information as calorie count, carbohydrate count, mineral content, vitamin content, and so forth. The list is as follows: onions, apples, steak, string beans, oranges, cheese, lamb chops, milk, corn flakes, lemons, bread, butter, hamburger, cupcakes, and carrots.

If you try to comment on each item in turn as it appears on the list, you will write a chaotic essay. You will repeat yourself far too often. Many of the things you will say about milk will be the same things you say about cheese. To avoid this repetition and chaos you need to classify the list, to move up the abstraction ladder seeking groups that look like the following:

Food

I. Vegetables
   A. Onions
   B. String beans
   C. Carrots

II. Fruit
   A. Apples
   B. Oranges
   C. Lemons

III. Meat
   A. Steak
   B. Lamb chops
   C. Hamburger

IV. Cereal
   A. Corn flakes
   B. Bread
   C. Cupcakes

V. Dairy
   A. Milk
   B. Cheese
   C. Butter

By following this procedure, you can use the similarities and dissimilarities of the different foods to aid your organization rather than have them disrupt it.

In division you move down the abstraction ladder. Suppose your problem now to be the reverse of the above. You are a dietician and someone asks you to list examples of foods that a healthy diet should contain. In this case you start with the abstraction, "food." You decide to divide this abstraction into smaller divisions such as vegetables, fruit, meat, cereal, and dairy. You then subdivide these into typical examples such as cheese, milk, and butter for dairy. Obviously, the outline you could construct here might look precisely like the one already shown. But in classification we arrived at the outline from the bottom up; in division, from the top down.

Very definite rules apply in using classification and division.

Keep all headings equal. In the example above, you would not have headings of Meat, Dairy, Fruit, Cereal, and Green Vegetables. "Green Vegetables" would not take in a whole class of food as the other headings do. Under the heading of "Vegetables," however, you could have subheadings of "Green Vegetables" and "Yellow Vegetables."

Apply one rule of classification or division at a time. In the example above, the classification is done by food types. You would not in the same outline include headings equal to the food types for such subjects as "Mineral Content" or "Vitamin Content." You could, however, include such subheadings under the food types.

Make each division or classification large enough to include a significant number of items. In the example above, you could have many equal major headings such as "Green Vegetables," "Yellow Vegetables," "Beef Products," "Lamb Products," "Cheese Products," and so forth. In doing so, however, you would have overclassified or overdivided your subject. Some of the classifications would have included only one item.

Avoid overlapping classifications and divisions as much as possible. In the example above, if you had chosen a classification that included "Fruits" and "Desserts," you would have created a problem for yourself. The listed fruits would have to go in both categories. You cannot always avoid overlap. (No one, for example, has devised a classification for the rhetorical modes that completely avoids overlap). But keep it to a minimum.

As long as they observe the rules, writers are free to classify and divide their material in any ways that best suit their purposes. An accountant, for example, who wished to analyze the money flow for construction of a state's highways might choose to classify them by source of funding: federal, state, county, city. An engineer, on the other hand, concerned with construction techniques, might choose to classify the same group of roads by surface: concrete, macadam, asphalt, gravel.

In a fact sheet about controlling the insects that prey on trees, the author chose to classify the insects according to damage produced:

Types of Damage

Leaf Chewers
A number of insects, mostly caterpillars and beetles, damage foliage by consuming all or parts of leaves or needles. This may take the form of skeletonizing, leaf mining, or free feeding. The skeletonizers consume the soft leaf tissues leaving a lacy pattern of the veins and sometimes the upper or lower leaf surface. The miners work between the leaf surfaces causing brownish or papery blotches or winding narrow trails. The free feeders eat the complete leaf and sometimes fold or roll the leaves or web leaves together.

Broad-leafed trees in otherwise good condition can withstand a complete defoliation without serious injury. They will produce a second crop of leaves the same season if defoliation occurs before August. Extensive defoliation for several successive seasons can kill the branches (called die-back) and possibly the tree.

Most evergreens, however, will die if completely defoliated. Any branch that is stripped of all its needles should be pruned out.

Sap-Sucking Insects and Mites
Aphids, plant bugs, scales, thrips, and mites extract the sap from buds, leaves, twigs, or stems. Their attacks frequently cause curling, spots, galls, yellowing, mottling, or deformed leaves and flowers. A healthy, well established tree will withstand most infestations without permanent damage. However, some die-back may follow heavy infestations and ornamentals are frequently disfigured. This detracts from their value in the landscape.

Some aphids and scale insects also secrete a sweet sticky material called honeydew. During periods of heavy infestations, honeydew coats the leaves and branches of the trees as well as sidewalks, lawn furniture, and other objects under the trees. A black sooty-mold fungus grows on this honeydew, giving the trees a sooty appearance.

Disease Carriers
Some insects are vectors of disease organisms. The most common of these is Dutch elm disease, which is spread by elm bark beetles. Chemicals are sometimes needed to control these insects, which carry diseases but which may not otherwise be harmful to trees.

Borers
Trees and shrubs may be attacked by several kinds of insects which bore or tunnel inside trunks, branches, or twigs. Borers generally infest weakened or dying trees rather than healthy, vigorous ones. Proper pruning, fertilizing, and watering will help prevent borer problems. There are few practical chemical controls for borers. Some species may be controlled in individual specimen trees by injecting the burrows or tunnels

with carbon tetrachloride and then plugging the treated burrows with clay or putty.[7]

Using this classification scheme, the author teaches his readers how to recognize whatever pests are plaguing their gardens at any given moment. The rest of the article explains how to deal with each type of pest. Obviously, insects can be classified in many ways; the types of damage they produce suggested the best classification scheme for this author, in view of the article's purpose.

Comparison and Contrast

In comparison you show how something is like something else. In contrast you show how it is unlike something else. You frequently use comparison and contrast. Someone asks you, "What is Charlie Jones really like?"

You answer, "Well he's a pretty unusual guy; he's a fine athlete. He's as good a football player as Pete Smith. But he's different from Pete. Pete is strictly an athlete. He doesn't care whether school keeps or not. But Charlie is a good student. He really hits the books—B's and A's all the time."

You assume here that the listener is familiar with Pete Smith. By comparing and contrasting the unfamiliar Charlie Jones to Pete, you aid the listener in understanding Charlie.

In technical writing you will use comparison and contrast in two different ways. First, you will sometimes structure a whole paper on comparison and contrast. Second, you will often use simple analogies—comparing the unknown to the known—in definitions and explanations. We will examine the more complete use of comparison and contrast first.

Frequently you will have to compare or contrast two or more processes or mechanisms. It is essential that you first list and explain the points of comparison and contrast. Without this listing your report will ramble with little or no coherence. After the listing, you have your choice of two different organizational approaches. We can best illustrate these approaches in outline form.

Assume that you are comparing two desalination processes: freezing, and flash evaporation. Your points of comparison are (1) cost of installation, (2) cost of producing water, (3) ease of maintenance, and

[7] John A. Lofgren, *Controlling Insect Pests of Shade and Ornamental Trees*, Agricultural Extension Service of the University of Minnesota Fact Sheet no. 28 (St. Paul, Minn., 1973).

(4) purity of the water produced. You might organize your material this way:

I. Points of Comparison and Contrast
   A. Cost
      1. Installation
      2. Production
   B. Ease of maintenance
   C. Purity
II. Freezing
   A. Cost
   B. Ease of maintenance
   C. Purity
III. Flash Evaporation
   A. Cost
   B. Ease of maintenance
   C. Purity

In this organizational approach, you take one process at a time and run through each of the points of comparison and contrast. This method has the advantage of giving you the whole picture for each process as you discuss it. The emphasis is on the processes. If your purpose was argumentative, you would call the points of comparison and contrast *criteria* or *standards*.

In the second approach, you discuss each process point by point:

I. Points of Comparison and Contrast
   A. Cost
      1. Installation
      2. Production
   B. Ease of maintenance
   C. Purity
II. Cost
   A. Freezing
   B. Flash Evaporation
III. Ease of maintenance
   A. Freezing
   B. Flash Evaporation
IV. Purity
   A. Freezing
   B. Flash Evaporation

The second point-by-point method has the advantage of sharper contrast or comparison. Often, in an argument, it is the better of the two methods.

In addition to organizing whole papers around comparison and contrast you will frequently use comparison and contrast (particularly comparison) as an aid in defining and explaining. These comparisons, called analogies in this use, may be quite short or fairly extended. Here is a famous extended analogy by Sir James Jeans that explains why the sky is blue:

> Imagine that we stand on an ordinary seaside pier, and watch the waves rolling in and striking against the iron columns of the pier. Large waves pay very little attention to the columns—they divide right and left and reunite after passing each column, much as a regiment of soldiers would if a tree stood in their road: it is almost as though the columns had not been there. But the short waves and ripples find the columns of the pier a much more formidable obstacle. When the short waves impinge on the columns, they are reflected back and spread as new ripples in all directions. To use the technical term they are "scattered." The obstacle provided by the iron columns hardly affects the long waves at all, but scatters the short ripples.
>
> We have been watching a sort of working model of the way in which sunlight struggles through the earth's atmosphere. Between us on earth and outer space the atmosphere interposes innumerable obstacles in the form of molecules of air, tiny droplets of water, and small particles of dust. These are represented by the columns of the pier.
>
> The waves of the sea represent the sunlight. We know that sunlight is a blend of many colors—as we can prove for ourselves by passing it through a prism, or even through a jug of water, or as Nature demonstrates to us when she passes it through raindrops of a summer shower and produces a rainbow. We also know that light consists of waves, and that the different colors of light are produced by waves of different lengths, red light by long waves and blue light by short waves. The mixture of waves which constitutes sunlight has to struggle through the obstacles it meets in the atmosphere just as the mixture of waves at the seaside has to struggle past the columns of the pier. And these obstacles treat the light waves much as the columns of the pier treat the seawaves. The long waves which constitute red light are hardly affected, but the short waves which constitute blue light are scattered in all directions.
>
> Thus the different constituents of sunlight are treated in different ways as they struggle through the earth's atmosphere. A wave of blue light may be scattered by a dust particle, and turned out of its course. After a time a second dust particle again turns it out of its course, and so on, until finally it enters our eyes by a path as zigzag as that of a flash of lightning. Consequently the blue waves of sunlight enter our eyes from all directions. And that is why the sky looks blue.[8]

---

[8] Sir James Jeans, *Stars in Their Courses* (Cambridge: Cambridge University Press, 1931), pp. 23–24. Reprinted by permission of the publisher.

Sir James has used analogy in a very imaginative way. By comparing light waves (an unfamiliar concept) to ocean waves (a familiar concept) he has made it easy for the reader to grasp his meaning. Such extended analogies are most useful when you are writing for laymen.

You should frequently use short, simple analogies when you are writing. Many people have difficulty in understanding the immense power released by nuclear reactions. A completely technical explanation of $E = mc^2$ probably would not help them very much. But suppose you tell them that if one pound of matter—a package of butter, for instance—could be converted directly to energy in a nuclear reaction, it would produce enough electrical power to supply the entire United States for thirty-six hours (that is, over eleven billion kilowatt hours). Such a statement reduces $E = mc^2$ to an understandable idea.

Analogies serve particularly well in definitions or descriptions. If, after describing a transistor, you tell a lay audience that it is similar to a water faucet in that you can use it to control the flow of electrons or shut them off completely, you make the concept more understandable.

Throughout your writing, use analogy freely. It is one of your best bridges to the uninformed reader.

Causal Analysis

In causal analysis, you state and defend the proposition that x causes y; or the converse proposition: y is the effect of x. Of course, x can represent more than one cause and y more than one effect.

Inductive and deductive reasoning are necessary ingredients in any causal analysis, so we will take them up first.

In inductive reasoning you move from the particular to the general. You present a series of facts and from them draw some general inference. To illustrate: Fifty people out of a hundred fall ill after a church picnic. An investigation shows that the fifty people who became ill ate potato salad. The fifty who did not become ill did not eat potato salad. A reasonable inference would be that the potato salad caused the illness. Final confirmation of the inference would depend upon an analysis of the potato salad that proved it was contaminated.

In deductive reasoning you move from the general to the particular. You start with some general principle, compare it to a fact, and draw a conclusion concerning the fact. Although you will seldom use the form of a syllogism in writing, we can best illustrate deductive reasoning with a syllogism:

1. All men are mortal.
2. Aristotle is a man.
3. Therefore, Aristotle is mortal.

Most often the general principle itself has been arrived at through inductive reasoning. For example, from long observation of lead, scientists have concluded that it melts at 327.4C. They have arrived at this principle inductively from many observations of lead. Once they have inductively established a principle, scientists can use it deductively as in the following syllogism:

1. Lead melts at 327.4C.
2. The substance in container A is lead.
3. Therefore, the substance in container A will melt at 327.4C.

Often you will argue both inductively and deductively within the same paper. First, you will examine a body of facts and establish a generalization. Then, in turn, you will use that generalization to arrive at some conclusions about another body of facts.

Causal analysis may go forward or backward. When we examine an effect and want to determine its cause, we move backwards. When we want to forecast, we examine existing conditions and try to predict what effects these conditions will cause. In discussing cause and effect we are normally dealing with time. Therefore, causal analysis often takes the form of a narrative.

In causal analysis we may be involved with more than one cause; that is, with complex causes. In this case you will, in your paper, list and explain all the causes. Often you will attempt to show a necessary relationship among them. Sometimes you will attempt to prove that one or another is the main cause.

The following paragraphs illustrate both the narrative movement of causal analysis and the interrelationship among complex causes:

> Most experts advise parents to deal with their children's fears—of strangers, loud noises, darkness, animals or imaginary monsters—by helping them to feel secure and by not demanding that these fears be overcome quickly. But parents also must learn more about these excessive and unnecessary fears: how they develop, and how they either can be cured or prevented.
>
> Some of your baby's fears begin during infancy. If he remains afraid of new people, strange places or noises, he will not grow independent enough to explore, and will miss out on many of life's extremely important experiences.

There are basic fear reactions that are quite normal for a helpless infant. Certain unexpected events, such as loud noises, abrupt changes in surroundings or sudden pain, may startle an infant, causing him to cry, gasp, become rigid and thrash his arms. However, those objects or events that naturally cause fears to develop may lead to other unnecessary fears called *learned* fears. Three processes are involved in the development of a learned fear:

Association: Previously neutral events become associated with experiences that babies naturally fear. This can occur within the first few months of life. For example, the infant who is placed in overly hot bath water one time may learn to fear bath time in general.

Imitation: Many fears are acquired through watching and imitating the reactions of adults or other children. This process usually begins at around two or three years of age. Parents can easily teach their children to fear spiders, darkness or thunderstorms, if they themselves consistently show fear in response to these things.

Symbolic Representation: As children grow in their ability to imagine, frightening fairy tales or television programs provide new sources of fear. Children are particularly susceptible between the ages of three and five. They may see scary witches, bears or monsters in shadows on the wall or hear ghostly moans when a branch grazes the side of the house.

Although most children's fears are perfectly understandable, the origins of others are not as easy to pinpoint. However, knowing the original cause of a fear is not always necessary in order to diminish it.[9]

The following two paragraphs illustrate exemplification. But the relationship between them is that of causal analysis. The first paragraph presents the cause, the second the effect.

As most of us know, the farm population of the United States has declined steadily for two decades, reaching a low of 10 million in 1965, slightly less than five percent of the nation's total population. Using a different yardstick, the total number of farms in the U.S. declined to just over 3 million in 1968, and commercial farms, those grossing over $10,000 per year (which, incidentally, produce 85 percent of the product sold), declined to approximately 1 million. There is no indication that this dwindling process is coming to an end: The annual rate of outmigration from farms stood at 6.3 percent during the period 1965–68—the highest rate in U.S. history.

With these dwindling numbers, of course, has come diminished political power. The term "farm bloc" has been forgotten. Neither candidate in the 1968 presidential election gave a major farm speech. Farm people and rural areas still maintain a respectable degree of power in the

---

[9]V. Thomas Mawhinney, "Mommy, Can I Leave the Lights On?" *Family Health*, October 1978, pp. 33–34, 47, 53.

Senate, but the political strength of farm people in the House of Representatives has declined to an extremely low level. In 1969, for example, Congressional districts claiming 25 percent or more farm population stood at only 31: the East contained none, the West 1, the South 14, and the Midwest 16. And this handful of farm Congressmen (by this generous definition) are concentrated in two tiny, embattled enclaves: the Committee on Agriculture and the Subcommittee on Agricultural Appropriations.[10]

Both paragraphs begin with a generalization as a central statement. In the first paragraph, the writer says, "As most of us know, the farm population has declined steadily for two decades. . . ." He then gives statistics illustrating the trend. The paragraph is unified because all the supporting data relate to the central statement. Unity is achieved in the second paragraph in the same way.

Many traps exist in causal analysis for the unwary writer. Avoid a rush to either conclusion or judgment. Take your time. Don't draw inferences from insufficient evidence. Don't assume that just because one event follows another, the first caused the second—a fallacy that logicians call *post hoc, ergo propter hoc* (that is: *after this, therefore, because of this*). You need other evidence in addition to the time factor in order to establish a causal relationship.

For example, in the sixteenth century tobacco smoking was introduced into Europe. Since that time, the average European's life span has increased severalfold. It would be a fine example of the *post hoc* fallacy to infer that smoking has increased the life span, which in fact probably stems from better housing, nutrition, and medical care.

Another common error is applying a syllogism backwards.

The following syllogism is valid:

1. All dogs are mammals.
2. Jock is a dog.
3. Therefore, Jock is a mammal.

But if you reverse statements (2) and (3) you have an invalid syllogism:

1. All dogs are mammals.
2. Jock is a mammal.
3. Therefore, Jock is a dog.

---

[10] Willard W. Cochrane, "U.S. Farm Policy, Collision in the 70's," *Minnesota Science,* Fall 1970, p. 4.

Jock, of course, could be a cat, a whale, a Scotsman, or any other member of the mammal family.

You can often find flaws in your own reasoning or that of others if you break the thought process down into the three parts of a syllogism.

Effective Exposition

Exposition is a common approach to a technical report. You will constantly be called upon to explain aspects of your work. Writing good expository prose is no easy chore. But you will be well on your way if you remember that persuasion, clarity, and completeness are your chief goals. You will be better equipped to achieve these goals if you master all the expository devices: topical arrangement, exemplification, definition, classification and division, comparison and contrast, and causal analysis.

Master also the ability to move up and down the abstraction ladder. The writer who stays high on the ladder produces little more than windy generalizations unrelated to reality. On the other hand, the writer who deals in facts and facts alone leaves the reader with too many unresolved questions. What is the point of all this information? What does it mean? What can I do with it? It's the writer's task to draw the needed inferences and generalizations for the reader. Just be sure that your inferences and generalizations are tied to fact, not floating like hot-air balloons that have slipped their cables.

Rich and subtle variation in use of the expository devices is one of the chief distinguishing marks of the good writer. Below we have reproduced a portion of a report called "Satellites of the Outer Planets."[11] The entire report has a simple topical arrangement consisting of an introduction, called "Planets and Satellites," followed by four sections: "Satellites of Jupiter," "Satellites of Saturn," "Satellites of Uranus," and "Satellites of Neptune." To develop the exposition, the writer uses several rhetorical devices within the overall pattern. To illustrate the interlocking nature of rhetorical devices used for support and development, we have identified the devices used in the introduction.

---

[11] Jet Propulsion Laboratory, "Satellites of the Outer Planets," *Mission to Jupiter/ Saturn* 2 (November 1977): 1. Published by National Aeronautics and Space Administration.

A satellite is a world that revolves around a planet as planets revolve around the Sun. Earth has a single large satellite, the Moon; Mars has two tiny satellites, Deimos and Phobos. But most of the satellites of the Solar System orbit the outer planets beyond Mars: 30 known objects, compared with the three satellites of Earth and Mars.

Definition

Comparison

Learning about these many other satellites is vital to our understanding of the origin, development, phenomena, and processes of the Solar System and its planets, including the Earth. However, satellites of the outer Solar System are so far away (and so hard to study) that scientists neglected them until recent years when new ways to observe them and new means to analyze the observations became available. NASA's close views of the Martian satellites, information about the Moon, and some NASA spacecraft pictures of Jupiter's large satellites created a surge of interest in the many satellites of the giant outer planets.

Causal Analysis

Satellites can be grouped into three types: objects about the size of the Earth-like planets, those about the size of larger asteroids, and small, irregularly shaped objects. Examples of each type are

Classification

(1) Saturn's largest satellite, Titan, a world larger than the planet Mercury.
(2) Uranus' satellite, Miranda, about the size of the asteroid Ceres.
(3) Leda, a recently discovered satellite of Jupiter that is about the size of the Martian satellite Deimos.

Exemplification

Satellites can also be grouped in terms of the shape and inclination of their orbits and their orbital motion.

Classification

Regular satellites are those with orbits that are nearly circular, close to the equatorial plane of the planet, and *posigrade* in direction; i.e., the satellite moves around the planet in the same direction the planets move around the Sun. Irregular satellites have orbits that are highly elliptical or are inclined to the plane of the planet's equator, or are in a *retrograde* direction (a direction opposite to that of the planet's revolution).

Definition

Regular satellites probably originated from the same processes that formed the planets. Irregular satellites were probably captured later.

Causal Analysis

EXERCISES ▊▊▊▊▊▊▊▊▊▊▊▊▊▊▊▊▊▊▊▊▊▊

1. Reproduced below is a portion of a report called *The Planet Venus*.[12] The portion reproduced explains some of the facts currently known about the planet and speculates about some of its mysteries. It's a good piece of expository writing. Write an analysis of it that shows which rhetorical devices the author has used, and where and why he has used them.

Venus as a Planet

The mass, diameter, and density of Venus are all only slightly less than those of the Earth. But its surface is hotter, its atmosphere much denser, and its rotations much slower than those of the Earth. Venus is one of the three planets of the solar system that do not have satellites; the others are Mercury and Pluto. Also Venus does not have a significant magnetic field.

The diameter of Venus is 12,100 km (7519 mi) and its average density is 5.25 times that of water. This high density may imply that Venus has a core of nickel and iron like the Earth.

The surface of Venus cannot be observed visually from Earth because Venus is shrouded in a thick blanket of clouds. However, radar waves from Earth stations penetrate to the surface and astronomers have used them to map the surface of Venus. In addition, the surface emits radio waves—microwaves—which penetrate the clouds and can

---

[12] U.S., National Aeronautics and Space Administration, *The Planet Venus* (Washington, D.C.: U.S. Government Printing Office, 1978) pp. 2–4.

be received on Earth. Astronomers use these radio waves to estimate the temperature of the planet's surface.

In the early 1960's when these measurements were first made, astronomers were surprised to discover that the surface of Venus is extremely hot. The microwaves carry the message that it is about 430°C which is hot enough to melt zinc. They also told us that the temperature is the same both day and night on the surface of Venus.

The surface temperature of Venus is much higher than that of the Earth. This difference is attributed to the dense atmosphere of Venus which traps incoming solar radiation and prevents heat from being re-radiated into space. This is termed the "greenhouse effect."

A telescope reveals no details in the yellowish clouds of Venus. However, if these clouds are photographed in ultraviolet light, dark shadings can be seen that rotate in a period of about 4 days. These shadings have also been seen more clearly in ultraviolet pictures of Venus taken from a passing spacecraft.

But the planet itself rotates much more slowly than its cloud tops. Again the discovery was made by radio waves. Transmitted from the Earth by a big antenna, these waves are reflected by Venus. Radar astronomers can tell from the form of the radio echo that Venus rotates once in 243.1 days, in a direction opposite to the rotation of the Earth. This is called a retrograde direction. A day (the time from one sunrise to the next) on Venus lasts about 117 Earth days [see figure].

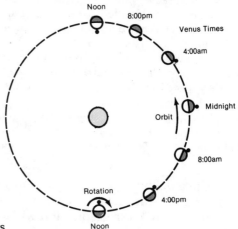

Axial Rotation: 243.1 Earth days
Orbital period: 225 Earth days (= Venus year)
Noon to Noon: 116.8 Earth days (= Venus day)
Venus year: 1.93 Venus days

A combination of retrograde rotation about its axis in 243.1 days and orbital revolution about the Sun in 225 days make a day on Venus 116.8 Earth days long, as shown in this diagram.

Another peculiar fact about the rotation of Venus is that it seems to be linked to the Earth. Each time Venus passes closest to the Earth at inferior conjunction,* the same face of Venus is turned towards the Earth. Most astronomers believe this is just a coincidence, but it may not be.

Astronomers do not know why Venus rotates so slowly. The other planets (except Mercury) rotate quickly on their axes. Mercury has been slowed by tidal friction because of its closeness to the Sun; but Venus is not only rotating slowly, it is also rotating in the "wrong" direction. The Sun could not have caused this. One possibility, proposed some years ago, is that Venus once had a big satellite that revolved around it in the retrograde direction. This satellite crashed into Venus and stopped the planet's rotation, or perhaps even pushed it a little into rotating the opposite way.

Astronomers looked at the spectrum of light reflected from the clouds of Venus and were able to find that its atmosphere contains much (97%) carbon dioxide. This again is a contrast to Earth which only has a small amount (0.03%) of carbon dioxide in its atmosphere. The reason there is so much carbon dioxide in the atmosphere of Venus is thought to be because Venus is a dry planet. The presence of oceans of water on Earth allowed Earth's early atmosphere of carbon dioxide to react with the rocks and form carbonates. Today, most of Earth's carbon dioxide is bound up as carbonates in rocks. Also living things on Earth used carbon dioxide from the atmosphere to provide carbon for their bodies. Oxygen was released by plants to form Earth's oxygen-rich atmosphere of today.

Another big difference between Earth and Venus, as mentioned earlier, is that Venus does not have a magnetic field of any significance. As a consequence its atmosphere has no protection from the blizzard of high-energy particles—protons and electrons—known as the solar wind. These particles are held away from the Earth by our magnetic field. But they speed directly into the upper atmosphere of Venus.

---

*Defined earlier in this way: "When Venus passes between Earth and Sun, and thus comes closest to the Earth, it passes through what astronomers call inferior conjunction. On the far side of the Sun it passes through superior conjunction."

2. Write a definition of some term in your academic discipline. Extend the definition with such devices as exemplification and comparison. Use a graphic if it will aid the reader.

3. Look at the list of topics on page 42. Choose one of the topics and devise an organizational plan for it. Use one of these rhetorical devices for your organizing principle: topical arrangement, classification or division, or comparison and contrast.

4. Write a paper in which you find, explore, and explain thoroughly some trend in the contemporary world. The possibilities are

limitless, but you may be happier exploring some trend in your own field. For example, are you in food science? Then you might be interested in the latest trends in packaging or in the percentage of meals eaten away from home. Are you in forestry? Are there trends in the use and kinds of wood products? Your best sources for this paper are books of facts such as almanacs and yearbooks. Examples of such books are *Information Please Almanac, The New York Times Encyclopedic Almanac, The World Almanac and Book of Facts, The Statistical Abstract of the United States, Canada Yearbook, Agricultural Statistics, Americana Yearbook, The Yearbook of Labour Statistics, The Municipal Yearbook, The United Nations Demographic Yearbook, The United Nations Statistical Yearbook.*

5. Write a causal analysis of some trend or trends, perhaps using the same sources and facts that you used for Exercise 4. Your causal analysis can take various forms:

    Trend A caused event B.
    Both trends A and B are caused by event X.
    If trends A and B continue, surely Y will result.

    Your exploration of causality may be a matter of showing the significance of some trend to a specific audience. For example, of what significance is the rising average age of the United States population to clothing manufacturers, restaurant owners, educators, or architects? Be very factual and don't push your conclusions beyond what is warranted by your evidence.

    Either Exercise 4 or 5 could well be written as a memorandum to an executive. See pages 303–306.

CHAPTER 6

Technical Narration, Description, and Argumentation

In the last chapter, we discussed technical exposition. In this chapter, we turn to narration, description, and argumentation. You will have frequent use of all four modes of discourse, sometimes in pure form, but more often in combination. You might, for example, use argumentation as the dominant organizational plan for a report supporting the feasibility of some proposed research. But within your argument you would probably use such expository devices as definition, exemplification, and causal analysis. Further, you might describe equipment to be used and narrate the history of past research to support your argument. The modes are means to ends, not ends in themselves. The ends, always, are to support your purpose persuasively and to satisfy the needs of your material and audience.

Narration

Narration tells a story—in technical writing a factual story. Narration relates a series of events in an ordered sequence, usually chronological. The second paragraph of the following passage, an introduction to an article that describes a nuclear reaction that took place over two billion years ago, is a typical example:

> In 1942, when Enrico Fermi and his associates started up their nuclear-fission reactor at Stagg Field in Chicago, there was every reason to believe it was the first such reactor on the earth. The record book must

now be corrected. In an open-pit uranium mine in the southeastern part of the Gabon Republic, near the Equator on the coast of West Africa, are the dormant remains of a natural fission reactor. Within a rich vein of uranium ore the natural reactor once "went critical," consumed a portion of its fuel and then shut down, all in Precambrian times. The experiment at Stagg Field had been anticipated by almost two billion years.

The history of the natural reactor is an extraordinary sequence of seemingly improbable events. First, uranium from an entire watershed accumulated in concentrated local deposits, including one at a place now called Oklo. Then the conditions necessary to sustain the fission chain reaction were established; these included constraints on the concentration of uranium in the ore, on the size and shape of the lode and on the amount of water and other minerals present. After the reactor had shut down, the evidence of its activity was preserved virtually undisturbed through the succeeding ages of geological activity. Finally, the discovery of the reactor involved an investigative tour de force worthy of the best sleuths in detective fiction.[1]

Narration is frequently used in technical writing to provide a historical overview that will aid the reader in understanding some exposition that is to come. The passage quoted above serves such a use. For the most part, such narration is related in a businesslike way. Sometimes, however, the writer may choose to use narration more dramatically—perhaps in brochures or advertisements for a lay audience—to simplify complex ideas and provide human interest. The following introduction to a National Aeronautics and Space Administration brochure written for the general public fulfills both these purposes:

When Galileo Galilei looked through his first primitive telescope in 1610, the surprise that greeted him—four hitherto unknown moons orbiting Jupiter—rocked the world and started a new revolution in scientific inquiry that continues today.

Men had wondered for millenia about the stars that lit the night sky. They puzzled over a few objects, pinpoints of light that didn't obey the rules, that meandered among the fixed stars, refusing to settle down in regular orbits. Men named them "planets" or "wanderers." Violent arguments—even inquisitions—raged over the question, what were these wandering stars?

But each new century found men taking small painful steps along the road called technology. New instruments were invented, first to measure the position of the wanderers: then, at last to magnify their images. Man realized the wanderers were other worlds, perhaps like his own.

---

[1] George A. Cowan, "A Natural Fission Reactor," *Scientific American*, July 1976, pp. 36–47.

Once man learned to break free of Earth's gravity and the murky atmosphere that clouded his vision, he reached out to those wandering worlds. He photographed them from close range, studied their atmospheres and surface markings and chemistry, probed beneath their skins so that he might understand what went on deep within.

As he explored, man came to understand that those other worlds relate to his home planet. They provided clues to Earth's origin and distant history, clues to how man might manage his home planet in the future.

Now one of the most exciting of those voyages to other planets—by two advanced Mariner-class spacecraft—the Voyager Project—will examine Galileo's Jovian moons in ways he couldn't have dreamed possible and will observe mighty Jupiter from close range. Then they will probe even farther to study yet another object that Galileo wondered about—the ringed planet Saturn and its satellites.[2]

Narrative overviews don't have to be restricted to past events. They can also be used to forecast future events. The writer of the above passage followed it with a narration of the projected flight plan of the Voyager Project:

The National Aeronautics and Space Administration will send the two spacecraft to advance our knowledge of the outer solar system—Jupiter, Saturn, their satellites, and perhaps even the farther reaches of space.

The spacecraft will be launched in the late summer of 1977, and will fly past Jupiter in 1979. Using the big planet's immense gravity to boost them on their way, the Voyagers will be accelerated toward Saturn, reaching the ringed planet in 1980 and 1981. There is also the possibility that one of the spacecraft will visit Uranus in 1986.

And then what, beyond Saturn? Both spacecraft will continue farther from the Sun, probing, studying and searching as they go. Far from Earth they will penetrate into galactic space, beyond the influence of the Sun, where they will cruise for eternity.[3]

Narration also plays an important role in relating the details of an experiment. Often the events of an experiment may best be described in chronological order: "Placing the bunsen burner under the test tube, we brought the solution to 98°C," and so forth. Notice the narrative movement in this excerpt from an account of an experiment with a new haying system.

The baling chamber of a standard 14 by 18-inch baler was reduced to 12 by 12-inch size. Other parts of the baler were modified to produce

---

[2]U.S., National Aeronautics and Space Administration, *Voyager—Journey to the Other Planets*, prepared by the Jet Propulsion Laboratory (Pasadena, Calif., 1977), p. 1.
[3]Ibid., p. 2.

well-formed, cube-shaped bales. Few problems were encountered in baling hay with high moisture content. The bale thrower functioned well with the small cubic bales. Most experimental bales were produced at a rate of 4 to 6 tons of dry hay per hour. Higher rates caused greater variation in the length and density of bales.

Whenever possible, bales were handled by mechanical means throughout the haying operation: A bale thrower loaded bales into trailing wagons. Hay was then unloaded from the rear of the self-unloading forage wagon into the hopper of an ordinary chain-and-flight farm elevator. The elevator delivered bales to a distributing conveyor above the drying compartment of the bin. Here, a plow that moves the length of the distributing conveyor discharged bales alternately to either side, uniformly filling the drying compartment. The handling rate of bales in and out of the drier was normally about 12 tons per hour.

Hand labor was required to rake bales from the trailing wagons into the elevator hopper. Some experimental work was done on a mechanical device to perform this function.

After the bales have been dried, the front wall of the bin can be swung open. The conveyor inside the bin then moves the bales to the front, where a man rakes them directly into an elevator hopper. At the University's Rosemount Station, dry bales were removed from the bin, conveyed to self-unloading wagons, and hauled to a storage area. In a farm setup, the drier should be located near storage areas so the bales can be conveyed directly from the drier to storage.[4]

In describing the experiment, the writer is actually describing process, and process description is probably the chief use of narration in technical writing. By process we mean a sequence of events that in chronological order progresses from a beginning to an end and results in a change or a product. The process may be humanly controlled, such as the manufacture of an automobile. Or it may be natural—the metamorphosis of a caterpillar to a butterfly, for example. In either case the key phrase is *chronological order*. It's the chronological order of the events that provides the organizational structure for process description. Figure 6-1 illustrates the idea nicely.

Process descriptions are written in one of two ways:

- *For the doer*—to provide instructions for performing the process.
- *For the interested observer*—to provide an understanding of the process.

A cake recipe provides a good example of instructions for performing a process. You are told when to add the milk to the flour,

---

[4] John Strait, "Haying Systems Using Cubic Bales," *Minnesota Science* 27, no. 4 (1971): 11–12.

Figure 6-1   A Verbal and Pictorial
Description of a Process

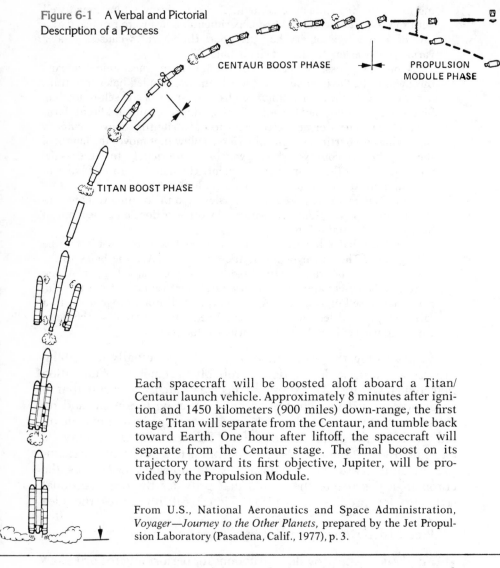

CENTAUR BOOST PHASE   PROPULSION
                             MODULE PHASE

TITAN BOOST PHASE

Each spacecraft will be boosted aloft aboard a Titan/
Centaur launch vehicle. Approximately 8 minutes after igni-
tion and 1450 kilometers (900 miles) down-range, the first
stage Titan will separate from the Centaur, and tumble back
toward Earth. One hour after liftoff, the spacecraft will
separate from the Centaur stage. The final boost on its
trajectory toward its first objective, Jupiter, will be provided by the Propulsion Module.

From U.S., National Aeronautics and Space Administration,
Voyager—Journey to the Other Planets, prepared by the Jet Propulsion Laboratory (Pasadena, Calif., 1977), p. 3.

when to reserve the whites of the eggs for later use. You are instructed
to grease the pan *before* you pour the batter in, and so forth. Writing
good performance instructions is such an important application of
technical writing that we have devoted all of Chapter 14 to it. In this
chapter we'll explain only the second type—providing an understanding of the process.

Insofar as you have grasped the proper order of events, the organization of a process description will not present any particular

problem to you. You simply follow the order of events as they occur. However, the amount of detail you include is likely to be a problem.

In writing a set of instructions, your goal is to have the reader perform the process. Therefore, you must be complete in every detail. In writing process descriptions to provide understanding, you'll find that extensive detail is not always necessary or even desirable. When *Time* magazine, for example, publishes an article about open-heart surgery, its readers do not expect a complete set of instructions on how to perform such an operation. Rather, they expect their curiosity to be satisfied in a general way. In selecting detail, be guided by your purpose and audience.

Our next example comes from a book called *Wood Handbook*. Its subtitle is *Wood as an Engineering Material*, and its preface makes clear that its primary audience is composed of architects and engineers. One small section of it deals with shipworms, marine organisms that attack wood immersed in salt water:

> Shipworms are the most destructive of the marine borers. They are mollusks of various species that superficially are wormlike in form. The group includes several species of *Teredo* and several species of *Bankia*, which are especially damaging. These are readily distinguishable on close observation but are all very similar in several respects. In the early stages of their life they are minute, free-swimming organisms. Upon finding suitable lodgement on wood they quickly develop into a new form and bury themselves in the wood. A pair of boring shells on the head grows rapidly in size as the boring progresses, while the tail part or siphon remains at the original entrance. Thus, the animal grows in length and diameter within the wood but remains a prisoner in its burrow, which it lines with a shell-like deposit. It lives on the wood borings and the organic matter extracted from the sea water that is continuously being pumped through its system. The entrance holes never grow large, and the interior of a pile may be completely honeycombed and ruined while the surface shows only slight perforations. When present in great numbers, the borers grow only a few inches before the wood is so completely occupied that growth is stopped, but when not crowded they can grow to lengths of 1 to 4 feet according to species.[5]

Notice that no attempt is made to give the full information about the shipworm that an entomologist might desire. We don't learn, for example, how the shipworm reproduces, nor do we even learn very clearly what it looks like. For the intended readers of the *Wood Handbook*, such information is not needed. They do need to know what is

---

[5] U.S., Department of Agriculture, *Wood Handbook*, prepared by the Forest Products Laboratory (Washington, D.C., 1974): p. 17.

presented—the process by which the shipworm lodges on the wood and how it bores into it. Knowing this, they can be alert to the potential for damage represented by the shipworm. Also, they will understand the need for the kinds of preventive measures, such as chemical treatment and sheathing, explained later in the text.

In another section of the book, the process of heat transfer through a wall is described:

Heat seeks to attain a balance with surrounding conditions, just as water will flow from a higher to a lower level. When occupied, buildings are heated to maintain inside temperature between inside and outside. Heat will therefore be transferred through walls, floors, ceilings, windows, and doors at a rate that bears some relation to the temperature difference and to the resistance to heat flow of intervening materials. The transfer of heat takes place by one or more of three methods—conduction, convection, and radiation (see figure).

Conduction is defined as the transmission of heat through solid materials; for example, the conduction of heat along a metal rod when

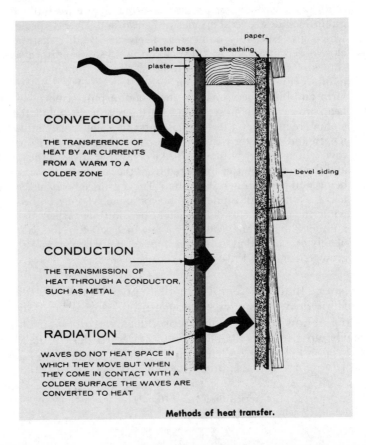

plaster base　plaster　sheathing　paper

CONVECTION

THE TRANSFERENCE OF
HEAT BY AIR CURRENTS
FROM A WARM TO A
COLDER ZONE

bevel siding

CONDUCTION

THE TRANSMISSION OF
HEAT THROUGH A CONDUCTOR,
SUCH AS METAL

RADIATION

WAVES DO NOT HEAT SPACE IN
WHICH THEY MOVE BUT WHEN
THEY COME IN CONTACT WITH A
COLDER SURFACE THE WAVES ARE
CONVERTED TO HEAT

Methods of heat transfer.

one end is heated in a fire. Convection involves transfer of heat by air currents; for example, air moving across a hot radiator carries heat to other parts of the room or space. Heat also may be transmitted from a warm body to a cold body by wave motion through space, and this process is called radiation because it represents radiant energy. Heat obtained from the sun is radiant heat.

Heat transfer through a structural unit composed of a variety of materials may include one or more of the three methods described. Consider a frame house with an exterior wall composed of gypsum lath and plaster, 2- by 4-inch studs, sheathing, sheathing paper, and bevel siding. In such a house, heat is transferred from the room atmosphere to the plaster by radiation, conduction, and convection, and through the lath and plaster by conduction. Heat transfer across the stud space is by radiation and convection. By radiation, it moves from the back of the gypsum lath to the colder sheathing; by convection, the air warmed by the lath moves upward on the warm side of the stud space, and that cooled by the sheathing moves downward on the cold side. Heat transfer through sheathing, sheathing paper, and siding is by conduction. Some small air spaces will be found back of the siding, and the heat transfer across these spaces is principally by radiation. Through the studs from gypsum lath to sheathing, heat is transferred by conduction and from the outer surface of the wall to the atmosphere, it is transferred by convection and radiation.[6]

The readers are furnished far more detail in this description than they were in the description of the shipworm. Even an illustration is provided to make the process more understandable. The process of heat transfer, while still background information, is more useful to the architects and engineers than is the life cycle of a shipworm. Many factors of design depend upon a complete understanding of the process. Again, audience and purpose have determined the amount of detail presented.

Nothing we have said makes the selection of the content for any process description easy. You must exercise a good deal of judgment in the matter. But we have given you some guidelines to aid your judgment. As in all writing, you must decide what your audience needs to know to satisfy its purpose and yours.

The matter of sentence structure is also of some importance in process description. In writing performance instructions, you will commonly use the active voice and imperative mood—sentences like "Clean the threads on the new section of pipe" and "Add pipe thread compound to the outside threads." In following your instructions, it's the reader, after all, who is the doer. With its implied *you*, the imperative voice directly addresses the reader. But in the generalized

---

[6] Ibid., pp. 20–24.

process description for understanding, where the reader is not the doer, the use of imperative mood would be inappropriate and even misleading.

In writing a process for understanding, therefore, you'll ordinarily use the indicative mood in both active and passive voice:

- *Active voice*—The size of the cover opening controls the rate of evaporation.
- *Passive voice*—The rate of evaporation is controlled by the size of the cover opening.

In active voice the subject does the action. In passive voice the subject receives the action. Use of the passive emphasizes the receiver of the action while deemphasizing or removing completely the doer of the action. (Incidentally, as the above examples illustrate, neither doer nor receiver has to be a human being or even an animate object.) In situations where the doer is unimportant or not known, you should choose passive voice. Conversely, in situations where the doer is known and important, you should choose the active voice. Because the active is usually the simpler, more direct statement of an idea, choose passive voice only when it is clearly indicated. We have more to say on this subject on pages 171–172.

Description

You may have had instruction in writing description before. But probably it was geared to imaginative writing in which you were asked to evoke a feeling for some place or object. Technical description may be imaginative, but it rarely has anything to do with evoking feeling. Rather, technical description is used to provide a rational understanding of the thing described, whether that thing is a mechanism, a place, or even, for example, a book.

The physical description of some mechanism is perhaps the most common kind of technical description. It is a commonplace procedure with little mystery attached to it. For example, we see a friend smoking a pipe and endangering his thumb by using it to tamp down hot tobacco coals. When he finishes his smoke, he risks cracking his pipe by banging it on an ashtray to shake out the ashes.

"Look," we say to him, "you ought to buy a pipe tamper and cleaner for yourself."

"What's that?"

"Well, it's a tool about the size of a penknife. In fact, it's made like a penknife. It's a narrow steel case with a dull knife blade and a thin pick folded into it that can be pulled out like penknife blades. One end of the case is round and flattened, about the size of a dime. That end of the case is the tamper. You use that to pack down the tobacco so that it burns more evenly.

"The pick is a thin piece of steel about the diameter of a toothpick and about half-an-inch longer. You use the pick to loosen the ashes in your pipe and turn them out into an ashtray. That way you don't have to bang your pipe and get it all dented."

"That so? What's the dull knife blade for?"

"You use the blade to ream out the cake that builds up in your pipe after you've smoked it awhile. The blade has a round tip, rather than a sharp one like most knife blades, so that you don't dig into the bottom of your pipe. The blade is sharp enough to cut through the cake but not sharp enough to cut into your pipe."

"Sounds like a handy gadget. I'll have to get one."

In this brief passage, we have used most of the techniques of good technical description. Our purpose has been to make you *see* the object and understand its function. We have described the overall appearance of the pipe cleaner and tamper and named the material with which it is made, steel. We have broken it down into its component parts. We have described the appearance of the parts and given their functions. We have used comparisons to familiar objects such as penknives, dimes, and toothpicks to clarify and shorten the description.

The use of comparison to familiar objects and the use of familiar terminology in general will greatly aid your audience in visualizing the thing described. We visualize things in essentially five ways—by shape, size, color, texture, and position—and have access to a wide range of comparisons and terms to describe all five.

For *shape*, you can describe things with terms such as cubical, cylindrical, circular, convex, concave, square, trapezoidal, or rectangular. You can use simple analogies such as C-shaped, L-shaped, Y-shaped, cigar-shaped, or spar-shaped. You can describe things as threadlike or pencillike or as sawtoothed or football shaped.

For *size*, you can give physical dimensions, but you can also compare objects to coins, paper clips, cigarette packages, bread boxes, and football fields.

For *color*, you can obviously use familiar colors such as red and yellow and also, with some care, such descriptive terms as pastel, luminous, dark, drab, and brilliant.

For *texture*, you have many words and comparisons at your disposal, such as pebbly, embossed, pitted, coarse, fleshy, honeycombed, glazed, sandpaperlike, mirrorlike, and waxen.

For *position*, you have opposite, parallel, corresponding, identical, front, behind, above, below, and so forth.

As we pointed out before, technical description is rational, but nothing bars it from being imaginative so long as you apply your imagination accurately.

The following brief passage shows you how a combination of dimensions and a few words indicating shape and position can help you accurately visualize a canal in cross section:

> What was a barge canal like? Engineering experience had settled the overall geometry of canals well before the end of the 18th century. They were trapezoidal in cross section; at the bottom their width was 20 to 25 feet and at water level it was 30 to 40 feet. The water in them was only three to four feet deep, so that the angle of the sides was quite flat. This gradual slope helped to minimize erosion due to wave action and slumpage due to seepage.[7]

A description of a mechanism can be as short as several sentences, as in this description of a cement mason's "darby."

Darby
A Darby (Fig. 15) is a long, flat, rectangular piece of wood, aluminum or magnesium from 30 to 80 in. long and from 3 to 4 in. wide with a handle on top. It is used to float the surface of the concrete slab immediately after it has been screeded, to prepare it for the next step in finishing. This tool should be used to eliminate any high or low spots or ridges left by the straightedge. It should also sufficiently embed the coarse aggregate for subsequent floating and troweling.[8]

Mechanism descriptions can also be lengthy pieces of writing. They grow in relation to the complexity of the thing described. In their longer form, they often follow a fairly standard pattern consisting of an introduction and a body.

In the *introduction* you give the function of the mechanism. You either describe its overall appearance or provide a graphic to fulfill that function. You divide the mechanism into its component parts and give an overview of how they work together. The following introduction to an encyclopedia description of a basic camera is typical:

---

[7] John S. McNown, "Canals in America," *Scientific American*, July 1976, pp. 116–124.
[8] *Cement Mason's Manual for Residential Construction* (Skokie, Ill.: Portland Cement Association, 1960), p. 12. Reprinted by permission.

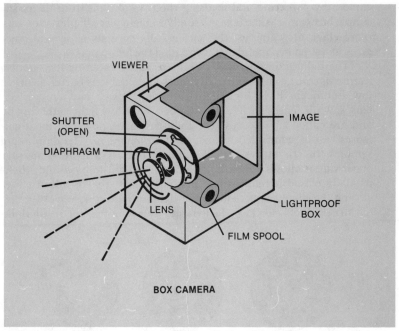

VIEWER

SHUTTER
(OPEN)

DIAPHRAGM

IMAGE

LENS

LIGHTPROOF
BOX

FILM SPOOL

BOX CAMERA

Camera, an instrument for taking photographs. Every camera has four basic parts: a light-tight box which is called the camera body, a lens, a shutter, and a film holder. When an object is photographed, the shutter opens to admit rays of light from the object, the lens gathers the rays and focuses them to produce a latent image on a light-sensitive film, and the shutter then closes. For an explanation of the process by which a photograph is produced from the latent image on the exposed film, see PHOTOGRAPHY: Photographic Process.[9]

In the *body* of the description, you describe the parts and give their function. (If the parts are complex, they, in turn, may be divided into *their* component parts.) You again use graphics when appropriate. Here is how the same encyclopedia article describes the body and lens of the basic camera:

Body. The camera body serves as a rigid framework on which all other parts of the camera are mounted, and it also serves to protect the film from exposure to light except at the instant a photograph is taken. The bodies of most modern cameras are made of plastic or metal.

Lens. The lens is mounted over an opening at the front of the camera body. Its function is to produce an image on the film at the back of the

[9] *Collier's Encyclopedia,* "Camera." Reprinted with permission from Collier's Encyclopedia © 1969 Crowell-Collier Educational Corporation.

camera by gathering and focusing the rays of light from the object. In simple box cameras the lens is usually a single piece of plastic or glass. In more elaborate cameras, the lens usually consists of several pieces of glass mounted in a metal cylinder called the lens barrel.

The light-gathering power of a lens is indicated by its *f-number*, or *relative aperture*, which is usually marked on the lens barrel. Lenses with low f-numbers have high light-gathering power and are called fast lenses, while lenses with high f-numbers have low light-gathering power and are called slow lenses. The light-gathering power of most camera lenses can be controlled by means of a *diaphragm*. The diaphragm resembles the iris of the eye. (See illustration.) When the diaphragm is open wide it allows a maximum amount of light to pass, and when it is partially closed it reduces the amount of light that is admitted. Since closing the diaphragm reduces the light-gathering power of the lens, it has the effect of "slowing" the lens and thus of increasing its f-number.

DIAPHRAGM.   In many cameras the amount of light that strikes the film can be controlled by opening or closing the diaphragm. The diaphragm illustrated here is designed for an f/4 lens. It is shown at its smallest aperture, f/22, opened three stops, f/8, and opened two more stops to f/4, the diaphragm's largest aperture.

Usually, no ending is necessary in a mechanism description. You stop when you have described the last part. The camera description we've been working with, for example, continues on to describe the shutter and the film holder and then simply stops. Sometimes, however, you may want to round off your description with a summary or a complimentary close (see pages 197–201).

The camera description we have shown you appears in a general encyclopedia and is intended for a lay audience. Therefore, it has the attributes of being simply written and fairly complete in its description of parts and functions. Its purpose is primarily instructional. In writing for an audience more knowledgeable about a mechanism, you might allow your description to have more professional terminology and to be more abbreviated. And if your purpose is to point out to a potential customer the good features of your mechanism, your description of the functions might be sales oriented. All these differences are evident in the catalog description of Trane Process Air Handlers shown in Figure 6-2.

Figure 6-2   Trane Catalog Description

PROCESS AIR HANDLERS

Trane Process Air Handlers are designed to meet the most stringent indoor and outdoor applications in clean or contaminated environments. Units are also designed to operate under high and low temperature applications. Air handlers provide cooling and heating of air and air mixtures, air filtering, vapor condensation, heat recovery and humidification. Spark proof and/or noncorrosive construction is available when required.

Some of the common features available on Trane Process Air Handlers include:

• Dampers — face and bypass, leak-tight, high pressure, noncorrosive materials, operators as required and special bearings.
• Filters — inertia, roll, bag, cartridge and charcoal.
• Heat Exchangers — plate fin, fin and tube and electric coils.
• Fans — centrifugal, with FC, BI, AF wheels, and vane axial.
• Access doors and panels.
• Bird screens.
• Motors and electrical controls as required.
• Humidifiers.

From *Trane Process Machinery, Heat Transfer and Combustion Equipment* (La Crosse, Wisc.: The Trane Corporation, 1976). Copyright © The Trane Corporation, 1976, used by permission.

Up to this point, we have dealt with static mechanisms, that is, mechanisms that have limited or no movement. Obviously, many mechanisms—think of an internal combustion engine—are dynamic. When you describe the functions of a dynamic mechanism and its components, you're describing motion. The following example concerning a four-stroke-cycle engine illustrates this point:

The Four-Stroke Cycle. In the four-stroke cycle an intake valve is opened as the piston is at the top of the cylinder, and the charge of fuel and air is drawn into the cylinder by the partial vacuum created by the piston's traveling down the cylinder. When the piston reaches the bottom of its stroke, the intake valve is closed and the piston, traveling upward, compresses the charge. When the piston reaches the top again, the mixture is ignited, and the resulting expansion of the hot gases forces the piston down. When the piston is at the bottom of its stroke again, the exhaust valve opens and the rising piston forces out the exhaust gases, leaving the cylinder clear for the introduction of the next charge of fuel and air. The entire process requires two complete down-and-up motions of the piston or two revolutions of the crankshaft. A flywheel stores up enough energy from the power stroke to carry the piston through the other three trips before the next power stroke.[10]

When you describe a dynamic mechanism, you move into process description and the use of narration. It's a good example of what we have several times emphasized about the manner in which the modes of discourse are mutually supportive rather than mutually exclusive. For a more elaborate example of a description of a dynamic mechanism, complete with illustrations, see pages 319–322.

In mechanism description as in all writing you must consider your purpose and the needs of your audience. Therefore, the organization and content of mechanism descriptions are variable. But they are all likely to include statements about function, some sort of division of the mechanism into its component parts, and physical descriptions in words or graphics or some combination of the two.

Many of the principles found in mechanism description have nonmechanical applications. For example, we don't normally think of wood as a mechanism, but the same rational principles we have been discussing are found in the following passage about wood. The subject is divided, objective physical details are described, and function is discussed:

Dry wood is made up chiefly of the following substances, listed in decreasing order of amounts present: Cellulose, lignin, hemicelluloses, extractives, and ash-forming minerals.

Cellulose, the major constituent, comprises approximately 50 percent of wood substance by weight. It is a high-molecular weight linear polymer that, on chemical degradation by mineral acids, yields the simple sugar glucose as the sole product. During growth of the tree, the linear cellulose molecules are arranged into highly ordered strands called fibrils, which in turn are organized into the larger structural ele-

---

[10] Ibid., "Engine."

ments comprising the cell wall of wood fibers. The intimate physical, and perhaps partially chemical, association of cellulose with lignin and the hemicelluloses imparts to wood its useful physical properties. Delignified wood fibers have great commercial value when reconstituted into paper. Moreover, they may be chemically altered to form synthetic textiles, films, lacquers, and explosives.

Lignin comprises. . . .[11]

The author completes the description by describing in order lignin, hemicelluloses, extractives, and ash-forming minerals.

The same techniques can even be applied to something as nonmechanical as a book. In the following passage, a public-relations device known as a capabilities booklet is described. The booklet is broken down into its component parts, and each part is described in terms of content and function:

> No single list of contents can possibly apply to all capability books. Here, however, are some of the major subjects you may find helpful in building your own booklet. This list is not necessarily in sequential order.

> *Brief corporate profile:* 200 to 600 words, usually up front, defining the company's new situation.
> *Company creed:* Short philosophic credo—if meaningful.
> *Short history:* Growth, expansion, scope, etc. with much more emphasis on the past five years than on the previous 20.
> *Photos of facilities:* Brief captions with outstanding photographs explaining advantages of using this equipment or this procedure or this service.
> *Photos or drawings of new products:* (Where applicable.) Range, versatility, or unique results of the facilities illustrated.
> *Proofs of performance:* Photos of current installations or applications with brief comment of where, what, and the *results* if available.
> *Differences or advantages:* Diagrams or copy underscoring major differences between your current capabilities compared with former methods or with those of competitors (without mentioning names.)[12]

Writers of technical material must often describe places as well as mechanisms and other objects. Place descriptions appear in research reports when the location of the experiment has some bearing on the research. Highway engineers must describe the locations of

---

[11] Department of Agriculture, *Wood Handbook*, pp. 2–5.
[12] *The Mead Short Course in the Graphic Arts, Vol. 4, Capabilities Booklets* (Dayton, Ohio: Mead Paper, 1974), p. 10. Copyright © 1974 by Mead Paper. All rights reserved.

projected highways. Surveyors have to describe the boundaries of parcels of property. Police officers have to describe the scene of an accident. In the following passage, an archaeologist reconstructs through imagination and scientific knowledge a Roman outpost in second-century Britain:

> A new fort, built of stone, rose on the eastern part of a level plateau. The enclosure was oblong; at the north and south ends the fort's massive walls, broken by central gates, run 85 meters; on the east and west sides the length of the walls is just under 150 meters.
>
> The overgrown ruins of the last wood fort at Vindolanda, abandoned 40 years earlier, stood just beyond the west wall of the new fort. Here the garrison engineers must have dumped many cartloads of clay in the process of covering up the earlier structures and preparing a level site for the stone foundations of what we call Vicus I. On both sides of the main road leading from the west gate were erected a series of long barrackslike buildings; the masonry was dressed stone bound with lime mortar. The largest barracks was nearly 40 meters long, and all of them seem to have been about five meters wide. We deduce that the structures were one story high and probably served as quarters for married soldiers. They were divided into single rooms, one to a family, by partitions spaced about seven meters apart. There was probably storage space under the roof, and a veranda outside each room would have provided cooking space. The four barracks we have located so far could have housed 64 families. The British army in India offered almost identical housing to the families of its Indian soldiers. Smaller buildings, possibly housing for the families of noncommissioned officers, stood nearby.[13]

When describing a place, you must be aware of your point of view—that is, where you are positioned to view the place being described. In the above passage, the author begins with an aerial point of view. He looks down upon the fort and describes it in general outline—its walls, roads, and buildings. That job done, he returns to ground level, and his point of view is now that of a well-informed guide walking you through the barracks. Notice the masonry, he says. Here are the rooms. One family lived in each. See how the rooms were partitioned off and where the storage space and cooking areas were. The writer uses the descriptive language of size, shape, and position. Physical description and function are gracefully combined.

When describing a place, if you, in your mind's eye, position yourself physically before you begin to write, you'll avoid confusing and annoying your readers with an inconsistent jumping about in

---

[13] Robin Birley, "A Frontier Post in Roman Britain," *Scientific American*, February 1977, pp. 39–46.

your description. You can shift your point of view as our archaeologist has done, but be aware when you are making the shift. Have a good reason for doing it, and don't do it too often. Most often the reason will be to allow yourself to first give an overview and then proceed to a closeup of selected details. That is evident in the above passage.

The following passage concerns the details photographed by two spacecraft on Venus. The scene is visualized from the point of view of the spacecraft's vantage points on the surface:

> Details on the surface of Venus were virtually unknown until two Soviet Venera spacecraft landed on it in 1975 and returned pictures. Earlier speculations had ranged from steaming swamps to dusty deserts, from carbonated seas to oceans of bubbling petroleum. All were wrong, though the desert concept seems closest to what is now known. Photographs from the Venera spacecraft revealed a dry rocky surface fractured and changed by unknown processes. There are rocks of many different kinds and a dark soil. The two spacecraft landed about 2000 km (1200 mi) from each other. One landed on an ancient plateau or plains area. At this site there are rocky elevations interspersed with a relatively dark, fine-grained soil. This soil seems to have resulted from a weathering of the rocks, possibly by a chemical action. The rocky outcrops are generally smooth on a large scale, with their edges blunted and rounded. The dark soil fills some of the cavities in the rocks. By contrast the other Venera landed at a site where there are rocks which look younger and less weathered, with not much evidence of soil between them.[14]

In this passage, the use of textured language is striking: *dry rocky surface fractured and changed by unknown processes; fine-grained soil; weathered; smooth; blunted and rounded.* Color is not mentioned, but the shade *dark* is a dominant note.

Remember that in any description you are selecting details. You would find it quite impossible to describe everything. Very often, the details chosen support a thesis, perhaps an inference or chain of inferences. Notice how the selected details in the following passage about craters on the moon support the final inference:

> In profile, the Surveyor V crater is dimple-shaped; it lacks a raised rim, and the slope of the crater wall increases gradually toward the center of the crater. It has a distinct small concave floor, however, about 2.4 meters wide. Both in plain view and in profile, the Surveyor crater resembles other craters, observed in Ranger pictures and Lunar Orbiter

---

[14] U.S., National Aeronautics and Space Administration, *The Planet Venus* (Washington, D.C.: U.S. Government Printing Office, 1978), p. 4.

photographs, which are inferred to have been formed by drainage of surficial debris into subsurface fissures. We infer that the Surveyor V crater has been formed by drainage of surficial fragmental debris into a fissure that passes beneath the center of the crater and extends for some distance both northwest and southwest beneath the crater chain.[15]

To summarize place description: Choose one or several points of view. Your choices should help to support whatever you are trying to do. To help your reader visualize the place, use the descriptive language of size, shape, color, texture, and position. Select details that support your purpose.

Argumentation

In argumentation, you deal with propositions that lie somewhere between the bounds of verifiable fact on one hand and pure subjectivity on the other. Your purpose is to convince your audience of the high probability that the propositions you are putting forth are correct. Let us illustrate.

Imagine for the moment that you are the planning engineer for a new housing subdivision called Hawk Estates. Hawk Estates, like many such new subdivisions, is being built close to a city, but not in a city. The problem at issue is whether Hawk Estates should build its own sewage disposal plant or tap into the sewage system of the nearby city. (You have already ruled out individual septic tanks because Hawk Estates is being built on nonabsorbent clay soil.) The city will allow the tap-in. You have investigated the situation and have decided that the tap-in is the most desirable alternative. The heads of the company planning to build Hawk Estates are not convinced. It is their money, so you must write a report to convince them.

First, you must state your major proposition: Hawk Estates should tap in to the sewage system of the city of Colorful Springs.

Now you must support your major proposition. Your support normally will consist of a series of minor propositions, which will in turn be supported by verifiable facts and statements from recognized authorities. In this case your support may consist of the following minor propositions and facts.

- *Minor proposition:* The city sewage service can handle Hawk Estates' waste. *Support:* To support this proposition you give the

---

[15] U.S., National Aeronautics and Space Administration, *Surveyor V, A Preliminary Report* (Washington, D.C.: Government Printing Office, 1967).

estimated amount of waste that will be produced by Hawk Estates, followed by a statement from the Colorful Springs City Engineer that the city system can handle this amount of waste. In an appendix put the actual figures that support the City Engineer's statement. This first minor proposition is your most important one. Questions of cost, convenience, and so forth would be irrelevant if Colorful Springs could not furnish adequate service.

- *Minor proposition:* The overall cost to Hawk Estates taxpayers will be only slightly more if they are tapped in to the city rather than having their own plant. *Support:* For facts you would list the initial cost of the plant versus the cost of the tap-in. You would further state the yearly fee charged by the city versus the yearly cost of running the plant. You might anticipate the opposing argument that the plant will save the taxpayers money. You could break the cost down per individual taxpayer, perhaps showing that the tap-in would cost an average taxpayer only an additional $4 a year, a fairly nominal amount.

- *Minor proposition:* The proposed plant, a sewage lagoon, will represent a nuisance to the homeowners of Hawk Estates. *Support:* Because the cost for the tap-in is admittedly higher, your argument will probably swing on this minor proposition. State freely that well-maintained sewage lagoons do not smell particularly. But then point out that authorities state that sewage lagoons are difficult to maintain. Furthermore, if not maintained to the highest standards, they emit a distinct odor. To clinch your argument, you show that the only piece of land in Hawk Estates large enough to handle a sewage lagoon is upwind of the majority of houses, during prevailing winds. With the tap-in, of course, all wastes are carried away from Hawk Estates and represent no problem of odor or unsightliness whatsoever.

In your conclusions you point out that in cost and the ability to handle the produced wastes, the proposed plant and the tap-in are essentially equal. But, you point out, the plant will probably become an undesirable nuisance to Hawk Estates. Therefore, you recommend that the builder should choose the tap-in over the plant.

Throughout any argument you appeal to reason. In most technical writing situations an appeal to emotion will make your case immediately suspect. Never use sarcasm in an argument. You never know whose toes you are stepping on or how you will be understood. Support your case with simply stated verifiable facts and statements from recognized authorities. In the example above, a statement from potential buyers that "sewage lagoons smell" would not be adequate.

But a statement to the same effect from a recognized engineering authority from a nearby university would be acceptable and valid.

The intention of argument is to convince. Its major technique is the use of a major proposition followed by a series of supported minor propositions. Within your propositions, anticipate and answer opposing arguments when you can.

Because their conclusions present probabilities rather than verifiable facts, the discussion sections of many scientific and technical reports are often rather subtle and understated arguments. Such is the case in the following discussion section from a report on an experiment with a tree-harvesting process known as "strip thinning."

About 200,000 acres of red pine 30–100 years old in northern Minnesota are expected to be available for thinning during the next 10 years. Strip thinning is a method of making mechanized harvesting adaptable to the management of these stands. The first cut, using tree length or full tree harvesting, can be made with minimum disturbance to the forest. These systems provide for efficient operation and removal of the harvest from the forest as tree length logs for processing at the mill. This strip thinning will leave the forest in a condition adaptable to future thinnings by individual tree selection or additional strips.

The layout of strips and landings can be used for the life of this stand, thus reducing future operation costs. If carefully planned, strips and landings can make the forest more adaptable to multiple use. Landings have potential use for wildlife openings, overflow campgrounds, and hunting sites. Strips can be oriented to increase snow accumulation and delay snowmelt for watershed purposes.

Although a systematic layout of straight strips was used, the system has enough flexibility to meet the varied conditions in the forest. For example, the 16-foot width is not mandatory for maneuverability of skidders on the strip. A 12-foot width would have been acceptable over a large part of the area. Strips need not be perfectly straight: slight curving can be used to take advantage of openings. Some selective cuttings may be possible in the area adjacent to strips.

Mechanized harvesting that permits removal of full trees, coupled with strip cutting that limits disturbance to 25 percent or less of the forest area, can be used for harvesting timber in sensitive areas, such as scenic zones and recreational areas. Careful harvesting under winter conditions would cause less disturbance than summer logging.

Further trials are planned to determine the full potential of these harvesting systems. Different age classes and species mixtures as well as additional harvesting patterns will be included.

These systems offer the forest manager a powerful tool in controlling forest conditions. Besides prescribing the trees to be cut, he also prescribes the removal method, slash placement, and clearing of landings. He can control forest conditions for tree growth, esthetics, wildlife, wa-

ter, or any special use for which required forest conditions can be pre-
scribed. Of course, skill in selecting landings and laying out strips that
can be effectively used in all future thinnings is required.[16]

Notice how the authors move easily from their major
proposition—*Strip thinning is a method of making mechanized harvesting adaptable to the management of these stands*—through a series
of minor propositions.

Within the framework of argument, you can use all the other
modes: exposition, narration, and description.

Persuasive Strategy

In the beginning of the chapter on technical exposition, we spoke of
the modes of discourse as persuasive strategies. We'll close this chapter by exploring briefly the proper choice of a rhetorical strategy. The
strategy or combination of strategies chosen depends in large part
upon the objectivity or subjectivity of the information you are working with. All information ranges along a continuum from complete
objectivity to complete subjectivity. To illustrate this continuum in
the classroom, we often conduct an experiment. We choose a dozen
students and tell them that we are going to give them three situations
and three corresponding questions to which they will respond with
one-word answers. In the first situation, a piece of chalk is held up
without identification and each student is asked to write one word
that will identify the object observed.

In the second situation, we ask each student to write down one
word that describes his or her reaction to modern art, the work of
Picasso or Jackson Pollock, for example.

In the third situation, each student is asked to write a one-word
reaction to the way the president of the United States is handling
whatever happens to be a controversial issue at the moment—the
domestic economy, for example.

Objectivity	Area	Subjectivity
	of	
	Reasonable Argument	

---

[16] Z. A. Zasada and John W. Benzie, *Mechanized Harvesting for Thinning Sawtimber Red
Pine*, Miscellaneous Report 99, Forestry Series 9 (St. Paul, Minn.: University of Minnesota Agricultural Experimental Station, 1970), pp. 13–14.

Then the answers for each situation are tallied. In the first situation, all the students inevitably—well, almost inevitably—write "chalk." In the second situation, we'll get a variety of one-word answers, such as "dislike," "junk," "interesting," "indifferent," "innovative," and "exciting." The reactions to the president's handling of the nation's economic problems may include "satisfactory," "unimpressive," "antiquated," "poor," and so forth.

The first situation, with the chalk, illustrates a response to a totally objective situation. Language is being used to record something that is completely verifiable: close examination and analysis of the object shown would clearly demonstrate whether it was a piece of chalk or not. Most of us understand how verifiable facts are applied, both in science and technology and in everyday life. If someone tells you that a room measures 25' × 60', and you disagree, you don't argue philosophically to prove your point. Rather, you get a tape measure and verify which of you is right or most nearly right. In this objective zone, then, you may describe, narrate, and explain, but you rarely need to argue in order to be persuasive.

The second situation, the reaction to modern art, takes us to the subjective end of the continuum—from the single answer "chalk" to many answers. The answers, in turn, range from favorable words like "interesting" and "innovative" to unfavorable words like "dislike" and "junk." In the objective zone, there is no reason to argue because we can verify the correct response. But subjective reactions to modern art are considerably less verifiable. Can we *prove* that the person who says "junk" is more correct than the person who says "interesting," or vice versa? Obviously, some works of art (regardless of their period) are superior to others, for fairly objective reasons. But one's *reaction* to any artistic trend—primitive, realistic, abstract, and so on—is purely a matter of individual taste: "I don't care if it's the greatest piece of sculpture in the world; to me, it looks like a hard-boiled egg."

In this subjective zone, when we do not share an opinion, its merits cannot be tested by objective scientific means. Although you can use description, narration, and exposition in an effort to challenge another's viewpoint, direct argument would not likely be persuasive because it is likely to butt up against deeply held personal values. You can, however, argue indirectly—"You should eat your spinach. I know you don't like its taste, but it's good for you."

In the last situation, the president's handling of the economic situation, the responses again range from favorable to unfavorable. But the differing opinions here are not purely subjective, as in matters of taste. They are based on the students' best analysis of whatever hard evidence they have concerning the president's actions. Many of our communications fall in this zone of the continuum, between pure objectivity and pure subjectivity. In this zone, argument is permissible—in fact, inevitable. And when we argue, we have at our command all the other modes: exposition, description, and narration.

Although scientific evidence is available on issues that arise in this middle zone, it does not provide indisputable answers. Therefore, we have to evaluate whatever evidence we find and use it as a basis for the conclusions on which we act. If you were an agronomist who had planted corn with 30 inches of space between rows in one field and 40 inches in another, and you found that the first method produced the best yield, what would you conclude? Assuming that other variables in the experiment (such as moisture, fertilizer, and drainage) were carefully controlled, you would probably infer that the 30-inch spacing *caused* the greater yield. But your scientific caution would not allow you to state your findings with any great certainty until you had repeated the experiment several times with the same results. And even then you would present the causal connection of spacing to yield as a conclusion: that is, an opinion, and not as a fact.

Henry David Thoreau once observed that if you find a trout in the milk can you can infer that someone has been watering the milk. He was right. But don't rush into conclusions. Gather your evidence. Weigh all the evidence and decide how much of it *really* supports your conclusions. Do not twist evidence to suit a preconceived result.

The whole process of gathering information, weighing evidence, drawing conclusions, and testing conclusions is, of course, the scientific method at work. Science depends upon the collection of facts and the projection of accurate conclusions based upon those facts. You, as a writer reporting on science and technology, must accurately record both facts and conclusions, using all four modes of discourse to do the job. Locating a particular topic on the objective-subjective continuum helps you to select the most appropriate and, therefore, the most persuasive strategy.

EXERCISES

1. Write a narration of two or three paragraphs that is intended to serve as a historical overview for some larger exposition. Write the overview for a lay audience. Choose as a subject some significant event in your own professional field such as Mendel's work on heredity, Salk's work on polio vaccine, the building of the Panama Canal, the establishment of the structure and function of DNA, the laying down of the first successful transatlantic cable, Nightingale's development of modern nursing.

2. Write two versions of a narration intended to provide an understanding of a process. The first version is for a lay audience whose interest will be chiefly curiosity. The second version is for an expert or technical audience that has a professional need for the knowledge. The process might be humanly controlled such as buying and selling stocks, writing computer programs, fighting forest fires, giving cobalt treatments, or creating legislation. It could be the manufacture of some product—paint, plywood, aspirin, digital watches, maple syrup, fertilizer, extruded plastic. Or you might choose to write about a natural process—thunderstorm development, capillary action, digestion, tree growth, electron flow, hiccuping, the rising of yeast dough.

3. Choose three common household tools such as a can opener, vegetable scraper, pressure cooker, screwdriver, carpenter's level, or saw. For each write a one-paragraph description such as you might find in an encyclopedia. Then write a second version of each paragraph that could serve as catalog description for a particular brand of the product, such as a General Electric can opener or a Stanley saw.

4. Write a description of a dynamic mechanism. You may choose either a manufactured mechanism—such as a farm implement, electric motor, or seismograph—or a natural mechanism, such as a human organ, an insect, or a geyser. Consider your readers to have a professional interest in the mechanism. They use it or work with it in some way. Divide the mechanism into its component parts, and, using both words and graphics, describe its functions and physical appearance.

5. Describe some place that could be of importance to you in your professional field—a river valley, a power plant, a nurse's station, a business office, a laboratory. Have a definite purpose and audience for your description, such as supporting some inference about the place for an audience of your fellow professionals. Or you might emphasize some particular aspect of the place for a lay audience. Use at least two points of view, one that allows you to give an overview of the place and one that allows you to examine it in some detail.

6. Argue pro or con for a proposition such as the following:

    •   Camera X is superior to camera Y.
    •   Organic foods are worth the additional cost.
    •   Nurses should be allowed to prescribe therapy and medication.

- Health care should be nationalized.
- Agricultural chemicals are necessary and safe.
- Nuclear power plants are the answer to power needs in the United States for the next 30 years.

State your major proposition and support it with three to five minor propositions that are, in turn, supported by factual evidence. Anticipate opposing arguments and answer them. Your audience is composed of college-educated lay people.

CHAPTER 7

Organizing Information

In a recent survey, a thousand engineers were asked what bugs them the most about writing. Leading the list of answers at 28 percent was "organizing and outlining."[1] Most writers would agree that organizing is a critical and difficult step. Poor organization leads to many other writing problems such as a lack of clarity and conciseness. Good organization smooths the entire process.

By stressing organization in this chapter we don't mean to imply that it is a static process performed in separation from all the other steps in writing such as gathering information and audience analysis. Rather, organization, as we indicate frequently, is a dynamic process that begins the moment you start thinking about a writing project. In Chapter 3, "Gathering Information," we give you ten steps to follow (page 46). The first one, "Calling Upon Your Memory," suggests that you begin by jotting down information you already possess on a subject. This "jotting down" is not only a start on gathering subject matter; it is also a start on the organizational process. The jotted-down thoughts, random though they may be, begin to give form to the paper you will ultimately write.

At the step labeled "Generalizing from Particulars and Particularizing from Generalities" (page 51) we urge you to "lean back in your chair and take a long, hard look at what you have so far learned." In your long hard look you apply the reasoning processes of induction and deduction. Both are organizational processes that may develop an entire report for you, or at least major sections of it. The

---

[1] Terry C. Smith, "What Bugs Engineers Most About Report Writing," *Technical Communication* 23, no. 4 (1976): 2–6.

last step of the ten is "Reviewing the Information Already Gathered." In this step we urge you to take stock frequently of your material as you gather it. We suggest that this periodic checking will keep you from going off on a tangent and will lead ultimately to a writing outline.

In Chapter 4, "Analyzing Your Audience," we advise you to be always thinking about your audience's purposes as well as your own. Audience factors—what readers already know, what they don't know, what they want to know, what you want them to know—govern organization in many ways. And, of course, the four modes of discourse explained in Chapters 5 and 6—exposition, narration, description, and argumentation—are all methods of organization.

Organization, then, is a continuing, dynamic process involving all the prewriting steps; by the time you reach the moment of getting a plan on paper, you will often find that the basic problem has already been solved. For example, your task may be to give an account of some happening—perhaps an accident. You see at once that a narrative approach is the best. In a rough outline, you line up your facts in chronological order, and you're ready to write. Or perhaps you're trying to persuade someone to action. The argumentation mode is called for. You have your major premise well in hand. You line up your minor premises in a way that best presents your case. You gather the appropriate supporting facts under each minor premise and again your organizational plan is simply done.

But, unfortunately, sometimes the right organization is not immediately obvious. On those occasions, most practicing writers would agree that a period of meditation about the material on hand must precede not only the writing of a rough draft but even the making of an outline. An outline constructed before you have thought about a wide range of possibilities may restrict you so tightly that you will either have to ignore it later or mutilate your material to fit it. This period of meditation can take many forms. For some writers it involves a lot of pacing around; for others, long periods of staring at pads of paper, typewriters, or walls. However it is done, what is involved is taking stock of the information you possess. We suggest for such moments a technique we have personally used and have also taught our students for many years—a modified form of brainstorming.

Brainstorming

What exactly is brainstorming? True brainstorming is a process in which you uncritically, without thought of organization, jot down every idea about a subject that pops into your head. The key to successful brainstorming is that you do not attempt to evaluate or organize your material at the first stage. These processes come later. Evaluation or organization at the first critical stage may cause you to discard an idea that may prove valuable in the context of all the ideas that the brainstorming session produces.

You can apply a modified form of brainstorming to organizing your report. We use the term *modified* because at the organizational stage you are not starting from scratch. Rather, you are working with the large body of information you have already gathered.

Take all your research materials—your interviews, your letters, your notecards, your questionnaires, your experimental data—and sit down at a table, preferably a large table. Place your materials in a stack before you. Now take a large pad of paper, the bigger the better. Thumb through your materials and write on the pad condensed versions of all the facts and ideas in your material. Also write down any ideas that your research materials suggest to you. These ideas you gain through association may be facts you knew before researching; they may be organizational ideas; they may be analogies to explain difficult concepts; they may be additional thoughts about your audience; they may be almost anything. Do not evaluate them. Write them down. As you proceed, draw lines between ideas to separate them.

Obviously, this is no five-minute exercise. Brainstorming a report that will run 1500 words when finished may take you several hours. If you are dealing with a complex subject, you may spend all day or several days simply revolving ideas around in your head and on paper.

We illustrate this process with excerpts from a brainstorming session on an essay you know in finished form—Chapter 4, "Analyzing Your Audience." We have put the ideas down here exactly as they appear on the brainstorming pad—abbreviations, fragmented ideas, and all:

Purpose vs audience analysis

---

purpose of writing—purpose of reading

---

on the job vs in school

---

need long known
   Aristotle defined generation gap
   Roman water commissioner
   typographical experiments

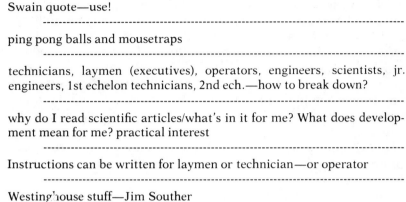

Swain quote—use!

ping pong balls and mousetraps

technicians, laymen (executives), operators, engineers, scientists, jr. engineers, 1st echelon technicians, 2nd ech.—how to break down?

why do I read scientific articles/what's in it for me? What does development mean for me? practical interest

Instructions can be written for laymen or technician—or operator

Westinghouse stuff—Jim Souther

The brainstorming session for Chapter 4 filled four sheets of 22-inch by 17-inch newsprint with closely packed writing. We paid no attention to organization at this point. Neither did we evaluate. Some of the ideas made it through to the finished product; some did not. (Some show up in other chapters.) All we tried to do was to put our ideas and facts out into the open where we could use them and think about them.

What do you do after the brainstorming session? First read over everything you have written and then, if possible, do something else for a while. Read a book, watch TV, work on another project, have a chat with your buddy down the hall, sleep on your ideas if you can. Let your subconscious mind work on your material. Many scientists and writers report that solutions to problems have come to them while they were asleep or doing something totally unrelated to solving the problem. Ernest Hemingway, telling about his early days in Paris, said that he had to learn not to think about his writing when he was not at work on it. In that way he allowed his subconscious mind to work on it.[2]

Probably you have had experiences, yourself, where a problem has solved itself overnight. Of course, the subconscious mind must have material to work on if it is to solve the problem. Your research and your brainstorming session supply that material.

Organizing Your Material

When you return to the material your brainstorming session produced, you must begin to organize and evaluate.

---

[2] Ernest Hemingway, *A Moveable Feast* (New York: Bantam Books, 1965), p. 12.

For your first step, reexamine your purpose. Usually, you will have had your purpose in mind while you were researching, but perhaps your research has modified it to some degree. At this point write it out in precise terms. Nothing clarifies fuzzy thought as quickly as having to express it on paper. State your purpose with your audience in mind. You are never merely writing, say, an essay on "How to build a hi-fi set," but rather, "How to build a hi-fi set if you are a high school student with a limited knowledge of electronics."

With your purpose clearly stated, examine the results of your brainstorming session. You are now about to bring order out of the hodgepodge before you.

Your first step should be to look for key ideas and organizational patterns that directly support your purpose. Find major ideas under which you can group less important ones. Let us use our audience analysis chapter again to illustrate. Looking at our four pages of information, we saw that types of audiences seemed to be the dominant idea. This suggested classification by audience type as the organizational scheme. First ideas may not always work, but in this case ours did. Once we had hit upon the proper rhetorical mode, our organization went well but not without a good deal of trial-and-error adjustments.

First, we had to decide how to classify our audience by type. We took a large sheet of paper and wrote the various audiences on it:

technicians
laymen (executives)
engineers
scientists
jr. engineers
operators

As an afterthought we added "history of audience analysis" to the list.

Next we took a large sheet of paper and wrote "Scientists" across the top of it. Under this heading we listed all the things from our four pages of ideas that seemed to apply to the scientist. Part of the list read like this:

definition by education
definition by experience
limited background information needed
OK for theoretical calculations

show all facts
OK to use math, formulas, scientific terms
tables OK
be honest

We did the same for "Engineers" and for "Laymen (executives)." We looked at our lists for a while. They didn't work as they should. Some of the things listed under "Scientists" fit just as well under "Engineers." We didn't want a lot of needless repetition. Some of the things we had to say about laymen and executives did not seem to belong in the same section. Some of the things we had to say did not fit under any of the headings.

We had to look for new headings. After several more trial-and-error runs, we ended up with these:

History of Audience Analysis
Laymen
Executives
Experts
Technicians
The Combined Audience

Under these headings we made a more complete outline, filling in facts and ideas under the major headings. No one but ourselves was to see the outline, so we skipped the formality of Roman numerals and other outline apparatus.

When we looked our outline over, we were pretty satisfied, but had a nagging doubt about beginning with the history of audience analysis. For one thing, we were violating our basic classification scheme. We could probably justify that, but were we also violating our own rules of audience analysis? Would you be as fascinated by Aristotle's ideas on audience analysis as we were? Or by the fact that Sextus Julius Frontinius, water commissioner of Rome in 97 A.D., invented the information report? Reluctantly, we decided that you wouldn't be. We scrapped that section. We kept just a few examples to illustrate the introduction to the chapter.[3]

When our rough but thorough outline was finished, we organized our research notes into piles corresponding to the headings and sub-headings of the outline and were ready to write our first draft of the

---

[3] If you *would* like to know more about this subject, we suggest you read Walter James Miller's "What Can the Technical Writer of the Past Teach the Technical Writer of Today?" *IRE Transactions on Engineering Writing and Speech* 4, no. 3 (1961): 69–76.

chapter. The planning of the chapter from brainstorming through organizing the research notes took 15 hours scattered over 4 days. In the hours not devoted to planning, we walked, read novels, did our other work, watched TV, slept, and tried not to think about the chapter.

The process we have just described is our way of organizing an extensive piece of writing. It may not work for everyone, although we know it does for many of our students. Other methods work for other people. Some people work directly from their research notes, laying them out on a large table and reading them over carefully. They then organize their material into piles, trying first one arrangement and then another. Some people begin by talking their papers into a tape recorder.

Some of these approaches and others we have not discussed are described by Dr. Blaine McKee in an article called, "Do Professional Writers Use an Outline When They Write?"[4] Dr. McKee's article, based on research conducted with members of the Society for Technical Communication, contains many interesting insights into the organizational practices of professional technical writers. Figure 7-1 is a table from his article showing the types of outlines used by the professionals.

Only 5 percent used no outline at all, and only 5 percent reported using the elaborate sentence outline. Most reported using some form of the topic outline or a mixture of words, phrases, and sentences. Most kept their outlines flexible and informal, warning against getting tied to a rigid outline too early. But all but 5 percent did feel the need to think through their material and to get some sort of organizational pattern down on paper before beginning the first draft of a report.

Figure 7-1   Types of Outlines Used

Type		Percent
Topic outline		60
Word	8.75 percent	
Word and phrase	51.25 percent	
Combination of word, phrase, and sentence		30
Sentence outline		5
No outline		5
		Total 100

[4]Blaine McKee, "Do Professional Writers Use an Outline When They Write?" *Technical Communication* 19, no. 1 (1972): 10–13.

However you approach the organizational process, keep these points in mind:

- You can save yourself a lot of initial chaos and hard work if you remember that certain kinds of reports (or sections of reports) have fairly standard organizations. For example, a description of a mechanism almost always involves focusing on each of its component parts. Comparison and contrast papers and physical research reports also follow a fairly definite pattern. We describe these organizational patterns in Chapters 5 and 6 and in Part 3, "Applications."
- Writing things down is a wonderful discipline. It clarifies thought, keeps your mind from wandering, and records key points.
- Stay relaxed. Don't rush the organizational process. As many times as need be, go over your information, objectives, and audience analysis. Be willing to try a trial-and-error approach until the right plan is found. It's much easier to rearrange and discard information at this stage than it is later.

Constructing the Formal Outline

An experienced writer seldom goes beyond the thorough but informal outlining procedure described in the previous section. Your instructors, however, will probably require you to submit formal outlines from time to time. They have several good reasons for this.

Reasons for Outlining

A formal outline is a good exercise in and of itself. You may be assigned to read an essay and outline it. By reading your outline, the instructor can quickly see if you have the ability to abstract the major ideas from a piece of writing—a necessary skill in research. Or the instructor may ask you to go through a body of research material and construct the outline for a paper that you will never actually write. This exercise allows both you and the instructor to concentrate on organizational skills without getting into writing skills.

In industry you would need an outline in cases of multiple authorship, such as a report prepared by a team or on a production-line basis. With a formal outline the writers know where their parts fit into the whole. And, of course, formal reports often require a table of contents, which closely resembles a formal topic outline (see pages 206–210).

Most often, however, instructors will ask for a formal outline of a paper that you are to write. Usually they will ask for the outline a week or two before the paper is due. They have two reasons for this early turn-in date. First, they get you to work on the paper early enough for you to think the paper through before you have to write it. Second, they allow themselves enough time to help you with your organization.

Writers often digress in reports. They have not always learned how to stay close to the main subject. Many writers hate to throw away any research material. They have worked hard to get their facts, and whether the facts really fit or not, they will attempt to shoehorn them in. Because of this tendency, writers sometimes include too much background material. A formal outline with its neat divisions will show a student writer and teacher alike when the student plans to develop irrelevant màterial at the expense of needed material. Look at the faulty outline in Figure 7-2. A paper written from this outline would never fulfill the stated purpose. The student read a good deal of the history of distillation and desalination and no doubt found it fascinating. But most of it is irrelevant to the purpose of the paper and should not be included.[5] The discussion of modern methods belongs in a prominent place. It is from these that the student must choose. But in the faulty outline, this discussion is the fifth subdivision under one of the subdivisions of the history of desalination—three times removed from prominence.

The student was correct in making the choice of a method a major division, but the outline as a whole is much too one-sided. In this instance, the senior officers designated as the audience do not need the ancient history. They are concerned with the application of modern desalination methods. They are concerned with theory only insofar as they need it to understand the application. The student has mislaid the purpose of the report and misjudged the needs of the audience. The outline clearly showed the teacher and the student where the organization failed.

The student's revised outline appears in Figure 7-3. In the revised outline, the student has discarded the irrelevant material. The first division of the outline states the problem, the solution of which is the purpose of the paper. The remaining divisions concentrate on the criteria that govern the choice and on the alternatives available. By the time division VII is reached, both the student and the audience have sound reasons for the choice. (Notice, also, that both outlines

---

[5] The student had not wasted time in absorbing all this history. The best reports, like the tips of icebergs, often suggest but do not display their underpinnings.

Figure 7-2   A Faulty Outline

DESALINATION METHODS FOR AIR FORCE USE

Purpose: To choose a desalination method for Air Force bases located near large bodies of salt water.

Audience: Senior officers.

I.   History of desalination

    A.   Distillation among ancient peoples

        1.   Greeks

            a.   Aristotle

            b.   Pliny the Elder

        2.   Arabians

    B.   Distillation in the Middle Ages

    C.   Distillation in modern times

        1.   France

            a.   Cellier–Blumenthal

            b.   Derosne

        2.   Gt. Britain

            a.   Fitzgerald

            b.   Hales

        3.   Shipboard equipment

        4.   The Middle East today

        5.   Modern methods

            a.   Electrodialysis

            b.   Reverse osmosis

            c.   Multistage flash

            d.   Long tube vertical

II.   Choice of a method

    A.   Theory of desalination

    B.   Long tube vertical

    C.   Justification of choice

Figure 7-3   A Revised Outline

DESALINATION METHODS FOR AIR FORCE USE

Purpose: To choose a desalination method for Air Force bases located near large bodies of salt water.
Audience: Senior officers.

    I.   Statement of the problem
        A. Need for a choice
        B. Choices available
        C. Sources of data
   II.   Explanation of criteria
        A. Cost
        B. Purity
        C. Quantity
        D. Ease of maintenance
  III.   Electrodialysis
        A. Theory of method
        B. Judgment of method
            1. Cost
            2. Purity
            3. Quantity
            4. Ease of maintenance
   IV.   Reverse osmosis
        A. Theory of method
        B. Judgment of method
            1. Cost
            2. Purity
            3. Quantity
            4. Ease of maintenance
[V, VI. Same as III and IV for remaining methods]
  VII.   Choice of a method

contain a statement of intended audience. Most conventional outlines do not require this step. We feel strongly that you should include it. Audience governs organization as much as purpose does.)

Outlining Conventions

As illustrated in Figure 7-3 and the following format, the outline has four parts: (1) title, (2) purpose statement, (3) audience statement, and (4) body.

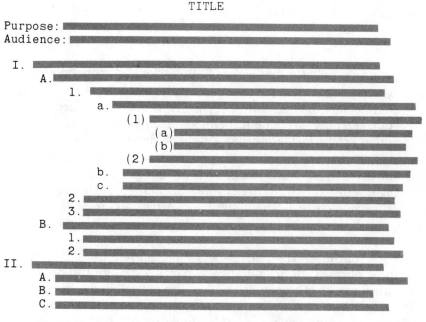

Observe the following conventions in the body of your outline:

Make all entries grammatically parallel. Do not mix complete sentences with incomplete sentences. Do not mix noun phrases with verb phrases, and so forth. A formal outline with a hodgepodge of different grammatical forms will seem to lack— and, in fact, may lack—logic and consistency.

Incorrect	*Correct*
I. The overall view	I. The overall view
II. To understand the terminal phase	II. The terminal phase
III. About the constant-bearing concept	III. The constant-bearing concept

- Never have a single division. Things divide into two or more; so, obviously, if you have only one division you have done no dividing. If you have a "I" you must have a "II." If you have an "A" you must have a "B," and so forth.

Incorrect

   I. Visual capabilities
      A. Acquisition
  II. Interception and
     closure rate
  III. Braking

Correct

   I. Visual capabilities
      A. Acquisition
      B. Interception and
        closure rate
        determination
  II. Braking

- Capitalize only the first letter of the entries and proper nouns.
- Do not have entries for your report's introduction or conclusion. Outline only the body of the report. Of course, the information in your purpose statement belongs in your introduction, and perhaps the information about audience belongs there as well.
- Use substantive statements in your outline entries. That is, use entries such as "Reverse osmosis" or "Judgment of method" that suggest the true substance of your information. Do not use cryptic expressions such as "Example 1" or "Minor premise."
- Use either a topic or a sentence outline.

Topic outlines. The entries of a topic outline are either single words or phrases, all grammatically parallel. Do not use periods or any other marks of punctuation after the entries in a topic outline. The two outlines on desalination in Figures 7-2 and 7-3 are both topic outlines. Students generally prefer topic outlines to sentence outlines, finding them easier to handle. Dr. McKee's research indicates that professional technical writers prefer them. They have two great advantages: (1) The table of contents that you must construct for more formal papers is essentially a modified topic outline. You are ahead of the game, then, when you have a good topic outline from the start. (2) Similarly, the headings in formal technical papers are lifted right from the outline. Like the entries in outlines, headings must be grammatically parallel.

Sentence outlines. In a sentence outline all entries are complete sentences with proper punctuation. It has one clear advantage over a topic outline. When a writer has constructed a sentence outline, most of the topic sentences for the paragraphs should be well in hand. Writing the paper then becomes a matter of filling in the sentences containing supporting details. The following is an example of a sentence outline:

SAFE MANNED INTERPLANETARY TRAVEL DURING SOLAR FLARES

Purpose: To demonstrate that manned interplanetary travel is safe during periods of solar flares providing NASA understands the problem and protects the astronauts with the latest prediction techniques and shielding methods.

Audience: Undergraduate college students with a scientific or engineering background are the intended audience.

I. Solar flares are associated with sunspots and produce dangerous radiation.

   A. Sunspots are regions of lower temperature than the surrounding areas of the sun's surface and appear as dark spots that are readily visible to an observer on the earth.

   B. Solar flares appear as sudden, intense brightenings of the areas around sunspots.

   C. Solar flares produce both corpuscular radiation and electro-magnetic radiation.

II. Biological effects constitute the greatest hazard to astronauts from solar radiation.

   A. Even small amounts of radiation will damage body cells.

   B. Depending on the radiation dose, the results can range from no obvious injury to death within 30 days.

III. Manned interplanetary travel will be possible with protection from solar flares.

   A. Predicting solar flares provides one means of protection.

      1. Short-range prediction is currently possible.

      2. Long-range prediction provides some protection but is not accurate enough for total protection.

   B. Shielding combined with short-range prediction will adequately protect the astronaut.

      1. Active shielding causes charged particles to miss the space capsule entirely.

      2. Passive shielding absorbs charged particles before they reach the astronauts.

EXERCISES

1. When asked to brainstorm about why the United States birth rate has been decreasing,* a group of students produced the reasons

---

*According to *The Statistical Abstract of the United States, 1978,* the birth rate per thousand population for the given years was as follows: 1940 (19.4), 1950 (24.1), 1955 (25), 1960 (23.7), 1965 (19.4), 1969 (17.8), 1970 (18.4), and 1975 (14.8).

listed below. Obviously, there are enough ideas on the list for several books. Looking at the list, see how many topics and subtopics you can break it down into. You may have to supply some "umbrella" topic headings. Select one of the topics (or subtopics). Decide what your purpose in handling the topic is and who your audience will be. Perhaps, for example, you might wish to show a lay audience how some development listed below is *one* of the causes for the declining birth rate. Using the material on the list and ideas of your own, organize a possible essay. If your instructor wishes you to, turn your organizational plan into a formal outline. If you were to write the essay, what facts would you need to support the opinions stated? Where would you look for such facts?

first child born later	inflation
divorce rate up	rising age of U.S. population
husbands and wives both working	more women in colleges and
increased family planning	universities
National Organization of Women	legalized abortion
End of '50s and '60s boom	higher costs for health care
world food shortages	better contraception
popular books on population problems	higher costs for food and shelter
	women in work force increasing
a mobile society	shift from rural to urban living
women's movement	changing sexual attitudes
average age at marriage increasing	the youth movement of the '60s

2. For about fifteen minutes, jot down everything you can think of about some technical or semi-technical subject. Choose a subject that you know well, so you can put down a good deal of information without further research—perhaps a subject related to a hobby or some school subject that you enjoy.

    Do not evaluate the material as you put it down. Brainstorm the subject. After you complete the fifteen-minute exercise, examine and evaluate what you have. Narrow down to the most specific subject and purpose that you can. Choose your audience. Keeping your specific subject, purpose, and audience in mind, organize your jottings into a rough outline. Do not worry overmuch about outline format, such as Roman numerals, parallel headings, and so forth.

3. Turn the informal outline you constructed for Exercise 2 into a formal outline. Use either a topic or sentence outline format. Your instructor may want to specify a format.

CHAPTER **8**

Achieving
a Clear Style

Examples of unclear writing style are all too easy to find, even in places where we would hope to find clear, forceful prose. Read the following sentence:

> While determination of specific space needs and access cannot be accomplished until after a programmatic configuration is developed, it is apparent that physical space is excessive and that all appropriate means should be pursued to assure that the entire physical plant is utilized as fully as feasible.

The murky sentence quoted comes from a report issued by a state Higher Education Coordinating Board. Actually, it's better than many examples we could show you. While difficult, it's probably readable. Others are simply indecipherable. When you have finished this chapter, you should be able to analyze a passage like the one above and show why it is so unclear. You should also know how to keep your own writing clear, concise, and vigorous. Here are the principles that underlie everything we will tell you in this chapter:

- Write as honestly and objectively as you can. Objectivity comes from facts correctly reported and honest inferences drawn from them.
- Do not be mysterious. Tell your readers as clearly as you can what you want them to understand, know, or do.

- Remember that you have invited your readers to come to you. The burden of being clear rests upon you.
- Do not strike a pose when you write. You are not trying to impress readers with your superiority. You are trying to get ideas from your brain to theirs by the shortest, clearest route.

We discuss paragraphs, lists, clear sentence structure, active verbs, active and passive voice, specific words, and avoiding pomposity. We have broken our subject into seven parts for simplicity's sake. But all the parts are closely related. All have but one aim—clarity.

The Paragraph

In Chapters 5 and 6, we discussed the rhetorical modes of exposition, description, narration, and argumentation. These modes may be used not only to develop reports but also to develop paragraphs within reports. Thus, paragraphs will vary greatly in organization and length, depending upon their purpose. In technical writing, however, the central statement more often than not appears at the beginning of the paragraph. This placement provides the clarity of statement that good technical writing must have. But in a paragraph aimed at persuasion the central statement may appear at the close, where it provides a suitable climax for the argument. No matter where a central statement is placed, unity is achieved by relating all the other details of the paragraph to the statement, as in this paragraph of inference and speculation:

Mariner 10 confirmed the earlier findings of Mariner 5 and Venera 4 on their flights to Venus in October 1967 and discovered a bow shock—a wave in front of the planet like the bow wave of a ship. Somehow the ionosphere of Venus forms this bow shock in the solar wind and stops the solar wind from plunging directly into the atmosphere of the planet. How and why this bow shock occurs is not fully understood. Certainly the effect is very different from that of the Earth, Moon, Mars, Mercury and Jupiter. The Venus bow shock might be a direct interaction of the solar wind with the atmosphere of Venus, or with just the ionosphere. It may alternatively arise because the solar wind

We have italicized the central statement concerning a "bow shock" in front of Venus. The rest of the paragraph presents facts and analysis in support of the central statement.

induces magnetic fields and produces thereby a pseudo or false magnetopause as though Venus had a magnetic field like the Earth. Or the solar wind might act on the atmosphere of Venus as it does on a comet to produce the tail. Any of these explanations might be confirmed or disproved by another mission to Venus.[1]

How long should a paragraph be? Examination of well-edited magazines such as *Scientific American* reveals that their paragraphs seldom average more than 100 words in length. The editors of magazines know that paragraphs are for the reader. Paragraphing breaks the materials into related subdivisions for the reader's better understanding. When paragraphs run on too long, the central statements that provide the generalizations needed for reader understanding are either missing or hidden in the mass of supporting details.

In addition to considering the need of the reader for clarifying generalizations, editors also consider the psychological effect of their pages on the reader. They know that large blocks of unbroken print have a forbidding appearance that intimidates the reader. If you follow the practices of experienced editors, you will break your paragraphs whenever your presentation definitely takes a new turn. As a general rule, paragraphs in reports and articles should average 100 words or under. In letters and memorandums, because of their page layout, you should probably hold average paragraph length to under 60 words.

How do you provide transition? Most often, a paragraph presents a further development in a continuing sequence of thought. When this is true, the paragraph's opening central statement will be so closely related to the preceding paragraph that it provides most of the transition you need. When a major transition between ideas is called for, consider using a short paragraph to guide the reader from one idea to the next.

The following five paragraphs provide an excellent example of paragraph development and transition:

Hyperactivity is essentially a symptom, which may be the result of a child's basic personality, a temporary state of anxiety, or subclinical seizure

Hyperactivity is the key word. It or *hyperactive* appears six times. Its repetition provides a major transitional device. The second

---

[1] U.S., National Aeronautics and Space Administration, *The New Frontier: Linking Earth and Planets*, Issue 5 (Pasadena, Calif.: Jet Propulsion Laboratory, 1974), p. 2.

disorders; or it may reflect a true hyperkinetic state. It may also, according to the Council on Child Health of the American Academy of Pediatrics, be strictly "in the eye of the beholder" (AAP Council on Child Health, 1975).

Descriptions of hyperactivity are generally given in behavioral terms, such as motor activity, attention span, frustration tolerance, excitability, impulse control, irritability, restlessness, and aggressiveness. Although these behaviors are measurable, they may not adequately reflect the kind of problems that different children have. Objective measures of attention span have been developed and have been used in some studies to help make the diagnosis, usually during tasks requiring continuous performance by the child. Standard questionnaires have also been designed, to be used by parents and teachers as they make observations at different times; such questionnaires provide useful evaluation information.

Researchers, however, acknowledge the difficulties of measuring these attributes and of interpreting the measurements. For example, in a study at the University of North Carolina designed to measure the motor activity and interest span of children diagnosed as hyperkinetic, Routh (1975) reported that of 78 referrals from physicians, teachers and parents, only 47% of the children were judged to be overactive—despite the fact that all of the children were considered to be "problem children" by those referring them to the testing service. In a study of teacher ratings of hyperactivity conducted at the University of Iowa, older teachers rated more children hyperactive than did young teachers (Johnson, 1974).

Clinical experience also indicates that many factors alter the activity patterns of such children or the perception

sentence leads into the central theme of the five paragraphs—that hyperactivity may sometimes be "in the eye of the beholder."

The central statement in the second paragraph shows its relationship to the first paragraph through the use of the words *hyperactivity* and *descriptions*. The second half of the statement leads into the subject of the paragraph—the behavioral activities that provide a measure of hyperactivity. Many central statements provide transition by looking both backward and forward.

The word *Researchers* provides the new element that will be considered in this paragraph. The words *measuring* and *interpreting* alert the reader that this paragraph continues the central theme of all five paragraphs.

Facts to support the difficulty of measurement are given.

The phrase *Clinical experience* shifts the reader from *researchers* as a source of information to a new source.

of the patterns by parents. Such factors range from the presence or absence of breakfast, weather conditions and resultant seasonal cycles that alter activities during cold seasons, to the interpersonal family relationships or existence of disruptive family problems.

The word *also* shows that the central idea of the previous paragraph is still being pursued. The clause beginning *many factors* introduces the subject matter of the paragraph.

Age of the child appears to be another determinant of detection of the syndrome of hyperactivity. The majority of cases are noted at about the age the child enters school, and there is usually a gradual diminution in the hyperactivity at puberty or slightly thereafter. While some observers cite isolated cases of identifiable hyperkinesis in older individuals, it is unusual. Also, in some areas of the world, the syndrome is not recognized as a problem by school authorities—again, an example of differences in perception or of occurrence.[2]

Age of the child introduces the central subject of this paragraph. The words *another, determinant, detection,* and *hyperactivity* all announce the relationship of the age of the child to the central theme of the five paragraphs.

The final sentence rounds off the five paragraphs by returning, in more formal language, to the central theme that hyperactivity may sometimes be "in the eye of the beholder."

The five paragraphs illustrate that you will develop paragraphs coherently when you keep your mind on the central theme. If you do so, the words needed to provide proper transition will come rather naturally. More often than not, your transitions will be repetitions of key words and phrases supported by such simple expressions as *also, another, of these four, because of this development, so, but,* and *however.* When you wander away from your central theme, no amount of artificial transition will really wrench your writing back into coherence.

Listing

One of the simplest things you can do to ease the reader's chore is to break down complex statements into lists. Visualize the printed page. When it appears as an unbroken mass of print, it intimidates readers and makes it harder for them to pick out key ideas. Get important ideas out into the open where they stand out. Do not bury them in a thick tissue of subordinate clauses and prepositional phrases. Lists help to clarify introductions and summaries. You may list (1) by

---

[2]*Diet and Hyperactivity: Any Connection* (Chicago: Institute of Food Technologists, 1976), pp. 1–2.

starting each separate point on a new line, leaving plenty of white space around it, or (2) using numbers within a line as we have done here. Examine the following summary from a student paper, first as it might have been written and then as it actually was:

> The exploding wire is a simple-to-perform yet very complex scientific phenomenon. The course of any explosion depends not only on the material and shape of the wire but also on the electrical parameters of the circuit. In an explosion the current builds up and the wire explodes, current flows during the dwell period, and "post-dwell conduction" begins with the reignition caused by impact ionization. These phases may be run together by varying the circuit parameters.

Now, the same summary as a list:

> The exploding wire is a simple-to-perform yet very complex scientific phenomenon. The course of any explosion depends not only on the material and shape of the wire but also on the electrical parameters of the circuit.
>
> An explosion consists primarily of three phases:
>
> 1. The current builds up and the wire explodes.
> 2. Current flows during the dwell-period.
> 3. "Post-dwell conduction" begins with the reignition caused by impact ionization.
>
> These phases may be run together by varying the circuit parameters.

The first version is clear, but the second version is clearer, and readers can now file the information in their minds as "three phases." They will remember it longer.

Some writers avoid using lists even when they should, so we hesitate to suggest any restrictions on the technique. But, obviously, there are some subjective limits. Lists break up ideas into easy-to-read, easy-to-understand bits, but too many can give your page the appearance of a laundry list. Also, journal editors sometimes object to lists where each item starts on a separate line. Such lists take space, and space costs money. So, use lists when they clarify your presentation but use them discreetly.

Clear Sentence Structure

The basic English sentence appears in two patterns, subject-verb-object (SVO) or subject-verb-complement (SVC), as in these two sentences:

John (s) hit (v) the ball (o).
The baby (s) cried (v) lustily (c).

Around such simple sentences as "John hit the ball" the writer can hang a complex structure of words, clauses, and phrases that serve to modify and extend the basic idea. He may say, "In his first Little League game, played on the Fourth of July in 1976, the bicentennial year, little John, the coach's son, hit the ball weakly to the pitcher's mound, retiring the side, and justifying in the minds of all present the thought that he was in the lineup merely because he was the coach's son."

Imagine such a sentence dealing with a complex technical subject. Because of the richness and flexibility of English, people do write such sentences. The ultimate English sentence could be infinitely long. No grammatical rule prevents this book from being one 90,000-word sentence.

Sentence Length

How long should a sentence be? Research cited by Dr. Rudolph Flesch[3] indicates the following scale of length versus reading ease:

very easy	8 words or less
easy	11
fairly easy	14
standard	17
fairly difficult	21
difficult	25
very difficult	29 words or more

Another study notes that professional writers' sentences average in the low 20s.[4]

A great many formulas to measure the readability of writing are available, some exceedingly complicated. However, in an article reviewing the use of such formulas, Dr. George R. Klare of Ohio University concludes "that counts of the two simple variables of word length and sentence length are sufficient to make relatively good predictions of readability."[5] That is, the use of shorter words and shorter sen-

---

[3] *The Art of Plain Talk* (New York: Harper & Brothers, 1946), p. 38.
[4] Porter G. Perrin and George H. Smith, *Handbook of Current English* (New York: Scott, Foresman and Company, 1955), p. 211.
[5] George R. Klare, "Assessing Readability," *Reading Research Quarterly* 10, no. 1 (1974–1975): 62–102.

tences usually correlates positively with ease in reading. But Dr. Klare cautions that shorter sentences and words simply *indicate* that a piece of writing will be more readable. Whether they *cause* the increased readability is another and more complicated question.

Dr. Klare's research suggests that while present readability studies do not give anyone the right to be arbitrary about formulas for writing, they do furnish us with some useful guides. And probably the best ones are the simplest. Hold your sentence length down. Technical concepts are hard enough to grasp even when described in simple sentences. Bring your skills in audience analysis to bear on the problem. Suppose you have an audience that includes a good many people with only an eighth-grade reading ability. Then, sentences that average 14 words, "fairly easy" on the Flesch scale, would be suitable. For most business correspondence, Flesch's "standard" 17-word average would be a good choice. For people with high reading skill—many executives, for example—a 20-word average would be appropriate. Stay away from the "very difficult" average of 29 words or more no matter who is in your audience. Remember, too, that you're dealing with *average* sentence length. Don't write sentences all of one length. For an average of 14 words, you would probably range from 5- to 25-word sentences. Dr. Klare also mentions word length as a factor. We will say more about that in a moment.

Sentence Order

What is the best way to order a sentence? Is a great deal of variety in sentence structure the mark of a good writer? One writing teacher, Dr. Francis Christensen of the University of Southern California, looked for the answers to those two questions. He examined large samples from twenty top writers, among them John O'Hara, John Steinbeck, William Faulkner, Ernest Hemingway, Rachel Carson, and Gilbert Highet. In his samples, he included ten fiction writers and ten nonfiction writers.[6]

What Dr. Christensen discovered seems to disprove any theory that good writing requires extensive sentence variety. The professionals examined depend mostly on basic sentence patterns. They write 75.5 percent of their sentences in plain SVO or SVC order, as in these two samples:

Doppler radar provides a tremendous increase in capability over conventional radar. (SVO)

Doppler radar can be tuned more rapidly than conventional radar. (SVC)

---

[6] Francis Christensen, "Notes Toward a New Rhetoric," *College English*, October 1963, pp. 7–18.

Another 23 percent of the time, the professionals begin a sentence with a short adverbial opener:

> As with any radar system, Doppler does have problems associated with it.

These adverbial openers are most often simple prepositional phrases or single words such as *however, therefore, nevertheless,* and the other conjunctive adverbs. Generally they provide the reader with a transition between thoughts. Following the opener, the writer usually continues with a basic SVO or SVC sentence.

These basic sentence types—SVO(C) or adverbial + SVO(C)—are used 98.5 percent of the time by the professional writers in Dr. Christensen's sample. What do the writers do with the remaining 1.5 percent of their sentences? For 1.2 percent, they open the sentence with verbal clauses based upon participles and infinitives such as "*Breaking* ground for the new church," or "*To see* the new pattern more clearly." The verbal opener is again followed most often with an SVO or SVC sentence, as in this example:

> Looking at it this way, we see the radar set as basically a sophisticated stopwatch that sends out a high energy electromagnetic pulse and measures the time it takes for part of that energy to be reflected back to the antenna.

Like the adverbial opener, the verbal opener serves most of the time as a transition.

The remaining .3 percent of the sentences (1 sentence in 300) are inverted constructions in which the subject is delayed until after the verb, as in this sentence:

> No less important to the radar operator are the problems caused by certain inherent characteristics of radar sets.

What can we conclude from this valuable study? Simply this: the professionals are interested in getting their content across, not in tricky word order. They convey their thoughts in a clear container, not clouded by extra words. You should do the same.

Active Verbs

The verb determines the structure of an English sentence. Many sentences in technical writing falter because the finite verb does not function properly; that is, (1) comment upon the subject, (2) state a relationship about the subject, or (3) relate an action which the subject performs. Look at the following sentence:

> *Sighting* of the ground was accomplished by the pilot at 7 A.M.

English verbs can easily be changed into nouns, but sometimes—as above—the change can lead to a faulty sentence. The writer has put the true action into the subject and subordinated the pilot and the ground as objects of prepositions. The sentence should read:

> At 7 A.M. the pilot *saw* the ground.

The poor writer can ingeniously bury the action of a sentence almost anywhere. With the common verbs *make, give, get, have,* and *use,* the writer can bury the action late in the sentence in an object:

> The punch card operator *has* the job of translating language symbols into machine symbols.

or

> The speaker did not *give* a satisfactory explanation of his technique.

Properly revised, the sentences read:

> The punch card operator *translates* language symbols into machine symbols.
> The speaker did not adequately *explain* his technique.

The poor writer can even bury the action in an adjective:

> A new discovery produces an *excited* reaction in a scientist.

Revised:

> A new discovery *excites* a scientist.

There is constructions may trap the writer into inefficient sentences. For example:

> There are some technical editors who prefer the passive voice.

Revised:

> Some technical editors prefer the passive voice.

When writing, and particularly when rewriting, you should always ask yourself: "Where's the action?" If the action does not lie in the verb, rewrite the sentence to put it there, as with this sample:

Music therapy is the scientific *application* of music to accomplish the *restoration, maintenance,* and *improvement* of mental health.

Rewritten to put active ideas into verb forms, the sentence reads this way:

Music therapy *applies* music scientifically *to restore, maintain,* and *improve* mental health.

The rewritten sentence defines "music therapy" in one-third less language than the first sentence, without any loss of meaning or content.

Active and Passive Voice

We discuss active and passive voice sentences on pages 127–128, but let us quickly review the concept again here. In an active voice sentence the subject performs the action and the object receives the action, as in "Mary hit the ball." In a passive voice sentence, the subject *receives* the action, as in "The ball has been hit." If you want to include the doer of the action, you must add this information in a prepositional phrase as in "The ball has been hit *by Mary*."

We urge you to use the active voice more than the passive, but we do not wish to imply that you should ignore the passive altogether. Passive voice is often useful. You can use the passive voice to emphasize the object receiving the action rather than the agent doing the action. The passive voice in "Influenza may be caused by any of several viruses" emphasizes *influenza*. The active voice in "Any of several viruses may cause influenza" emphasizes the *viruses*.

Often the agent of action is of no particular importance. When such is the case, passive voice is appropriate because it allows you to drop the agent altogether:

Edward Jenner's work on vaccination was published in 1796.

Be aware, however, that inappropriate use of passive voice can cause you to omit the agent when knowledge of the agent may be vital. Such is often the case in giving instructions: "All doors to this building will be locked by 6 P.M." may not produce locked doors until it is rewritten in active voice: "The night manager will lock all doors to this building by 6 P.M."

While passive voice has its uses, most editors feel that too much of it produces lifeless and wordy writing. They advise, therefore, against using it except when it is clearly appropriate. The *Council of Biology Editors Style Manual* succinctly expresses the reasons for this advice:

> Use the active voice except where you have good reason to use the passive. The active is the natural voice, the one in which people usually speak and write, and its use is less likely to lead to wordiness and ambiguity. Avoid the "passive of modesty," a device of writers who shun the first-person singular. "I discovered" is shorter and less likely to be ambiguous than "it was discovered." When you write "Experiments were conducted," the reader cannot tell whether you or some other scientist conducted them. If you write "I" or "we" ("we" for two or more authors, never as substitute for "I"), you avoid dangling participles, common in sentences written in the third person passive voice.[7]

As the biology editors point out, passive voice does, indeed, cause many dangling participles, as in "While conducting these experiments, the chickens were seen to panic every time a hawk flew over." Chickens conducting experiments? Not really. Active voice straightens the matter out: "While conducting these experiments, we saw that the chickens panicked every time a hawk flew over." (See also "Dangling Modifier," pages 460–461.)

Write in the active voice whenever appropriate. When rewriting, find and revise inappropriate passive voice sentences into the active.

Specific Words

Remember the semanticists' abstraction ladder that we described for you on page 103: a ladder composed of rungs that ascend from very specific words such as *table* to abstractions such as *furniture*, *wealth*, and *factor*. The human ability to move up and down this ladder enabled us to develop language, on which all human progress depends. Because we can think in abstract terms, we can call a moving company and tell it to move our furniture. Without abstraction we would have to bring the movers into our house and point to each object we wanted moved. But like many helpful writing techniques, abstraction is a device you should use carefully.

---

[7]CBE Style Manual Committee, *Council of Biology Editors Style Manual: A Guide for Authors, Editors, and Publishers in the Biological Sciences*, 4th ed. (Council of Biology Editors, 1978), p. 21.

Stay at an appropriate level on the abstraction ladder. Do not say "inclement weather" when you mean "rain." Do not say "over-whelming support" when you mean "62 percent of the workers sup-ported the plan." Do not settle for "suitable transportation" when you mean "a bus that seats 32 people."

Writing that uses too many abstractions is lazy writing. It re-lieves writers of the need to observe, to research, and to think. They can speak casually of "factors," and neither they nor their readers really know what they are talking about. Here is an example of such lazy writing. The writer was setting standards for choosing a desali-nation plant to be used at Air Force bases.

- The cost must not be prohibitive.
- The quality of water must be sufficient to supply a military estab-lishment.
- The quality of the water must be high.

The writer here thinks he has said something. He has said little. He has listed slovenly abstractions when with a little thought and research he could have listed specific details. He should have said:

- The cost should not exceed $2.00 per thousand gallons.
- To supply an average base with a population of 5000, the plant should purify 750,000 gallons of water a day (AFM 88-10 sets the standard of 150 gallons a day per person).
- The desalinated water produced should not exceed the national health standard for potable water of 500 parts per million of dissolved solids.

Abstractions are needed for generalizing, but they cannot replace specific words and necessary details. Words mean different things to different people. The higher you go on the abstraction ladder, the truer this is. The abstract words—*prohibitive, sufficient,* and *high*—could have been interpreted in as many different ways as the writer had readers. No one can misinterpret the specific details given in the rewritten sentences.

Abstractions can also burden sentences in another way. Some writers are so used to thinking abstractly that they begin a sentence with an abstraction and *then* follow it with the specific word, usually in a prepositional phrase. They write,

The problem of producing fresh water became a problem at overseas bases.

instead of,

Producing fresh water became a problem at overseas bases.

or,

The circumstance of the manager's disapproval caused the project to be dropped.

instead of,

The manager's disapproval caused the project to be dropped.

We do not mean to say you should never use high abstractions. A good writer moves freely up and down the abstraction ladder. But when you use words from high on the ladder, use them properly—for generalizing and as a shorthand way of referring to specific details you have already given.

Pomposity

When writing, state your meaning as simply and clearly as you can. Do not let the mistaken notion that writing should be more elegant than speech make you sound pompous. Writing *is* different from speech. Writing is more concise, more compressed, and often better organized than speech. But elegance is not a prerequisite for good writing.

A sign at a service station where one of us gets his gasoline reads, "No gas will be dispensed while smoking." Would anyone in that service station speak that way? Of course not. He would say, "Please put out that cigarette" or, "No smoking, please." But the sign had to be elegant, and the writer sounds pompous, and illiterate as well.

If you apply what we have already told you about the active voice, active verbs, specific words, and clear sentence structure, you will go a long way toward tearing down the fence of artificiality between you and the reader. We want to touch on just four more points: empty words, use of the first-person point of view, elegant variation, and pompous vocabulary.

Empty Words

The easiest way to turn simple, clear prose into elegant nonsense is to throw in empty words, like these phrases that begin with the impersonal "it": "It is evident," "It is clear that," or, most miserable of all, "It is interesting to note that." When something is evident, clear, or interesting readers will discover this for themselves. If something is not evident, clear, or interesting, rewrite it to make it so. When you must use such qualifying phrases, at least shorten them to "evidently," "clearly," and "note that." Avoid constructions like "It was noted by Jones." Why not simply say, "Jones noted"?

Avoid excessive use of phrases and clauses introduced by "of" and "which." Such phrases bring in many empty words.

Compare,

The cooling of the mixture makes it easier to handle.

with,

Cooling the mixture eases its handling.

In the second version we have eliminated the prepositional phrase and put the action in the verb where it belongs.

"Which" is used awkwardly here:

Evolution, a theory which was controversial sixty years ago, is now generally accepted.

You save two words with,

A controversial theory sixty years ago, evolution is now generally accepted.

Many empty words are simply old stock phrases writers throw in out of sheer habit. You see them often in business correspondence. A partial list follows:

to the extent that	with reference to
in connection with	relative to
with regard to	with respect to
as already stated	in view of
inasmuch as	with your permission
hence	as a matter of fact

We could go on, but so could you. When such weeds crop up in your writing, pull them out.

Another way to produce empty words is to run the abstract word in tandem with the specific word. This produces such combinations as

twenty in number *for* twenty

> wires of thin size *for* thin wires
>
> red in color *for* red

When you have expressed something specifically, do not throw in the abstract term for the same word.

First Person Point of View

Once, many reports and scientific articles were written in the third person—"This investigator has discovered"—rather than first person—"I discovered." This practice is now much less common. Most style manuals for scientific journals now advise against the use of the third person (see page 172) on the grounds that it's wordy and confusing. We agree with this advice completely.

The judicious use of *I* or *we* in a technical report is entirely appropriate. Incidentally, such usage will seldom lead to a report full of *I*'s and *we*'s. After all, there are many agents in a technical report other than the writer. In describing an agricultural experiment, for example, researchers will report how *the sun shone, photosynthesis occurred, rain fell, plants drew nutrients from the soil,* and *combines harvested.* Only occasionally will they need to report their own actions. But when they must, they should be able to avoid such roundabout expressions as "It was observed by this experimenter."

Elegant Variation

Elegant variation will also make your writing sound pompous. (H. W. Fowler and F. G. Fowler invented the term *elegant variation* in their book, *The King's English.* H. W. Fowler used it again in his *Dictionary of Modern English Usage.* Writers should own both books and use them often.) Elegant variation occurs when a writer substitutes one word for another because of an imagined need to avoid repetition. This substitution can lead to two evils: (1) The substituted word may be a pompous one. (2) The variation may mislead the reader into thinking that some shift in meaning is intended. Look at the following example:

> The scientists used the Mars Mariners to study Mars' magnetic fields. The spacemen further utilized the far-reaching craft to investigate the red planet's radiation belts.

Confusion reigns. The writer has avoided repetition, but the reader may think that "scientists" and "spacemen" are two different groups. "Utilize" can be a synonym for "use," but it has the additional meaning of *convert*. The reader may well conclude that some conversion of the Mariner's equipment was needed before it could "investigate" the radiation belts. What happens if the reader doesn't know that "red planet" is a nickname for Mars? Are "Mars Mariners" and "far-reaching craft" one and the same?

In addition, in this context "spacemen," "utilized," "far-reaching craft," "investigate," and "red planet" sound pretentious.

The writer could have improved this passage in two ways. First, pronouns could be used to avoid some repetition, ignoring the other repetitions:

> The scientists used the Mars Mariners to study Mars' magnetic field. They further used them to study Mars' radiation belts.

Better yet, he could have compressed the two sentences into one:

> The scientists used the Mars Mariners to study Mars' magnetic field and its radiation belts.

In either of the rewritten passages, the writer could have avoided confusion and pomposity.

Remember also that intelligent repetition provides good transition. Repeating key words reminds the reader that you are still dealing with your central theme (see pages 163–165).

Pompous Vocabulary

Generally speaking, the vocabulary you think in will serve in your writing. Jaw-breaking thesaurus words and words high on the abstraction ladder will not convince readers that you are intellectually superior. They will merely convince them that your writing is hard to read. We are not really telling you here that you must forego your hard-won educated vocabulary. If you are writing for readers who would understand words like "extant" or "prototype," then use them. But use them only if they are appropriate to your discussion. Don't use them to impress people. Remember too the conclusion reached by Dr. Klare (see pages 167–168) that excessive word length is a significant factor in reducing reader understanding.

Nor are we talking about the specialized words of your professional field. At times these are quite necessary. Just remember to define them if you feel your reader will not know them. What we are talking about is the desire some writers seem to have to use pompous vocabulary to impress their readers.

Pompous writers will seek to bury you under the many-syllabled words they use to express one-syllable ideas, as in this example from the U.S. Department of Transportation:

> The purpose of this PPM [Policy and Procedure Memorandum] is to ensure, to the maximum extent practicable, that highway locations and designs reflect and are consistent with Federal, State, and local goals and objectives. The rules, policies, and procedures established by this PPM are intended to afford full opportunity for effective public participation in the consideration of highway location and design proposals by highway departments before submission to the Federal Highway Administration for approval. They provide a medium for free and open discussion and are designed to encourage early and amicable resolution of controversial issues that may arise.

We urge you to read as much good writing—both fiction and nonfiction—as time permits. Stop occasionally as you do and study the authors' choice of words. You will find them to be lovers of the short word. Numerous passages in Shakespeare are composed almost entirely of one-syllable words. The same holds true for the King James Bible. Good writers do not want to impress you with their vocabularies. They want to get their ideas from their heads to yours by the shortest, simplest route.

Summary

A final example will summarize much that we have said. Insurance policies have for so long been verbal bogs that most buyers of insurance have long since given up on finding one clearly written. However, the St. Paul Fire and Marine Insurance Company decided that perhaps it was both possible and desirable to simplify the wording of its policies. For a trial run, the company turned Dr. Rudolf Flesch loose on one of their policies, and he rewrote it using only words familiar to the average reader. He kept both the sentence length and syllable count down and used no long paragraphs listing conditions and exceptions. He used predominantly active voice and active verbs. The insurance company becomes "we" and the insured "you." Definitions are included where needed rather than segregated in a glossary.

The resulting policy is wonderfully clear. Compare a paragraph of the old with the new.[8]

---

[8] The excerpts from the St. Paul Fire and Marine Insurance Company's old and new Personal Liability Catastrophe Policy are reprinted with the permission of the St. Paul Companies, St. Paul, Minn. 55102.

Old

Cancellation

This policy may be cancelled by the Named Insured by surrender thereof to the Company or any of its authorized agents, or by mailing to the Company written notice stating when thereafter such cancellation shall be effective. This policy may be cancelled by the Company by mailing to the Named Insured at the address shown in this Policy written notice stating when, not less than thirty (30) days thereafter, such cancellation shall be effective. The mailing of notice as aforesaid shall be sufficient notice and the effective date of cancellation stated in the notice shall become the end of the policy period. Delivery of such written notice either by the Named Insured or by the Company shall be equivalent to mailing. If the Named Insured cancels, earned premium shall be computed in accordance with the customary short rate table and procedure. If the Company cancels, earned premium shall be computed pro rata. Premium adjustment may be made at the time cancellation is effected or as soon as practicable thereafter. The check of the Company or its representative, mailed or delivered, shall be sufficient tender of any refund due the Named Insured. If this contract insures more than one Named Insured, cancellation may be effected by the first of such Named Insureds for the account of all the Named Insureds; notice of cancellation by the Company to such first Named Insured shall be deemed notice to all Insureds and payment of any unearned premium to such first Named Insured shall be for the account of all interests therein.

New

Can This Policy Be Canceled?   Yes, it can. Both by you and by us.

If you want to cancel the policy, hand or send your cancellation notice to us or our authorized agent. Or mail us a written notice with the date when you want the policy canceled. We'll send you a check for the unearned premium, figured by the short rate table—that is, pro rata minus a service charge.

If we decide to cancel the policy, we'll mail or deliver to you a cancellation notice effective after at least 30 days. As soon as we can, we'll send you a check for the unearned premium, figured pro rata.

Examples that substitute specific, familiar words for the high abstractions of the original policy are used freely. For instance:

You miss a stopsign and crash into a motorcycle. Its 28-year-old married driver is paralyzed from the waist down and will spend the rest of his life in a wheelchair. A jury says you have to pay him $1,300,000. Your stand-ard insurance liability limit is $300,000 for each person. We'll pay the balance of $1 million.

Or:

We'll defend any suit for damages against you or anyone else insured even if it's groundless or fraudulent. And we'll investigate, negotiate and settle on your behalf any claim or suit if that seems to us proper and wise.

You own a two-family house and rent the second floor apartment to the Miller family. The Millers don't pay the rent and you finally have to evict them. Out of sheer spite, they sue you for wrongful eviction. You're clearly in the right, but the defense of the suit costs $750. Under this policy we defend you and win the case in court. The whole business doesn't cost you a penny.

Incidentally, there is no fine print in the policy. It is set entirely in 10-point type, a type larger than that used in most newspapers and magazines. Captions and even different colored print are used freely to draw attention to transitions and important information. You can clean up your own writing by following the principles discussed in this chapter and demonstrated in this new insurance policy.

If you exercise care, your own manner of speaking can be a good guide in writing. You should not necessarily write as you talk. In speech, you may be too casual, even slangy. But the sound of your own voice can still be a good guide—in this way. When you write something, read it over; even read it aloud. If you have written some-thing you know you would not speak because of its artificiality, re-write it in a comfortable style. Rewrite so that you can hear the sound of your own voice in it.

EXERCISES

1. You should now be able to rewrite the example sentence on page 161 into clear, forceful prose. Here it is again; try it:

   While determination of specific space needs and access cannot be ac-complished until after a programmatic configuration is developed, it is apparent that physical space is excessive and that all appropriate means should be pursued to assure that the entire physical plant is utilized as fully as feasible.

2. Below are some expressions that the Council of Biology Editors
    believes should be rewritten.[9] Using the principles you have
    learned in this chapter, rewrite them:

    an innumerable number of tiny veins
    at this point in time
    bright green in color
    we conducted inoculation experiments on
    due to the fact that
    during the time that
    fewer in number
    for the reason that
    goes under the name of
    if conditions are such that
    in the event that
    in view of the fact that
    it is often the case that
    it is possible that the cause is
    it would appear that
    lenticular in character
    oval in shape
    plants exhibited good growth
    prior to
    serves the function of being
    subsequent to
    the fish in question
    the treatment having been performed
    throughout the whole of this experiment
    the tube which has a length of 3 m
    a process for the avoidance of waste
    judging by present standards, these trees are

- If we interpret the deposition of chemical signals as initiation of
  courtship, then initiation of courtship by females is probably the
  usual case in mammals.
- A direct correlation between serum vitamin B_{12} concentration
  and mean nerve conduction velocity was seen.

---

[9] CBE Style Manual Committee, *Style Manual*, pp. 19–23. Reproduced with permission
from *Council of Biology Editors Style Manual*, 4th edition. CBE Style Manual Committee. Council of Biology Editors, 1978.

- It is possible that the pattern of herb distribution now found in the Chilean site is a reflection of past disturbances.

- Following termination of exposure to pigeons and resolution of the pulmonary infiltrates, there was a substantial increase in lung volume, some improvement in diffusing capacity, and partial resolution of the hypoxemia.

- Many biological journals, especially those that regularly publish new scientific names, now state in each issue the exact date of publication of the preceding issue. In dealing with journals that do not follow this practice, or with volumes that are issued individually, the biologist often needs to resort to indexes . . . in order to determine the actual date of publication of a particular name.

- Some in the population suffered mortal consequences from the lead compound in the flour.

3. Turn the following sentence into a paragraph of several sentences. See if listing might be a help. Make the central idea of the passage its first sentence.

   If, on the date of opening of bid or evaluation of proposals, the average market price of domestic wool of usable grades is not more than 10 percent above the average of the prices of representative types and grades of domestic wools in the wool category which includes the wool required by the specifications (see (f) below), which prices reflect the current incentive price as established by the Secretary of Agriculture, and if reasonable bids or proposals have been received for the advertised quantity offering 100 percent domestic wools, the contract will be awarded for domestically produced articles using 100 percent domestic wools and the procedure set forth in (e) and (f) below will be disregarded.

4. Write a short report, or rewrite one you did earlier, using the stylistic principles of this chapter.

PART 2

Creating the Professional Report

When you have researched and organized your information and familiarized yourself with good technical writing style, you are ready to create your report. Depending upon the complexity of your subject and the needs of your audience, you may find it necessary to include in your report such elements as a title page, table of contents, letter of transmittal, abstract, bibliography, notes, and so forth. A more complex report will also often contain graphic aids. Your purpose in including all these elements is to ease your reader's passage through your report. You should know how to handle them well. To assist you in these matters, we have included in this part chapters on prose elements, mechanical elements, and graphical elements. The final chapter in Part 2 tells you how to make appropriate use of these elements in planning, writing, and revising your report.

CHAPTER 9

Prose Elements

In this chapter we discuss the prose elements, such as the letter of transmittal, the preface, and the abstract, that make up the more formal report. In Chapter 10, we explain the mechanical elements— title page, table of contents, list of illustrations, and so forth. The formal elements of reports answer functional needs. For example, many professional reports are long, and the table of contents serves the functional purpose of sorting things out for the reader. Professional reports are likely to pass between people who do not know each other very well. The letter of transmittal serves as an introduction and breaks the ice, so to speak. Busy readers, who sometimes lack the time to read an entire report, must read professional reports selectively. They need an informative abstract to give them the major points in a hurry. To a large extent, then, the degree of formality needed in a report depends on situation and audience. You'll be better able to understand how the elements of reports work together after you see the function each element fulfills. We'll discuss this matter more fully in Chapter 12, "The Professional Report." To help you find needed elements, here is a list of them with reference to the page where the explanation of each begins:

Letter of Transmittal and Preface

We have placed the letter of transmittal and preface together because in content they are often quite similar. They usually differ in format and intended audience only. You will use the letter of transmittal when the audience is a single person or a single group. Many of your major reports in college will include a letter of transmittal to your professor, usually placed just before or after your title page. When on the job, you may handle the letter differently. Often it is mailed before the report, as a notice that the report is forthcoming. Or it may be mailed at the same time as the report but under separate cover.

Generally, you will use the preface for a more general audience where you may not know specifically who will be reading your report. The preface or letter of transmittal introduces the reader to the report. It should be fairly brief. Always include the following basic elements:

- Statement of transmittal or submittal. (This step is included in the letter of transmittal only.)
- Statement of authorization or occasion for report.
- Statement of subject and purpose.

Additionally, you may include some of the following elements:

- Acknowledgments.
- Features of the report that may be of special interest or significance.
- List of existing or future reports on the same subject.
- Background material.
- Summary of the report.
- Special problems (including reasons for objectives not met).
- Financial implications.
- Conclusions and recommendations.

How many of the secondary elements you include depends upon

the structure of your report. If, for example, the report itself contains an introductory summary, there may be no point in including such a summary in the preface or letter of transmittal. See Figures 9-1 and 9-2 for a sample letter of transmittal and a sample preface.

If the report is to remain within an organization, the letter of transmittal will become a memorandum of transmittal. This changes nothing but the format. See pages 303–306.

Abstracts and Introductory Summaries

Discussed below are two kinds of abstracts: informative and descriptive. Each kind is often set off by itself on a separate page or pages of a report. Sometimes the descriptive abstract is placed on the title page

Figure 9-1   Letter of Transmittal

Gatlin Hall
Weaver University
Briand, MA 02139

July 27, 1980

Dr. Ross Alm
Associate Professor of English
Weaver University
Briand, MA 02139

Dear Dr. Alm:

I submit the accompanying report entitled "Venus and Mercury as Planets" as the final project for English 430, Technical Writing.

The report discusses the characteristics of both Venus and Mercury, covering size, mass, density, physical appearance, and atmosphere.   I have made an effort to provide a base for understanding the significance of the space probes to Venus and Mercury carried on by NASA's Jet Propulsion Laboratory (JPL).   Recent information about Mercury obtained by the most recent probe, Mariner 10, is incorporated into the report.

I am indebted to Ms. Mary Fran Buehler of JPL who has allowed me to quote extensively from her unpublished work on Mariner 10.

Sincerely,

Anne K. Chimato

Anne K. Chimato
English 430

Figure 9-2   Preface

PREFACE

    In recent years the National Aeronautics Administration
(NASA) has explored the inner planets of our solar system,
Venus and Mercury, with unmanned space probes. This report,
part of NASA's educational series for high school students,
reports the information from the latest probe, Mariner 10.
Characteristics of both Venus and Mercury—including size,
density, physical appearance, and atmosphere—are discussed.

    Of particular interest in this report is the surprising
finding that Mercury, contrary to scientific expectation, has
a magnetic field. This finding may cause present theories
about the generation of magnetic fields within planets to be
revised.

    For lists of other reports on NASA's unmanned space
probes write to NASA, Jet Propulsion Laboratory, California
Institute of Technology, Pasadena, California 91125.

(see Figure 10-1, page 207). Because your full report contains complete documentation, you need not footnote or otherwise document
the information in abstracts.

Never use "I" statements in either kind of abstract. Report your
information impersonally as though it were written by someone else.
The informative abstract in Figure 9-3 shows the style. This is not an
arbitrary principle. If you were to publish an article, your abstract
would likely be reprinted in an abstracting journal where the use of
"I" would be inappropriate.

Frequently a report will contain both a descriptive and an informative abstract. When this is so, it is common practice to label the
descriptive abstract as *Abstract* and the informative abstract as *Introductory Summary*.

Informative Abstract

An informative abstract is a brief statement of the key information in your complete report. Good informative abstracts are difficult
to write. At one extreme, they lack adequate information; at the

Figure 9-3   Informative Abstract

ABSTRACT

   Venus has a diameter of 7520 miles and a density slightly
less than that of Earth.   Because of Venus' persistent cloud
cover, little is known about its surface features.   Venus'
day is the equivalent to 127 Earth days, and its year equals
225 Earth days.   Its cloud layer is approximately 20-miles
thick.   Venus' surface temperature is approximately 475 C,
and its surface pressure is equivalent to the pressure in
Earth's ocean at a depth of 2400 feet.   Mercury has a diameter
of 3000 miles and a density greater than Earth's.   Mercury
has a period of revolution around the sun equal to 88 Earth
days, and its day is equal to 176 Earth days.   Therefore, Mercury's day is 2-years long by its time.   Mercury's surface is
cratered like the surface of Earth's Moon.   Mercury has virtually no atmosphere but does show minute quantities of
helium.   Its surface temperature varies from 325 C to -125
C.   An unexpected discovery of Mariner 10 was that Mercury
has a magnetic field.   Because Mercury rotates so slowly,
scientists thought that it would have no magnetic
field.   This finding may cause present theories about the
generation of magnetic fields within planets to be revised.
The cratering of Mercury suggests that at some point in
Earth's early history it may have been similarly pock-
marked.   Evidence for this conclusion can also be found in
the most ancient of Earth's rocks.

other, they are too detailed. In the informative abstract you must pare down to material essential to your purpose. This can be a slippery business.

Suppose, for example, you are writing a report to explore the knowledge about the way the human digestive system absorbs iron from food. In the body of the report you discuss an experiment conducted on Venezuelan workers that followed isotopically labeled iron through their digestive systems. To enhance the credibility of the information presented, you include some details about the experiment. You report the conclusion that vegetarian diets decreased iron absorption. How much of all this should you put in your abstract? Given your purpose, the location and methodology of the experiment would not be suitable material for the abstract. You would simply report the fact that in one experiment vegetarian diets have been shown to decrease iron absorption.

When you are selecting material for your informative abstract, remember that the abstract should be able to stand alone. In fact, for the busy reader it may serve as a substitute for the entire report. Therefore, you must include enough information to satisfy the purpose of your report. Normally, you would state major conclusions and decisions, if any, in the informative abstract. Certain business report formats handle this matter differently, however. See, for example, pages 493–497.

A good rule of thumb for length, unless you have instructions otherwise, is that your abstract should be from five to ten percent of the length of your body. For business reports, this will normally be an acceptable limit. But most journals, because of rising publication costs, will set arbitrary limits of under 200 words for abstracts. The abstracts printed in Figure 2-12 on page 34 are good examples of such short abstracts.

Our advice for writing informative abstracts has necessarily been rather general. Depending upon where your abstract may appear, you may receive much more specific advice. For example, most professional journals or societies publish style books that include specifications about how to write an abstract. We have reprinted one such specification in Figure 9-4. Read it to gain more insight into abstract writing.

Descriptive Abstract

The main purpose of the descriptive abstract is to help busy readers decide if they need or want the information in the report enough to read it entirely. The descriptive abstract merely tells what the full report contains. It cannot serve as a substitute for the report itself. Many reports contain descriptive abstracts, and many abstracting

Figure 9-4   Instructions for Abstracting

1.4   An abstract is a brief summary of the content and purpose of the
Abstract   article. In APA journals the abstract is used in place of a concluding
summary and appears directly under the by-line. All APA journals
except *Contemporary Psychology* require an abstract.

The abstract allows readers to survey the contents of an article
quickly. Because, like the title, it is used by *Psychological Abstracts*
for indexing and information retrieval, the abstract should be self-
contained and fully intelligible without reference to the body of the
paper and suitable for publication by abstracting services without
rewriting. Information or conclusions that do not appear in the
main body of the paper should not appear in the abstract. Because
so much information must be compressed into a small space, authors sometimes find the abstract difficult to write. Leaving it until
the article is finished enables you to abstract or paraphrase your
own words.

An abstract of a *research* paper should contain statements of
the problem, method, results, and conclusions. Specify the subject
population (number, type, age, sex, etc.) and describe the research
design, test instruments, research apparatus, or data-gathering
procedures as specifically as necessary to reflect their importance
in the experiment. Include full test names and generic names of
drugs used. Summarize the data or findings, including statistical
significance levels, if any, as appropriate. Report inferences made
or comparisons drawn from the results.

An abstract of a *review* or *theoretical* article should state the
topics covered, the central thesis, the sources used (e.g., personal
observation, published literature, or previous research bearing on
the topic), and the conclusions drawn. It should be short but informative. For example, "The problem was further discussed in
terms of Skinner's theory" is not an informative statement. The
abstract should tell the reader the *nature* or *content* of the theoretical
discussion: "The discussion of the problem centered on Skinner's
theory and the apparent fallacy of determinism."

An abstract for a research paper should be 100–175 words;
one for a review or theoretical article, 75–100 words. General style
should be the same as that of the article.

Remember, to the degree that an abstract is succinct, accurate,
quickly comprehended, and informative, it increases your audience.

From *Publication Manual of the American Psychological Association*, 2nd ed. (Washington, D.C.:
American Psychological Society, 1974), p. 15. Copyright © 1974 by the American Psychological
Association. Reprinted by permission.

journals print them. Here is a descriptive abstract based on the information in Figure 9-3:

> This report describes the characteristics of both Venus and Mercury covering size, rotation, revolution about the sun, density, physical appearance, and atmosphere. New findings about Mercury's magnetic field are presented. Conclusions about Earth's early history are drawn from Mercury's cratered surface and Earth's ancient rocks.

Note that the descriptive abstract discusses the *report*, not the subject. After reading this abstract, you know that "New findings about Mercury's magnetic field are presented," but you do not know what those findings are. You must read the report for that information. Whether the report is ten or a thousand pages long, a descriptive abstract can cover the material in less than ten lines. In contrast, the informative abstract must reflect the length of the report's discussion section.

Thus, to do its job an informative abstract must grow longer as the discussion grows longer. For further examples of descriptive abstracts, see Figure 2-11 on page 33.

The Introduction

In your introduction you should announce four things immediately: (1) subject, (2) purpose, (3) scope, and (4) plan of development. Early in your paper, you also give any needed theoretical or historical background, and this is sometimes thought of as part of the introduction.

Subject

Never begin an introduction with a superfluous statement. The writer who is doing a paper on core memory in computers and begins with the statement, "The study of computers is a vital and interesting one" has wasted the readers' time and probably annoyed them as well. Announce your specific subject loud and clear as early as possible in the introduction, preferably in the very first sentence. The sentence, "This paper will discuss several of the more significant applications of the exploding wire phenomenon to modern science" may not be very subtle, but it gets the job done. The reader knows what the subject is. Often, in conjunction with the statement of your subject, you will also need to define some important terms that may be unfamiliar to your readers. For example, the student who wrote the above sentence followed it with these two:

> A study of the exploding wire phenomenon is a study of the body of knowledge and inquiry around the explosion of fine metal wires by a sudden and large pulse of current. The explosion is accompanied by physical manifestations in the form of a loud noise, shock waves, intense light for a short period, and high temperatures.

In three sentences the writer announced the subject and defined it. The paper is well under way.

Sometimes, particularly if you are writing for nonspecialists, you may introduce your subject with an interest-catching step. This step may be rather extended, as in this example:

> Traveling in orbit around the earth at an altitude of some 270 miles, the Skylab astronauts rotated their spacecraft to establish a bearing on one of their principal check points: a cluster of several hundred round green spots, each half a mile in diameter, arrayed in an orderly pattern on the earth's surface below them. What they were viewing was a dense concentration of circular irrigated fields in north-central Nebraska, a pattern easily identified from space. Passengers on commercial jet airliners increasingly notice the same sight over many other areas of the continental U.S., including eastern Colorado, central Minnesota, the Texas Panhandle, the Pacific Northwest and northern Florida. Now the proliferating green circles can be seen even in the middle of the Sahara. What is being observed is perhaps the most significant mechanical innovation in agriculture since the replacement of draft animals by the tractor.
>
> These circular green fields, most often found in arid or semiarid country, are being watered by the world's first successful irrigation machine.[1]

Or you may simply extract a particularly interesting fact from the main body of your theme. For example,

> Last year, the Civil Aeronautics Authority attributed 16 aircraft accidents to clear air turbulence. What is known about this unseen menace that can cripple an aircraft, perhaps fatally, almost without warning?

In this example, the writer catches your interest by citing the accident rate caused by clear air turbulence and nails the subject down with a rhetorical question in the second sentence. Interest-

---

[1] William E. Splinter, "Center-Pivot Irrigation," *Scientific American*, June 1976, p. 90.

catching introductions are used in brochures, advertisements, and magazine and newspaper articles. You will rarely see an interest-catching introduction in business reports or professional journals. If you do, it will usually be a short one.

Purpose

Your statement of purpose tells the reader *why* you are writing about the subject you have announced. By so doing you also, in effect, answer the reader's question, "Should I read this paper or not?" For example, a student who stated her subject as "The scientific principles of the Cockcroft-Walton Accelerator" continued with "The purpose of this report is to offer a complete explanation of the Cockcroft-Walton Accelerator to future undergraduates here at the University." Readers who do not need such an explanation will know there is no purpose in their reading the report.

Another way to understand the purpose statement is to realize that it often deals with the *significance* of the subject. A writer who had the application of human engineering to technical writing as his subject announced his purpose this way:

> Regardless of your writing specialty, however, you will be more effective as a technical writer if you become familiar with human engineering and learn to apply human-engineering principles.[2]

Scope

The statement of scope further qualifies the subject. It announces how broad and, conversely, how limited the treatment of the subject will be. Often it indicates the level of competence expected in the reader for whom the paper is designed. For example, the student who wrote, "The purpose of this report is to offer a complete explanation of the Cockcroft-Walton accelerator to future undergraduates here at the University" declared her scope as well as her purpose. She is offering a complete, not a partial, explanation; and the audience she is designing her report for is not composed of high school students or graduate physicists, but undergraduates. She further qualified her audience and scope by stating, "This explanation of the accelerator will be comprehensive enough to allow sophomore students of physics to use and operate the accelerator without further research."

---

[2] Max Weber, "Human Engineering and its Application to Technical Writing," *Technical Communication* 19 (Third Quarter 1972): 2.

Plan of Development

Part of the introduction is telling the reader how you plan to develop your paper. The principle of psychological reinforcement is at work here. If you tell your readers what you are going to cover, they will be more ready to comprehend as they read along. The following, taken from the introduction to a paper on iron enrichment of flour, is a good example of a plan of development:

> This study presents a basic introduction to three major areas of concern in regard to iron enrichment: (1) questions on which form of iron is best suited for enrichment use; (2) potential health risks from super enrichment—cardiovascular disease, hemochromatosis, and the masking of certain disorders; and (3) the inconsistencies in basic knowledge as they relate to the definitions, extent, and causes of iron deficiency.

You need not necessarily think of the announcement of subject, purpose, scope, and plan of development as four separate steps. Often, subject and purpose or scope and plan of development can be combined. In a short paper, perhaps two or three sentences might cover all four points, as in this example:

> Concern has been expressed recently over the possible presence in our food supply of a class of chemicals known as "phthalates" (or phthalate esters). To help assess the true significance of this concern as to the safety and wholesomeness of food, this article discusses the reasons phthalates are used, summarizes their toxicological properties, and evaluates their use in food packaging materials in the context of their total use in the environment.[3]

Also, introductions to specialized reports may have peculiarities of their own. These will be discussed in Part 3, "Applications."

Theoretical or Historical Background

When necessary theoretical or historical background is not too lengthy, you can incorporate it into your introduction. In fact, when handled properly such information may well catch the interest of the reader, as in the following example.

> Climatologists attribute the warming trend to the furnaces of civilization which have been spewing forth increasing loads of carbon dioxide

---

[3] *Phthalates in Food* (Chicago: Institute of Food Technologists, 1974), p. 1.

to the atmosphere. This colorless and odorless gas, exhaled by man and used by plants to make themselves green, restricts the escape into space of infrared radiation from the sun-warmed earth. Since increased CO_2 absorbs more of the infrared radiation than formerly, a larger amount of heat accumulates, causing a slight but significant increase in average global temperature. This impact of atmospheric CO_2 on climate— dubbed the "greenhouse effect"—has become more apparent in recent years because of the escalating rate at which power plants and industry throughout the world have burned coal, oil from shale, and synthetic oil and gas.[4]

If necessary background material is extensive, however, it more properly becomes part of the body of your paper.

The Discussion

The discussion, or body, will be the longest section of your report. Your intention and your content will largely determine the form of this section. Therefore, we can prescribe no set form for the discussion. In writing your discussion you will use one or more of the rhetorical forms described in Chapters 5 and 6 or the special techniques described in Part 3, "Applications." In addition to prose you will probably also use captions and various visual aids such as graphs, tables, and illustrations. These we describe in Chapters 10 and 11.

When thinking about the discussion, remember that almost every technical report answers a question or questions: What is the best method of desalination to create an emergency water supply at an overseas military base? How are substances created in a cell's cytoplasm carried through a cell's membranes? What is the nature of life on the ocean floor? How does a hydraulic pump work? To answer such specific questions, general questions must be asked and answered. Often the reporter's old standbys—Who? What? When? Where? Why? How?—are good starting points. However you approach your discussion, project yourself into the minds of your readers. What questions do they need answered to understand your discussion? What details do they need to follow your argument? You will find that you must walk a narrow line between too little detail and too much.

Too little detail is really not measured in bulk, but in missing links in your chain of discussion. You must supply enough detail to

---

[4]Carolyn Krause, "Carbon Dioxide and Climate," *Oak Ridge National Laboratory Review* 10 (Fall 1977): 40–41.

lead the reader up to your level of competence. You are most likely to leave out crucial details at some basic point that, because of your familiarity with the subject, you assume to be common knowledge. If in doubt about the reader's competence at any point, take the time to define and explain.

Many reasons exist for too much detail, and almost all stem from writers' inability to edit their own work. When you realize that something is irrelevant to your discussion, discard it. It hurts, but the best writers will often throw away thousands of words, representing hours or even days of work. You must always ask yourself questions like these: Does this information have significance, directly or indirectly, for the subject I am explaining or for the question I am answering? Does the information move the discussion forward? Does it enhance the credibility of the report? Does it support my conclusions? If you don't have a "yes" answer to one or more of these questions, the information has no place in the report, no matter how many hours of research it cost you.

The Ending

Depending upon what sort of paper you have written, your ending can be (1) a summary, (2) a set of conclusions, (3) a set of recommendations, or (4) a graceful exit from the paper. Frequently, you'll need some combination of these. We'll look at the four endings and at some of the possible combinations. It's also possible in executive reports that the "ending" may actually be placed at the front of the report. See pages 82, 256–258, and 493–497.

Summary

Many technical papers are not argumentative. They simply present a body of information that the reader needs or will find of interest. Frequently, such papers end with summaries. In a summary, you condense for your readers what you have just told them in the discussion. Each major section of your discussion should be restated in the summary. Sometimes you may wish to number the points for clarity. The following, from a paper of about 2500 words, is an excellent summary:

> The exploding wire is a simple-to-perform yet very complex scientific phenomenon. The course of any explosion depends not only on the material and shape of the wire but also on the electrical parameters of the circuit.

An explosion consists primarily of three phases:

(1) The current builds up and the wire explodes.
(2) Current flows during the dwell period.
(3) "Post-dwell conduction" begins with the reignition caused by impact ionization.

These phases may be run together by varying the circuit parameters.

The exploding wire has found many uses: it is a tool in performing other research, a source of light and heat for practical scientific application, and a source of shock waves for industrial use.

Summaries should be concise and they should introduce no material that has not been covered in the report.[5] You construct a summary as you construct an informative abstract. You read your discussion over, noting your main generalizations and your topic sentences. You smoothly blend these together into a paragraph or two. Sometimes you will represent a sentence from the discussion with a sentence in the summary. At other times you will shorten such sentences to phrases or clauses. The last sentence in the example above actually represents a summary of four sentences from the writer's discussion. The four sentences, themselves, were the topic sentences from four separate paragraphs.

Conclusions

Some technical papers work toward a conclusion. They ask a question, such as "Are nuclear power plants safe?" present a set of facts relevant to answering the question, and end by stating a conclusion: "Yes," "No," or sometimes, "Maybe." The entire paper aims squarely at the final conclusion. In such a paper, you argue inductively and deductively. You bring up opposing arguments and show their weak points. At the end of the paper, you must present your conclusions. Conclusions are the inferences drawn from the factual evidence of the report. They are the final link in your chain of reasoning. In simplest terms, the relationship of fact to conclusion goes something like this:

Facts	*Conclusion*
Car A averages 25 miles per gallon Car B averages 40 miles per gallon	On the basis of miles per gallon, Car B is preferable.

---

[5] Writers are often confused on this point. There is no prohibition against introducing new material immediately *after* a summary. For example, you may wish to indicate further research that needs to be done in the area you have discussed. Steering the reader to further research is often a stimulating way to end a paper.

Because we presented a simple case, our conclusion was not difficult to arrive at. But even more complicated problems present the same relationship of fact to inference.

In working your way toward a major conclusion, you ordinarily have to work your way through a series of conclusions. In answering the question of nuclear power plant safety, you would have to answer a good many subquestions concerning such things as security of the radioactive materials used, adequate control of the nuclear reaction, and safe disposal of nuclear wastes. The answer to each subquestion is a conclusion. You may present these conclusions in the body of the report, but it's usually a good idea to also draw them all together at the end of the report to prepare the way for the major conclusion.

Earlier (page 195) we showed you the introduction to a report that questioned whether the class of chemicals known as phthalates endangered public health. Here are the conclusions to that report:

> Based on the observations made thus far, there is no evidence of toxicity in man due to phthalates, either from foods, beverages, or household products as ordinarily consumed or used.
> These observations, coupled with the limited use of phthalate-containing food packaging materials and the low rate of migration of the plasticizers from packaging material to food, support the belief that the present use of phthalates in food packaging represents no hazard to human health.[6]

All the conclusions presented are supported by evidence in the report.

In larger papers or when dealing with a controversial or complex subject, you would be wise to precede your conclusions with a summary of your facts. By doing so, you will reinforce in your reader's mind the strength and organization of your argument. For an example of such a combination, see Appendix A, pages 494–496. In any event, make sure your conclusions are based firmly upon evidence that has been presented in your report. Few readers of professional reports will take seriously conclusions based upon empty, airy arguments. Conclusions are frequently followed by recommendations.

Recommendations

A conclusion is an inference. A recommendation is the statement that some action be taken or not taken. The recommendation is, of course, based upon the conclusions and is the last step in the process. You conclude that Brand X bread, for example, is cheaper per pound

---

[6]*Phthalates in Food*, p. 2.

than Brand Y and just as nutritious and tasty. Your final conclusion, therefore, is that Brand X is a better buy. Your recommendation is "Buy Brand X."

Many reports such as feasibility reports, environmental impact statements, and research reports concerning the safety of certain foods or chemicals are decision reports that end with a recommendation. For example, we are all familiar with the government recommendations that have removed certain artificial sweeteners from the market and that have placed warnings on cigarette packages. These recommendations were all originally stated at the end of reports looking into these matters.

Recommendations are simply stated. They follow the conclusions, ordinarily in a separate section, and look something like this:

> Based upon the conclusions reached, we recommend that our company
> - Not increase the present level of iron enrichment in our flour.
> - Support research into methods of curtailing rancidity in flour containing wheat germ.

Frequently, you may have a major recommendation followed by several implementing recommendations, as in the following:

> We recommend that the Department of Transportation build a new bridge across the St. Croix River at a point approximately three miles north of the present bridge at Hastings.
>
> - The Department's location engineers should begin an immediate investigation to decide the exact bridge location.
> - Once the location is pinpointed, the Department's right-of-way section should purchase the necessary land for the approaches to the bridge.

You need not support your recommendations when you state them. You should have already done that thoroughly in the report and in the conclusions leading up to the recommendations. It's likely, of course, that a full-scale report will contain, in sequence, a summary, conclusions, and recommendations. For more information about this subject and for examples of summaries, conclusions, and recommendations, see pages 408–412 and 494–497.

Complimentary Close

A short, simple, nonargumentative paper often requires nothing more than a graceful exit, a complimentary close. As you would not end a conversation by turning on your heel and stalking off without a "good-bye" or a "see you later" to cover your exit, you do not end a paper without some sort of close. In a short informational paper that has not reached a decision, the facts should be still clear in the readers' minds at the end, and they will not need a summary. One sentence, such as the following that might end a short speculative paper on lasers, will probably suffice: "Since lasers seem to have almost unlimited uses, they cannot fail to receive increasing scientific attention in the years ahead."

Sometimes an appropriate quotation can be used to get you out of a report gracefully, as in the following:

> As Roger Revelle and H. E. Seuss stated in 1957: "Human beings are now carrying out a large-scale geophysical experiment of a kind that could not have happened in the past nor be repeated in the future. Within a few centuries we are returning to the atmosphere and oceans the concentrated organic carbon stored in the sedimentary rocks over hundreds of millions of years. This experiment, if adequately documented, may yield a far-reaching insight into the processes determining weather and climate."[7]

EXERCISES

Reprinted below is the discussion section from a report titled *Mercury in Food*.[8] It was written both for food technologists and intelligent lay persons. Its purpose is to examine the possible dangers of mercury poisoning in the food we eat. It reaches conclusions but not recommendations. For the purpose of this exercise, pretend that you have written the report as a class assignment. Provide the following prose elements:

- Letter of transmittal to your writing teacher.
- Descriptive abstract.
- Introduction.
- Summary.
- Conclusions.

---

[7] Krause, "Carbon Dioxide and Climate," p. 47.
[8] Institute of Food Technologists' Expert Panel on Food Safety and Nutrition and the Committee on Public Information, *Mercury in Food* (Chicago: Institute of Food Technologists, 1973), pp. 1–2.

Mercury in Food

Mercury and the Environment

During normal growth process, plants absorb mercury from the soil and air; in some instances, plants even concentrate it to small droplets of the metal. Some bacterial organisms, when exposed to inorganic mercury, can convert it to organic mercury compounds. These microscopic organisms and the material they produce, generally called alkyl mercury, may be consumed by fish or animals. Some animals and vegetables have the ability to convert organic forms of mercury back to inorganic compounds.

This constant cycling of mercury from one form to another has gone on for eons without any recognizable toxic effect on the food supply of the world. It is extremely doubtful that the concentration of mercury in the oceans has increased significantly as a result of man's use of the metal.

Awareness of Potential Dangerous Effects Increasing

Until the last two decades, man had only been vaguely aware of the problems arising from the misuse of mercury. A series of isolated incidences tragically demonstrated the potential dangers.

In 1953, a veritable epidemic hit the fishermen and their families in the villages on Minamata Bay, Japan. A number of people who were highly dependent on seafood showed signs of brain damage. Some of these cases were fatal. An intense investigation revealed that a local chemical plant was discharging a waste stream containing organic mercury into the bay. This was absorbed by the fish in the area and eventually passed on to the villagers.

After finding the cause of the problem, authorities were able to eliminate the source of pollution. Mercury levels in the bay returned to normal and once again the local fish was safe to eat.

Mercury showed its insidious effect in Sweden when naturalists noted that certain species of birds were diminishing. Here a study disclosed that these birds had been feeding on grain seeds that had been treated with mercury as a protection against fungus. A similar product was also a favorite pesticide. In 1966 the Swedish government banned or restricted these uses of mercury compounds. The bird life is reviving and the awareness of the potential problem is very much alive.

The practice of treating seeds with mercury compounds has also had its impact on humans. In 1968, a New Mexico farmer fed treated grain seed to his hogs and he and his family subsequently ate the meat from these animals. As a result, three of the farmer's children were crippled and a fourth child was born blind and retarded. Bags containing mercury treated seeds carry a label warning that the contents are poisonous to animals and humans, and such seeds are dyed a bright pink to differentiate them from untreated seeds. Unfortunately, however, these warnings were ignored or misunderstood by those involved. As a result of this, and similar incidents which occurred in other parts of the world, the United States Department of Agriculture in March of 1969 banned the practice of treating seeds with organic mercury compounds.

Maximum Allowable Exposure Levels Set
Urine specimens indicate that man and other animals absorb some mercury from their food and water. People who work in industries using mercury may show urine levels ten or twenty times higher than those who are not directly involved with the metal. The United States Department of Labor has established maximum allowable exposure levels to protect workers. Studies have shown that it is possible for a worker to absorb approximately one milligram of mercury per day from industrial exposure when working within the established allowable limits. Periodic physical examinations, which emphasize special methods of evaluation for possible neurological damage, indicate that the allowable levels are safe for the employees' health.

Guidelines Established for Levels in Food
The United States Food and Drug Administration has conducted routine analysis of foods for mercury content for several years. Practically all foods tested showed levels of mercury concentration well within the norms for the natural environmental content of the element. Only fish and fishery products showed concentrations greater than could be considered to be normal. As a result of their investigations and augmented by the Japanese and Swedish experiences, the Food and Drug Administration in 1969 established a 0.5-parts-per-million (ppm) guideline as the maximum safe limit for mercury in fish.

Only Fish Products Exceed Limits
Swordfish and tuna are the only commercially popular fish that have shown a mercury content exceeding 0.5 ppm. These two species of fish accumulate organic mercury compounds as they grow larger because they consume vast quantities of smaller fish.

Since both tuna and swordfish are caught at sea, far from any possible sources of industrial pollution, the mercury in their systems must come from natural sources. Most probably man, for many years, has eaten tuna and swordfish with concentrations of mercury higher than the established limit without signs of any harmful effect. Analysis of museum specimens of tuna caught during the period 1879 to 1909 reveals that they contain levels of mercury as high as those in fish being caught today. Scientists must, therefore, conclude that mercury levels in tuna and very probably swordfish, have not appreciably changed in the past 90 years.

Why Man Hasn't Suffered from Eating Fish
Recent studies at the University of Wisconsin have discovered that some fish, including tuna, have a built in mechanism for blocking and reducing the toxicity of mercury in their tissues. This research may answer the question of how man has safely eaten fish containing mercury concentrations higher than those allowed by the Food and Drug Administration. Most experts agree that 0.5 ppm level for fish has a considerable margin of safety built into it. It is heartening to recognize there have been no reported cases of mercury poisoning in the U.S.A. from eating fish.

CHAPTER 10

Mechanical Elements

In Chapter 9, we discussed the basic prose elements of reports and gave you a list of all the elements. In this chapter we discuss the mechanical elements of reports. Our approach is descriptive, not prescriptive; that is, we describe some of the more conventional practices found in modern technical reporting. We realize fully, and you should too, that many colleges, companies, and journals call for different practices from the ones we describe. Therefore, we do not recommend that you must follow at all times the practices in this chapter. If you are a student, however, your instructor may, in the interests of class uniformity, insist that you follow this chapter fairly closely.

Whether you follow the set of conventions described in this chapter or some other set, you should realize the importance of following some consistent, conventional plan when you put your reports together. Inconsistency in format confuses the reader. It is just one more thing for your reader to worry about. And in the mechanical elements of your report, as well as in the more interesting elements of your actual discussion, you should put the reader's need to understand above all else.

In order, the mechanical elements included in this chapter are (1) covers, (2) title pages, (3) table of contents, (4) list of illustrations, (5) glossary and list of symbols, (6) captions, (7) numbering systems, (8) documentation, (9) pagination, and (10) appendixes and annexes.

Covers

Covers serve three purposes. The first two are functional and the third esthetic and psychological.

Common typewriter paper requires protection during handling and storage. Pages ruck up, become soiled and damaged, and may eventually be lost if they are not protected by covers. Because they are what readers first see as they pick up a report, covers are the appropriate place for prominent display of identifying information such as the report title, the company or agency by or for which the report was prepared, and security notices if the report contains proprietary or classified information. Incidentally, students should not print this sort of information directly on the cover. Rather, they should type the information onto gummed labels readily obtainable at the five-and-ten or the college bookstore, and then fasten the labels to the cover. The student label might look like this:

```
           THE GREAT RED SPOT OF JUPITER

                         by
                  Philip Q. Dowsing

   English 430              12 December 1979
```

Esthetically and psychologically, covers bestow dignity, authority, and attractiveness. They help to convert a bundle of manuscript into a finished work that looks and feels like a report, and has some of the characteristics of a printed and bound book.

Suitable covers need not be expensive and sometimes *should not be.* Students, particularly, should avoid being pretentious. All three purposes are frequently well served by covers of clear plastic or light cardboard, perhaps of thirty- or forty-pound substance. Students can buy such covers in a variety of sizes, colors, and finishes.

While you are typing your report, remember that when you fasten it into its cover about an inch of left margin will be lost. If you wish an inch and a half of margin, you must leave two and a half inches on your paper. Readers grow irritated when they must exert brute force to bend open the covers in order to see the full page of text.

Title Pages

Like report covers, title pages perform several functions. They dignify the reports they preface, of course, but far more important, they provide identifying matter and help to orient the report users to their reading tasks.

To give dignity, a title page must be attractive and well designed. Symmetry and balance are important, as are neatness and freedom from clutter. The most important items should be boldly printed; items of lesser importance should be subordinated. These objectives are sometimes at war with the objective of giving the report users all the data they may want to see at once. Here we have listed in fairly random order the items that sometimes appear on title pages. A student paper, of course, would not require all or even most of these items; we have starred the four that are usually sufficient for simple title pages.

*1. Name of the company (or student) preparing the report
*2. Name of the company (or instructor and course) for which the report was prepared
*3. Title and sometimes subtitle of the report
 4. Code number of the report
 5. Contract numbers under which the work was done
 6. List of contributors to the report (minor authors)
 7. Name and signature of the authorizing officer
*8. Date of submission or publication of the report
 9. Company or agency emblem and other decorative matter
10. Proprietary and security notices, if applicable
11. Abstract
12. Library identification number
13. Reproduction restrictions
14. Distribution list

Understandably, placing all of these items on an 8½ × 11 inch page would guarantee a cluttered appearance. The moral is plain: if you cannot put down all that you would like, put down—clearly and boldly—all that you must. See Figures 10-1 and 10-2 for typical title pages.

Table of Contents

A table of contents (TOC) performs at least three major functions. Its most obvious function is to indicate by number the page on which discussion of each major topic begins; that is, it serves the reader as a locating device. Less obviously, a TOC displays the extent and nature of the topical coverage and suggests the logic of the organization and the relationship of the parts. Still earlier, in the prewriting stage, provisional drafts of the TOC enable the author to "think on paper"; that is, they act as outlines to guide the composition.

Figure 10-1   Title Page

THE WORKING PERSONALITY

A Technical Memorandum

Prepared

by

Antonia M. Webster

for

English 418

Advanced Technical Writing

Abstract

This report proposes the adoption of a new course in the
Social Sciences designed to develop traits of character and
personality in students preparing for professions in law and
medicine.

December 23, 1979

Figure 10-2   Title Page

LIVESTOCK RESEARCH INCORPORATED
1020 Union Avenue
Whiting, Pennsylvania 15564

FINAL REPORT

ON

CROSSBRED FINNISH LANDRACE

MARKET LAMBS

Report 117–W–70–15–F

Prepared by

David V. Modic
_____

Approved by

_____

March 5, 1979

A system of numbers, letters, type styles, indentations, and other mechanical aids has to be selected so that the TOC will perform its intended functions. Figure 10-3 shows a TOC quite suitable for student reports.

As you see in Figure 10-3, the term CONTENTS heads and identifies the page. The titles of preliminary parts, chapters, and main terminal parts are given in capitals. Divisions and subdivisions of chapters are given less prominence by means of indention and a combination of capital and small letters. Beginning page numbers of parts are given at the right, lined up vertically on their right-hand digits. Spaced leader dots are sometimes used to carry the reader's eye from the end of each title to the beginning page and, at the same time, pull the page together visually. Number the titles of body sections.

It is important for you to guard against certain tendencies and temptations when you design your own TOC. As with the title page, avoid clutter. Seldom is there justification for listing parts subordinate to the subdivisions of sections; very shortly a point is reached where users have almost as much trouble locating items in the TOC as they have in locating them by flipping through the pages of the report. Also, make first-level items appear dominant. Finally, you are urged to prepare the final draft of the TOC by referring to the typed pages of the otherwise finished report. It is simply amazing to discover the differences between an early version of a TOC and the updated version. Whole sections may have dropped out or have been placed in a new order. Wording of captions may have been drastically changed. Old page numbers may no longer apply. Remember that the TOC entries and the captions on the text pages must be worded exactly the same. Every entry in the TOC must also be in the report. However, as we have already pointed out, every caption in the report need not be in the TOC. Captions used to mark divisions of subdivisions are probably best left out of the TOC.

List of Illustrations

If a report contains more than a few illustrations, say more than three or four, it is customary to list the illustrations either on a separate page or on the TOC page. Before we go into detail, however, it seems essential to explain what we mean by *illustrations*.

Illustrations are of two major types, tables and figures. A table is any array of data, often numerical, arranged vertically in columns and horizontally in rows, together with the necessary captions and notes. Any illustration that does not satisfy this definition of *table* is automatically a figure. Figures, therefore, include photographs,

Figure 10-3   Table of Contents

CONTENTS

maps, graphs, organization charts, and flow diagrams—literally anything that does not qualify as a table by the preceding definition. (For further details, see Chapter 11.)

If the report contains both tables and figures, it is customary to use the page heading ILLUSTRATIONS, a combining term. If only figures appear in the report, the page title becomes FIGURES, and correspondingly, if only tables appear, it becomes TABLES. If both figures and tables are present, it is the custom to list all the figures first and then all the tables.

The titles of all illustrations should be as brief and yet as self-explanatory as can be. Avoid the cumbersome expression "A Figure Showing Characteristic Thunderstorm Recording." Say, simply, "Characteristic Thunderstorm Recording." On the other hand, do not be overly economical and write just "Characteristic" or "A Comparison." At best, such generic titles are only vaguely suggestive.

Figure 10-4 shows a simple version that should satisfy most ordinary needs. Notice in the figure that Arabic numerals are used for figures and Roman numerals for tables. This practice is common, but by no means standard.

Glossary and List of Symbols

Reports dealing with technical and specialized subject matter almost invariably include abbreviations, symbols, and terms not known to the layman. Thus a communication problem arises. Technically trained persons have an unfortunate habit of assuming that what is well known to them is well enough known to others. This assumption is seldom justified. Terms, symbols, and abbreviations undergo changes in meaning with time and context. In one context, ASA may stand for American Standards Association; in another context, for Army Security Agency. The letter K may stand for Kelvin or some mathematical constant. The meaning given to Greek letters may change from one report to the next even though both were done by the same person. Furthermore, writers seldom have complete control over who will read their reports. A report intended by the author for an engineering audience may have to be read by members of management, the legal department, or sales. In doubtful instances, it is wise to play it safe by including a list of symbols or a glossary, or both. Readers who do not need these aids can easily ignore them; those who do need them will be immeasurably grateful.

Figure 10-5 illustrates a list of symbols. Figure 10-6 illustrates a glossary. In Figure 10-6 notice that the terms to be defined are set up to stand out plainly from the definitions. Notice also that the defini-

Figure 10-4 List of Illustrations

ILLUSTRATIONS

vi

Figure 10-5   List of Symbols

SYMBOLS

A	Mass number
A.W.	Atomic weight
c	Velocity of light (2.998×10^{10} cm/sec)
D	H^2 atom (deuterium)
E	Energy
e-	Electron
e	Electronic charge (1.602×10^{-10} abs. coulomb)
ev	Electron volt
F	Free energy
(g)	Gas phase
H	Heat content
h	Planck's constant (6.624×10^{-27} erg sec)
I_{sp}	Specific impulse
k	Boltzmann's constant (1.3805×10^{-16} erg/deg)
ln	Natural logarithm

*   *   *   *   *   *   *   *

α	Alpha particle
γ	Gamma ray
ζ	Bond energy
μ	Micro
ρ	Density

v

Figure 10-6   Glossary

GLOSSARY

Btu ------------------the amount of heat required to raise
the temperature of one pound of water
one degree Fahrenheit

degree day ------------a temperature standard around which
temperature variations are measured

design temperature ----the maximum reasonable temperature expected during the heating or cooling
season upon which the design calculations are based

heat transmission
coefficient ------the quantity of heat in Btu transmitted per hour through one square
foot of a building surface

infiltration ---------the air leaking into a building from
cracks around doors and windows

sensible heat ---------heat that the human body can sense

thermal conductivity - the quantity of heat in Btu transmitted by conduction per hour through
one square foot of a homogeneous material for each degree Fahrenheit difference between the surfaces of the
material

thermal resistance ----the reciprocal of thermal conductivity

vii

tions here are fragmentary sentences. Whatever kind of definition you decide to write—full or fragmentary sentences—you should be consistent throughout your glossary.

Captions

We use the term *captions* to refer to the various headings and titles that you may use to display your report's coordination and subordination to the reader. The caption itself is a phrase which describes what is discussed in the paragraph or paragraphs that follow it. Coordination and subordination are shown through your captions by a consistent use of various type faces and positions for different level captions.

You may be surprised by the amount of typographical variations that you can obtain from a standard typewriter:

1. You can set up a caption entirely in capital letters.
    NEMATODES
2. You can use a combination of capitals and small letters.
    Nematodes and Fungi
3. You can use "expanded capitals."
    N E M A T O D E S
4. You can center headlines as in 1, 2, and 3 above, or you can place them at the left margin.
    Pathogen

    Fungi
5. Of course you can also indent them five or more spaces from the left margin.
    Fungi
6. Finally, you can choose between underlining and not underlining.
    Pathogens
    Fungi

As illustrated, you can use the possibilities in various combinations, so that with ingenuity and foresight you can construct ten or more distinctive levels of captions using only a standard typewriter.

Because captions are used to show the organization of a paper, an obvious relationship exists between them and the TOC. All entries in the TOC must be reproduced verbatim as captions. In addition, you may also include captions to display lower levels of organization not included in the TOC. Rarely should a student paper include more than four levels of captions. Figure 10-7 illustrates a system of captions that might actually run over four or five pages.

Figure 10-7   A Four-Level System of Captions

CONTROLLING   SOIL-BORNE
PATHOGENS   IN   TREE   NURSERIES

TYPES OF SOIL-BORNE PATHOGENS AND THEIR EFFECTS ON TREES

The Soil-borne Fungi
     At one time it was thought that soil-borne fungi ....
....

     Basidiomycetes.   The basidiomycetes are a class of fungi
whose species ....

     Phycomycetes.   The class phycomycetes is a very diver-
sified type of fungi. It is the ....

The Plant Parasitic Nematodes
     Nematodes are small, nonsegmented ....
....

     TREATMENTS AND CONTROLS FOR SOIL-BORNE PATHOGENS
....
....

The caption is not a part of the sentence following it. It stands outside the text. An example of misuse would be

     Data systems.   These are designed for a specific
     command function.

Note that "Data systems" is the subject of the sentence through the pronoun "These." While it seems natural, avoid such use of the caption. In the interests of clarity, repeat the key term somewhere in the first sentence or two that follow the caption.

Numbering Systems

Quite often, in order to facilitate reference back and forth in a report or book, a numbering system is combined with the captions. The three systems in most common use are the traditional outline system, the century-decade-unit system (often called the Navy System), and the multiple-decimal system. The three schematics that follow illustrate the three systems.

Traditional outline system:

```
            T I T L E  C A P T I O N
    I.   FIRST–LEVEL CAPTION
         A.   Second–Level Caption
                 1.   Third–level caption.
                 2.   Third–level caption.
         B.   Second–Level Caption


    II.  FIRST–LEVEL CAPTION
         A.   Second–Level Caption
                 1.   Third–level caption.
                 2.   Third–level caption.
         B.   Second–Level Caption
```

Century-decade-unit system:

```
            T I T L E  C A P T I O N
    100  FIRST–LEVEL CAPTION
         110   Second–Level Caption
                 111   Third–level caption.
                 112   Third–level caption.
         120   Second–Level Caption


    200  FIRST–LEVEL CAPTION
         210   Second–Level Caption
                 211   Third–level caption.
                 212   Third–level caption
         220   Second–Level Caption
```

Multiple-decimal system:

```
                              T I T L E   C A P T I O N

        1   FIRST-LEVEL CAPTION

            1.1   Second-Level Caption

                  1.1.1   Third-level caption.

                  1.1.2   Third-level caption.

            1.2   Second-Level Caption


        2   FIRST-LEVEL CAPTION

            2.1   Second-Level Caption

                  2.1.1   Third-level caption.

                  2.1.2   Third-level caption.

            2.2   Second-Level Caption
```

Many companies or government agencies specify one of these systems; therefore it pays to be familiar with them all. If you use a numbering system with your captions, you must also repeat the numbers before the entries in the TOC. Often, numbers will be used only with the discussion elements of your report; that is, those elements currently marked with Roman numerals in Figure 10-3. However, it is also acceptable to use numbers with other major elements: Introductions, Conclusions, Recommendations, and References.

Documentation

There is a bewildering complexity of documenting systems from college to college, journal to journal, company to company. Therefore, we cannot claim a universal application for the instructions that follow. Use them barring conflicting instructions from your instructor, college, employer, or the style book of the journal or magazine in which you hope to publish.

Before we go into the mechanics of documentation, it might be wise to discuss why and when you need to document.

First of all, documenting fulfills your moral obligation to give credit where credit is due. It lets your reader know who was the originator of an idea or expression and where his or her work is found. Second, systematic documentation makes it easy for your readers to research your subject further.

When do you document? Well, to begin with, give credit when you quote directly from the work of another. Beyond documenting direct quotations, it is not always easy to determine what should and what should not be documented. You do not document general information or what might be called common knowledge. For example, even if you referred to a technical dictionary to find that creatinine's more formal name is methyglycocyamidine, you would not be obligated to show the source of this information. It is general information, readily found in many sources. If on the other hand you should include in your paper an opinion that certain quantum effects can be explained without inferring that atomic particles behave like waves, you would need to document the source of this opinion.

Generally, you should document if the material involves speculation or if it is found in only a few sources. Do not clutter your pages with references to information readily found in many sources. If in doubt as to whether to document or not, play it safe and document. If you can pinpoint your obligation sentence by sentence or paragraph by paragraph then document closely and specifically. If you owe a more general obligation, then indicate that general obligation as in this example:

> [1] Much of the information in this section was provided by Dr. James Blake, Associate Professor of Chemistry at the Arden School of Technology.

We will explain two systems of documentation. The first involves using footnotes, the second a bibliography or reference list.

Notes and Footnotes

Footnotes may be (1) displayed at the bottom of the page where the citation occurs,[1] or (2) gathered together at the end of the paper under the heading NOTES. The first method is used in this book. The second is more common in student reports and journal articles. In both cases the footnote form is the same. In both cases the citation is made by a superscript number; that is, a number placed above the line of print, as earlier in this paragraph.

In a list of notes, the title NOTES is written in solid capital letters without underscoring and is centered on the line. The first entry in the list begins one blank line below this title. (See Figure 10-8, page 220, for a sample list of notes.) Entries appear in the list of notes in the order that the citations occur in the paper. Each entry is numbered consecutively with superior Arabic numerals in the text. In the note itself the number is placed on-line, flush with the left margin,

---

[1] For examples, see footnotes to Chapter 8.

Figure 10-8   List of Notes Including Unpublished Sources

```
                            NOTES

1.   John Sterling Harris, "So You're Going to Teach Technical
Writing: A Primer for Beginners," Technical Writing
Teacher 2 (Fall 1974): 4.

2.   Paul L. Briand, "The Nonsense about Technical Writing,"
Journal of Engineering Education 57 (1967): 507.

3.   Donald H. Cunningham and Vivienne Hertz, "An Annotated
Bibliography on the Teaching of Technical Writing," College
Composition and Communication 21 (1970): 177-79.

4.   Ibid.

5.   United Nations, Economic and Social Council, Bibliography
of Publications Designed to Raise the Standard of Scientific
Literature (Paris, 1963), pp. 204-206.

6.   Briand, "Nonsense about Technical Writing," p. 508.

7.   A Manual of Style, 12th ed. (Chicago: University of
Chicago Press, 1969), p. 337.

8.   Private communication with Thomas L. Warren, Department
of English, University of South Dakota at Springfield,
Springfield, South Dakota, 20 March 1974.

9.   H. W. Fowler and F. G. Fowler, The King's English (New
York: Oxford University Press, 1940), p. 46.

10.   Juanita Williams Dudley, "Writing Skills of Engineering
and Science Students," IEEE Transactions on Engineering
Writing and Speech 14 (June 1971): 42.

11.   Harris, "Technical Writing," p. 5.
```

and followed by a period; the body of the note is then begun in the
third space to the right of the number.

Each line of an individual entry is single-spaced; double-spacing
is used between separate entries.

Generally, the correct form for an individual entry is as follows:
facts of composition followed by facts of publication. These facts,
however, will vary widely with certain types of entries; consult the
samples which follow and the samples in Figure 10-8 to find the
correct form for the most widely used types of entries.

Though few authorities agree on every detail of an individual
entry, all agree on the basic aspects: (1) the author's or editor's name,
(2) the title, and (3) the facts of publication. Following the widely
used Chicago *Manual*,[2] we recommend that you place a comma be-

---

[2]*A Manual of Style*, 12th ed. (Chicago: University of Chicago Press, 1969). Many of the
major technical and scientific journals generally conform to the citation style recommended by the *Manual*. In current practice, a journal article will usually be followed
by a bibliography or reference list rather than notes.

tween the three components of the footnote. (Some style books recommend a period between components 2 and 3. Try to be cheerful about such things and follow whatever style is dictated.)

To attempt to give a specific example of every type of individual entry you may come across would be impossible and is indeed unnecessary. If you conform to the basic format indicated above, you should have no difficulty. Each of the three components is discussed in detail below.

Author. In footnotes put the author's name in natural order, first name or initial, middle name or initial, and last name. Use the name as it appears on the title page. For multiple authors or editors, indicate first the name of the author or editor whose name appears first on the title page; if there are two or three authors or editors, list all names. If there are more than three authors or editors, list the first one and designate all authors or editors beyond the first one by "et al." (from the Latin *et alii*, "and others").

Examples:

> Van Wyck Brooks,
> (an author)
> P. L. Robinson and R. E. Dodd,
> (two authors)
> W. B. Yeats, ed.,
> (an editor)
> Ludovico Ariosto and Paul L. Briand, eds.,
> (two editors)
> Hobard H. Willard, Lynne L. Merritt, Jr., and John A. Dean,
> (three authors)
> Charles D. Hodgman et al.,
> (more than three authors)
> United Nations, Economic and Social Council, Commission on Human Rights,
> (author a named group)

Title. Underline the book titles in typed or handwritten papers. This corresponds to the italics used in print to designate a title. Anything separately published is underlined. The title of an article from a book, magazine, encyclopedia, collection, anthology, or newspaper is placed within quotation marks; this title is then followed by the underlined title of the publication from which it has been taken. Give the title in full exactly as it is on the title page, but do not include subtitles and long explanatory phrases.

Examples:

> *An Introduction to Automatic Computers,*
> (book title)

"Moths and Ultrasound," *Scientific American,*
    (title of article in a periodical)
Review of *A Model of the Brain* by J. Z. Young,
    (book review)
"More Demand for Burley Predicted," Louisville *Courier-Journal,*
    (newspaper headline with the city not part of the paper's name)
"Magnetism Problems and the Analog Computer,"
    (an unpublished thesis or dissertation)
"Landscape Architecture," *Encyclopaedia Britannica,* 11th ed.,
    (an encyclopedia article)
Interview Topic: The Air Force Role in Space,
    (an untitled interview)
"New Techniques in Making Nearly Perfect Mirrors," no. 2 in The
Kitzhaber Memorial Lectures in Geometrical Optics,
    (an unpublished lecture)

Facts of publication. Include those facts necessary to identify the source clearly, so the reader may verify your citation or pursue further the information taken from the source.

When citing a book, the title should be immediately followed by the number of the edition, if you are using an edition other than the first. Then, in parentheses, list place of publication, followed by a colon; the publisher's name, exactly as it appears on the title page or copyright page, followed by a comma; and the date of publication. Unless the city's location is obvious (as with New York or San Francisco), and it cannot be confused with another city of the same name, include the state (or country if applicable) when listing place of publication:

Middletown, Conn.: Wesleyan University Press
Washington, D.C.: American Chemical Society

When several places are listed on the title page, use only the first. For date of publication use the copyright date or the date of the latest edition. When no date is given, use the abbreviation "n.d." Place a comma after the end parenthesis, then cite the page or pages from which your information has come; for example, "p. 24" or "pp. 32–6."

For periodical citations it usually is not necessary to indicate the publisher and place of publication. The volume number is listed immediately after the name of the periodical (the periodical title is underscored, but the number is not). The year of publication follows, in parentheses; then a colon; then the inclusive page numbers of the particular citation.

If a journal renumbers its pages in each issue, the month or issue number must also be included. The month (or day, for weekly periodicals) is indicated before the year, and both are enclosed in parentheses. Alternatively, the issue number may be given after the volume number, preceded by "no." If the periodical is published in one or more series, give the series designation just before the volume number.

Yearbooks and conference proceedings usually may be cited in the same fashion as periodicals. For bulletins, circulars, and other monographs which appear in numbered series, follow the citation form for books, inserting the title of the series and the number of the monograph, if any, after the title of the monograph.

Government publications that are books or monographs written by individuals are handled like other books or monographs. It is much more difficult to cite official reports, government bulletins, pamphlets, contract reports, etc., with or without known authors. Certainly the citation should include the name of the government (nation, state, city, etc.), the name of the agency, the author if known, title of the report, and the date of publication. In the case of an anonymous document, the sponsoring body (such as a government agency) should be treated as the author. If the document has a number (for example, AFOSR TN–58–6, AERE–R–4705, AFM 10–4) this should be included. It is very helpful to indicate where a document can be located, for example, Office of Technical Services, U.S. Department of Commerce. You may have to consult a librarian for help in citing specific government publications.

Examples:

> *Introduction to Quantum Mechanics* (New York: McGraw-Hill Book Company, Inc., 1935), pp. 10–12.
> > (a typical book citation)

> *Economic Geography of South America,* 3d ed. (New York: McGraw-Hill Book Company, Inc., 1940), p. 40.
> > (third edition of a book)

> *Electrochemistry of Fused Salts,* trans. Adam Peiperl (Washington, D.C.: The Sigma Press, Publishers, 1961), p. 21.
> > (translated book)

> *J Animal Science* 18 (1959) : 815–19.
> > (a typical periodical citation, to be preceded by the title of the article)[3]

---

[3] It is common practice to abbreviate journal titles. You can find the abbreviations for most journals in the *American Standards for Periodical Title Abbreviations,* published by the American Standards Association. If you cannot easily obtain this source, you will not be wrong in most instances if you give the name of the journal in full.

Scientific American 213, no. 3 (1965) : 162–74.
    (periodical which repaginates each issue)
Scientific American (September 1965), pp. 162–74.
    (alternate way to cite a monthly periodical which repaginates each issue)
Can J Res, Sec F 26 (1948): 151–59.
    (periodical published in sections or series)
Southern Cooperative Series Bulletin 47 (1957).
    (monograph published in numbered series, to be preceded by the title of the monograph)
National Electronics Conf Proc 3 (1948) : 97–108.
    (annual proceedings of a society)

Describe any unpublished material as completely as you can.

Unpublished Master's thesis, University of Denver, 1980.
    (unpublished thesis)
Interview with Walter C. Work, Hq USAF, Washington, D.C., 20 April 1979.
    (interview)
Presented at the 36th annual meeting of the Colorado-Wyoming Academy of Science, Denver, Colorado, 30 April 1975.
    (lecture or paper presented at a meeting)
Private communication with John Gan, Colorado Springs, Colorado, 14 September 1967.
    (telephone conversation, brief meeting, or letter from an individual)

Subsequent references. If you refer to an entry more than once, you can choose among several ways of constructing your citation for the subsequent series. If you refer to the same entry for two or more footnotes in sequence with no intervening footnotes, the simplest citation is "Ibid." (from the Latin *ibidem*, "the same place"). If you are referring to the same entry, but to a different page of that entry, construct your citation as follows, "Ibid., p. 26."

When you are referring to the same entry more than once, but other footnotes intervene, your second and subsequent citations need repeat only the author's name and a shortened form of the title. When the page reference changes, follow the title with a comma and the proper page reference, for example, "Taylor, *Analysis*, p. 202."

Bibliographies and Reference Lists

Bibliographies can serve two purposes: they can be used in place of notes or footnotes to document sources, or they can serve as general lists of references relating to the subject of a report. A bibliographical list can include works used for general background in preparing a report, as well as works specifically cited. Or the bibliography may be intended primarily as a guide to additional information for interested readers.

In reports where the subject matter is primarily scientific, the bibliography is the most widely accepted way to cite specific sources. In this case, the title REFERENCES is used. In reports where the subject matter is drawn from the humanities, notes or footnotes are the accepted way to document sources, and a bibliography serves as a more general list of references.[4]

The Chicago *Manual of Style* recommends different formats for bibliographic entries in the sciences and humanities. We will discuss the scientific style here. For information on accepted style in the humanities, refer to the *Manual of Style*, pp. 372–384, or the latest *Handbook of the Modern Language Association*.

The format you follow for the entries in a scientific bibliography or reference list is similar to the format for footnotes, but with some significant differences:

1. Use periods instead of commas between the basic components of author, title, and facts of publication. Omit the parentheses around the facts of publication.

2. Transpose the authors' names: last, first, middle name or initial. If you must list several names, punctuation between names should follow the style of Figure 10-9.

3. The year of publication follows the author's name and precedes the title as in Figure 10-9.

4. Capitalize only the first letter of the title and subtitle and proper nouns. Do not enclose periodical titles with quotation marks.

5. Arrange the entries in alphabetical order rather than in order of appearance in the text. You need enter each entry only once, doing away with the need for "Ibid." and other subsequent entry forms. Determine alphabetical order by the author's last name or, if no author is listed, by title (disregarding *the, a,* or *an*).

6. Omit page numbers from book entries. You will see the reason for this in a moment. In periodical entries, give inclusive page numbers for each article.

With a reference list, source citation becomes very simple. At a point of reference in the text, place a parenthetical reference like one of these: (Brock 1978) or (Asher 1978, p. 93) or (Eisen 1978, pp. 93–96). The first (Brock 1978) refers the reader to the entry in

---

[4]Because this text is about writing, we have followed the accepted style for the humanities and used footnotes to cite our sources.

Figure 10-9   Bibliography Demonstrating Common Types of Scientific Citations

REFERENCES

Editors, <u>Astronomy</u>. 1979. Voyager approaches Jupiter.
    <u>Astronomy</u> 7, no. 4:14-15.

Groth, E. J. et al. 1977. The clustering of galaxies.
    <u>Scientific American</u> 237, no. 5:76-98.

Leaky, R. E., and Lewin, R. 1977. <u>Origins: What new discoveries</u>
    <u>reveal about the emergence of our species and its possible</u>
    <u>future</u>. New York: E. P. Dutton.

Schwartzenburg, D., and Burnham, R. 1979. The grand
    adventure. <u>Astronomy</u> 7, no.3:7-15,22-24.

The Viking Lander Imaging Team. 1978. The Martian landscape.
    NASA SP-425. Washington, D.C.: U.S. National Aeronautics
    and Space Administration.

the bibliography with author Brock and date 1978. You will use this form when the reference is to a periodical entry where the bibliography includes the inclusive pages of the article cited. The second form (Asher 1978, p. 93) refers the reader to page 93 of the reference authored by Asher. The third form (Eisen 1978, p. 93–96) refers the reader to pages 93 to 96 of the Eisen work. The second and third forms are used to refer to bibliographical entries, usually books, where the inclusive pages have not already been given. If the work has two or three authors, include the last name of each. For more than three authors, use *et al.*: (Ratti et al. 1979, pp. 45–47).

Parenthetical references should be on the line with the text and not above the line. The point of reference is normally at the end of the material used.

Example:

> As early as 1973, experts were predicting that "by 1990 we will be using continuous-thrust spacecraft for all space exploration projects" (Brock, p. 93). The development of continuous-thrust propulsion systems—which now appears to be a certainty—will greatly reduce flight times for all space missions (Ratti et al. 1979; Jones and Stein 1978).

Within the reference list or bibliography itself, it two or more publications are listed for the same author or authors, only the first entry gives the name. In immediately succeeding entries the author's name is replaced by a line followed by a period:

Leakey, R. E., and Lewin, R. 1977. *Origins: What new discoveries reveal about the emergence of our species and its possible future.* New York: E. P. Dutton.

——— . *People of the lake.* 1978. Garden City, N.Y.: Doubleday & Co., Anchor Press.

Copyright

Fairly stringent copyright laws protect published work. When you are writing a student report that you do not intend to publish, you need not concern yourself with this problem. If, however, you intend to publish a report, you should become familiar with copyright law. You can find a good summary of it in the University of Chicago's *Manual of Style* (1969), pp. 87–99.

Pagination

The numbering of pages, or pagination, is seldom a complicated problem. Brief manuscripts and reports having little prefatory matter are nearly always given Arabic numbers, running in one continuous series from first page to last. It is the convention to place these numbers at the bottom of the page or at the upper right-hand corner. When centered near the bottom of the page, each number may be given a hyphen on each side:

- 17-

As reports grow longer and more complicated, the page-numbering system may also need to be more complex. In such reports, the preliminary pages are often given small Roman numbers, as *i, ii, iii.* The pages of the body of the report are given Arabic numerals. Sometimes the introduction is considered part of the preliminary material, sometimes part of the body. In either case the Arabic series begins at 1 and usually continues through to the last page of the report. A number is not placed on the title page. However, a number is allowed for it; that is, the following page is number 2 or ii.

The Arabic series is not invariably retained throughout the report. Sometimes the appendixes or annexes, if any, are given capital letters, and then the pages of an appendix bear their distinctive letter plus their own series of Arabic numbers:

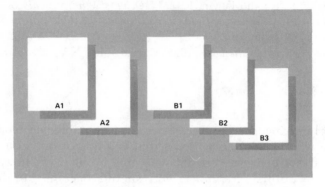

Using the letter-plus-number system for appendixes and annexes possesses two advantages. For one things, it is a distinctive system which clearly reveals that a page is part of an appendix or annex and not part of the body of the report. For another thing, use of a distinctive system makes it possible to print the appendixes and annexes before the pagination of the main report has been made firm. As a corollary, pages may be added to or removed from an appendix or annex without disturbing the pagination of the rest of the report.

What do you do if an appendix is a reprint of a document having its own page numbers? The answer, quite simply, is to retain the original page numbers and then beneath them add the new page numbers in brackets:

128

[426]

Clarity for the reader's use is always a paramount objective in pagination. In any case, you should never be more complicated than you have to be.

Appendixes and Annexes

Appendixes and annexes have become increasingly important in modern reporting practices. Many report users want to extract the heart of a report with a minimum expenditure of time and energy. This development has led to several changes in reporting practices within recent years, including greater economy of language and the adoption of appendixes and annexes to hold subordinate materials. The "main report" may now consist of twelve to twenty pages whereas the same content in earlier days might have required readers to plow through a hundred or more pages.

A knowingly designed TOC, brisk language in the prefatory parts, meaty summaries of two to five pages bringing well forward the conclusions and recommendations, the insertion of colored separator pages to "seal off" detail in appendixes and annexes—all have combined to make the report user's job simple and the report maker's job more difficult and hazardous.

Here, however, we are concerned with a limited aspect of the whole effort leading to "reader's ease"—the use of appended materials to clean out and clean up the oversized body of the voluminous report of earlier days. Therefore, you should know that detailed data, analyses, and supporting evidence are now being relegated to appendix position.

During the final stages of major reorganization, you should determine whether the following materials should be shifted to appendixes or annexes:

1. Case histories
2. Supporting illustrations
3. Detailed data
4. Transcriptions of dialogue
5. Intermediate steps in mathematical computation
6. Copies of letters, announcements, and leaflets mentioned in the report
7. Samples, exhibits, photographs, and supplementary tables and figures
8. Extended analyses
9. Lists of personnel
10. Suggested collateral reading
11. Anything else that is not essential to the sense of the main report

This tendency to reduce the body of the report to basic material has some disadvantages. In a drastic attempt to achieve conciseness and economy, some reports have been stripped of their circumstantial detail. Others have been made so brief and cryptic that they defy interpretation. Others have had their main parts excised and transferred beyond the colored separator page—the iron curtain of some modern reports. On the other hand, the motives behind these changes have a sound foundation. Reports *have* become too long and too detailed. Body discussions *have* become clogged with marginally relevant material. Reports *have* become tedious for the ordinarily busy and interested reader.

We therefore suggest that you be receptive to suggestions that certain front and body materials be shifted to the back. Consider each such suggestion on its merits, but consider also the effect on the whole. Do not be so eager to shift material to the appendixes and annexes that you prevent the reader from understanding the main body of the report.

EXERCISES

1. Compile a list of difficult and unusual terms you may have to introduce in the final report of your own selected project. Will the vocabulary problem be great enough to require you to include a glossary?

2. Prepare at least a tentative table of contents for your proposed final report. (See pages 206, 210.) Will you need a numbering system? If so, will the traditional system (page 217) be adequate or should you use one of the more complicated systems such as the multiple-decimal system (page 218)?

3. Decide what material, if any, you should place in the appendix of your final report. Why would you not include this material somewhere in the body of your report? If it is placed in the appendix, who would use the material and for what purpose?

4. How would you argue the point that reports generally profit if their format is conventional and the mechanical elements are given standard treatment?

5. In a published report or textbook locate as many of the mechanical elements discussed in this chapter as possible. What variations and departures do you find?

6. How does a bibliography (strictly speaking) differ from a list of notes. What are the virtues and limitations of each system?

7. The following sources are given in scrambled order and without attention to proper punctuation.
   a. List them, with punctuation, as they would appear in a scientific bibliography.
   b. List them as they would appear in a list of notes.

   Thomas F. Staton. How to Instruct Successfully. 1960. McGraw-Hill Book Company, Inc. New York, London. Page 12.

   Brown, James W., Richard B. Lewis, and Fred F. Harcleroad. AV Instruction, Media and Methods. McGraw-Hill Book Co. New York, 1969. Page 12.

Lalter Arno Wittich and Charles Francis Schuller, *Audiovisual Materials, Their Nature and Uses.* Harper & Row: New York, Evanston, London, 1967

The Severity of an Earthquake, by the Geological Survey, U.S. Department of the Interior. U.S. Government Printing Office, Washington, D.C., 1979, page 16.

"The Invisible Threat: The Stifled Story of Electron Waves." By Susan Schiefelbein. Saturday Review, September 15, 1979. Vol. 6, No. 18, pages 16–20.

S. Michael Halloran, "Technical Writing and the Rhetoric of Science," Journal of Technical Writing and Communication, pages 77–88. Vol. 8, Number 2, 1978.

Stress in Community Groups by Jerry W. Robinson, Jr., and others. Cooperative Extension Service, University of Illinois at Urbana-Champaign, 1975. Pages 6–10.

CHAPTER 11

Graphical Elements

From your reading of earlier chapters you should have concluded that report writing requires you to make many kinds of decisions. Some of these decisions, and highly important ones, involve the use of visual aids. When should you use prose to convey your ideas, and when should you turn to visual aids instead? The answers are not always clear, and sometimes you have to experiment with several means to find which will work best. But for your general guidance we can offer three principles:

1. Prefer visual aids if your ideas are primarily quantitative, structural, or pictorial.
2. Estimate whether your readers are primarily "eye minded" or "word minded." Generally, technically minded people are also eye minded.
3. Decide where your own skill lies—with words or with visual means.

Prose and visual aids are usually closely related. You may need to explain some aids in detail, while a brief mention will suffice for others. The amount of explanation will depend on the importance of the aid to your exposition, the complexity of the aid, and the technical expertise of your audience. Generally speaking, the more important and complex the aid, and the less expert your audience, the more prose explanation you need. In any event, always at least mention the visual aid to your reader. And mention it early in the discussion. Do not lead readers through a complicated prose explanation and then refer to the visual aid that simplifies the whole explanation. Send them to the aid immediately, and they can cut back and forth between the prose and the aid as necessary.

When a visual aid is important to your explanation, locate it as close to the explanation as you can. If it is a small aid, put it right on the text page. If the aid itself takes up a page, place it facing the prose explanation or at least on an adjoining page. When visual aids are less important to your explanation, they may be located further away, perhaps in an appendix.

Visual aids are generally divided into two categories: tables and figures. In a table the data are organized into columns (vertical) and rows (horizontal). Necessary numbers, titles, explanations, and so forth are provided. A figure is any visual aid that is not a table, a category that includes graphs, drawings, diagrams, and photographs. In a formal report, tables and figures are listed separately in a list of tables and a list of figures. In the rest of this chapter we will give you rather basic explanations of tables and figures.

Tables

A table is any arrangement of data set up in columns and rows, classified as either *formal* or *informal*.

Informal Tables

We "read" an informal table much as we would read normal prose conveying essentially the same information. Just as we usually read a paragraph only once, so do we scan through an informal table only once. It follows that informal tables must be relatively simple and immediately clear. As you may guess, the writer who inserts an informal table is actually exercising a choice between using prose and using the tabular display.

Informal tables have certain earmarks to distinguish them from the formal variety.

1. They are brief and simple.
2. They are not identified by table number and often have no title.
3. So that they will be seen as a continuation of the text, they have no ruled frame around them, and they have only the necessary minimum of internal ruled lines.
4. Lacking titles and table numbers, informal tables cannot be listed in the index or list of tables at the front of the report.

The example below shows a common variety of informal table. Portions of the surrounding text have been reproduced to illustrate how the sense of the table ties in with that of the prose preceding and following.

Figure 11-1   Informal Table

The relative effectiveness of these four nematocides can be seen in a study done at The West Virginia University where each of the nematocides used brought the following percentages of reduction in nematode population in red pine seedling beds:

Methyl bromide	99.2
D-D	85.0
Mylone	68.0
Vapam	64.0

Therefore, in this study methyl bromide was the most satisfactory material for controlling the plant parasitic nematodes in the red pine seedbed. In other studies as well, methyl bromide proved most effective in the control of plant parasitic nematodes.

Informal tables have certain virtues. Because they use minimal verbal machinery, they are space-savers, often conveying factual information that would require prose occupying several times the same space. They give readers relief by breaking up pages of solid prose. They signal readers that something different or special is being said. Readers can slow down and be sure of picking up the sense.

Formal Tables

Formal tables constitute a major problem in the use of graphic aids. For clarity of discussion here, we will first identify in Figure 11-2 the standard parts of such tables by name, location, and function.

Figure 11-2   Typical Table Format

TABLE NUMBER TITLE			
Stub Heading[a]	Column Heading[b]	Column Heading	
		Subheading	Subheading
Line Heading			
Subheading	Individual "cells" for tabulated data		
Subheading			
Line Heading			

[a]Footnote
[b]Footnote

You should notice several points in our skeleton table. First, the number and title of the table are traditionally placed *above* the table. Usually tables are numbered consecutively throughout a report, but sometimes they are numbered or lettered consecutively throughout each chapter: I-A, I-B; II-A, II-B, etc.; or 1-1, 1-2; 2-1, 2-2, etc.

Every column and subcolumn and every line or group of lines must have a heading that clearly identifies the data. All headings should have parallel grammatical form. Substantive headings consisting of nouns or noun phrases are usually most appropriate. The stub heading classifies the line headings only; it should not introduce the column headings. In other words, stub and column headings have vertical reference, whereas line headings have horizontal reference. Violations of these principles are shown in the very faulty table below:

Model Number	100	101	102	103
Year of origin	1964	1968	1969	1972
Lens f-number	5.6	3.5	2.8	1.4
Shutter speed (max)	1/300	1/500	1/800	1/1000
Retail cost new	$82	$89	$104	$117

Note how a few simple revisions make the table more logical and easier to read.

Descriptive Data	Model Numbers			
	100	101	102	103
Year of origin	1964	1968	1969	1972
Lens f-number	5.6	3.5	2.8	1.4
Shutter speed (maximum)	1/300	1/500	1/800	1/1000
Retail cost new (dollars)	82	89	104	117

However, even this arrangement may not be the ultimate. Assume that the main intent of the table is to compare the different models. In such a case most people would prefer the following table in which the data directly compared are arranged in columns rather than rows.

Model Number	Year of Origin	Lens f-number	Shutter Speed (maximum in seconds)	Retail Cost New ($)
100	1964	5.6	1/300	82
101	1968	3.5	1/500	89
102	1969	2.8	1/800	104
103	1972	1.4	1/1000	117

Notice that in these sample tables we have eliminated the vertical lines of the skeleton table. The modern trend is to eliminate as much clutter from tables as possible. Therefore, vertical columns are often separated simply by spacing rather than lines. Also, the horizontal double line between the headings and data is frequently eliminated and a single line substituted.

For economy's sake we now list additional conventions governing the construction of tables and their incorporation in the text of reports:

1. Every table should be as simple, clear, and logical as it can be made.
2. Every table must be keyed into the text before the table appears, either earlier on the page or on the preceding page.
3. Every reasonable effort should be made to insert a table in the text where the reader is expected to refer to it.
4. If it is necessary to refer to an earlier table or a later table, give its page number as well as table number.
5. Column headings and line headings should be arranged in some rational order—as alphabetical, geographical, temporal, or quantitative.
6. Where applicable, column headings and line headings should include units of measure, as miles, degrees, or percentages.
7. In columns, whole numbers are aligned on the right-hand digits. Fractional numbers are aligned on the decimal points.
8. The source of data appearing in a table must be indicated, whether by acknowledgment in the text, by a footnote, or both.
9. As indicated in our skeleton table on page 234, footnotes are internal to the table and are usually designated by superscript lower-case letters. The letter designation is used superscript in the text, but may be either superior or on-line in the footnote. If on-line it is followed by a period.

Figures 11-3 and 11-4 illustrate typical tables. Figure 11-3 shows the use of a footnote in a table. Figure 11-4 shows how to handle a

Figure 11-3   Sample Table with Footnote

TABLE II
AVERAGE LABORATORY WORKERS' WAGES

POSITION	WAGE[a]
Project Supervisor (Full-Time)	$10,000 & up/yr
Technical Aides	$3.00 – $3.75/hr
Students (Part-Time)	$1.50 – $2.00/hr

[a]These figures are very rough estimates obtained by interviews and correspondence with laboratory workers working in Pennsylvania.

table that runs over more than one page. Here we have included some of the text, so that you can get a feel for the way a table looks on a page. Notice, also, how the table is introduced.

Graphs

Our purpose in this section is to describe some of the basic kinds of graphs you can construct to illustrate your reports. Before you read the ground rules of graphing that follow, look at the diagram below to get a general idea of what distinguishes each kind of graph from the others. (Later we will discuss the different graphs in detail, with more precise illustrations.)

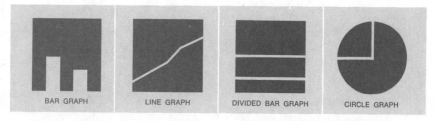

BAR GRAPH   LINE GRAPH   DIVIDED BAR GRAPH   CIRCLE GRAPH

1. Keep the graph simple. Provide no more detail than you absolutely need, no more ideas than the reader can grasp easily, only essential facts. Keep in mind the purpose of the graph.
2. Make the graph big enough. The answer to what is "big enough" is subjective, but also rather obvious. No squinting should be necessary.
3. Keep the graph orderly. In making comparisons, present the quantities in order from large to small or small to large or chronologically, according to your purpose.

Figure 11-4   Sample Table with Runover

10. Remove power from both instruments and disconnect the nanovolt source.

G. TROUBLE ISOLATION AND CORRECTION

Table 5-1 contains a check list of some of the more common malfunctions likely to be encountered with the recommended corrective measures. The instrument should provide years of trouble-free operation, if used in accordance with operating instruction.

Table 5-1. Troubleshooting Chart

INDICATION	POSSIBLE CAUSE	CORRECTIVE ACTION
1. Power light not illuminated with power switch ON	a. Jumper on battery terminals 1 and 2 not installed.	a. Replace jumper.
	b. Open fuse, F301.	b. Replace fuse and trace source of overload with ohmmeter before re-applying power.
	c. Defective indicator lamp.	c. Replace indicator lamp.
	d. Defective diode CR205.	d. Replace diode CR205.
2. Internal noise exceeds specified limits.	a. Defective component in AC amplifier.	a. Isolate source of noise to a specific AC amplifier by use of the

Table 5-1. Troubleshooting Chart (Cont'd.)

INDICATION	POSSIBLE CAUSE	CORRECTIVE ACTION
3. Drift exceeds specified limits.	a. Ambient temperature not within specified limits. (More likely to occur in rack-mounted instrument.)	a. Control ambient temperature to be within specified limits.
4. No meter indication with RANGE switch on any position and power light on.	a. Defective component in power supply.	a. Check power supply beginning with output of inverter transformer secondary, inverter and inverter driver. After locating approximate source of trouble, eliminate components by substitution.

5-3. REMOVAL OF MODEL 3990 FROM CASE

To remove the Model 3990 null detector from the case proceed as follows:

1. Remove the four corner screws on the front of the case; two screws at the top and two at the bottom.

Remove the two side screws on the rear of the case.

Courtesy of Honeywell Test Instruments Division, *Honeywell Technical Manual: Instructions for Guarded Null Detector, Model 3990.* Reproduced by permission of Penn Airborne Products Co.

4. Make the illustration attractive. A graph should be well balanced and blend into the page design.

5. Use any device possible to help the reader interpret your graph. Color, arrows, heavier lines, and annotation may be used to this end. Including a summary of the graph material either on the face of the graph or in a caption below the title will aid the reader's understanding.

6. Avoid two- and three-dimensional presentation of data. Areas and volumes may be appropriate on occasion but usually fail to make comparisons as clearly as a graph using one dimension. Graphs which attempt to depict volume are very difficult to interpret correctly.

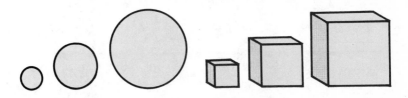

7. Make the units agree. Design your graphs so that the reader cannot possibly compare dollars with tons, or miles with hours. Make sure that the graph will be read so as to compare dollars with dollars, tons with tons, etc.

Multiple Bar Forms

Bar graphs are easily and accurately read. The reader does not come away with conflicting visual and verbal impressions. In general, it makes no difference whether the bars run vertically or horizontally, but there are some exceptions. One factor to be considered is page layout. Another is the nature of the data. People expect certain kinds of comparisons to be read up and down—altitudes for example—and others, like distances, to be read from side to side. The degree of reading precision you desire will determine whether you use a scale and, if you use it, how many grids to include on it.

Avoid, if possible, using keys to graphs. Put information on the bar itself or right next to it. Do not use a grid unless accuracy in reading the graph is of more importance than the dramatic quality of the graph. Figure 11-5 shows a bar graph with the information expressed vertically; the same facts could also be depicted using horizontal bars.

Line Graphs

Of all graphs used by scientists and engineers, the line graph is by far the most popular and useful. Certainly, it lends itself the best to purely technical information. The line graph can be used to plot the

Figure 11-5   Vertical Bar Graph

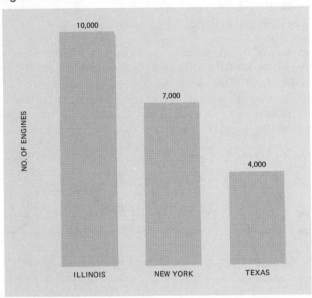

behavior of two variables: the *independent* variable and the *dependent* variable. The independent variable is normally plotted horizontally; that is, along the abscissa. The dependent variable is normally plotted vertically; that is, along the ordinate. The dependent variable is affected by changes in the independent variable. The rudimentary form of the line graph is dimensionless. That is, the *shape* of the relationship of the two variables is suggested, but is not spelled out quantitatively. Figure 11-6 shows the manner in which the plate current of a diode vacuum tube increases with increases in plate voltages. A particular diode may not behave exactly as shown, but diodes as a class tend to behave as illustrated.

Line graphs do, however, most often show values. In order to avoid a cluttered appearance, these values are often shown in hash marks on the abscissa and ordinate, as in Figure 11-7, rather than on a complete grid. Figure 11-6 also illustrates the use of the "suppressed zero." When the values plotted are so high that it would be impractical to start with zero, a scale starting at a point above zero is used. Whenever you must follow this practice be sure your reader understands clearly that the zero has been suppressed. One way to do this is to separate the scale, as in Figure 11-7.

When accuracy is of paramount importance, a complete grid must be used. Figure 11-8 illustrates a line graph plotted on graph paper with a fine grid that permits accuracy of construction and interpretation.

Figure 11-6   Simple Line Graph

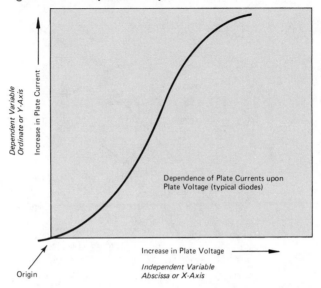

Figure 11-7   Line Graph with Hash Marks and Suppressed Zero

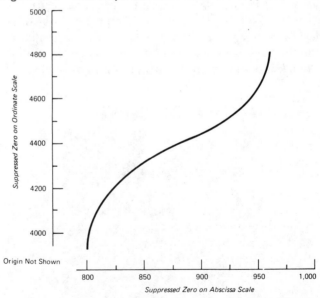

Divided Bar and Circle

Both the divided bar and the circle or pie graph make for ease and accuracy of reading. They both show proportions, the relative size of related quantities. Both are commonly found in business and finan-

Figure 11-8   Line Graph Requiring Fine Grid

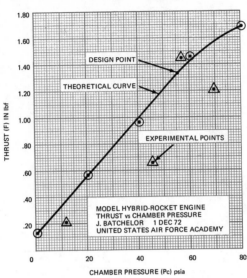

cial reports, but they are also useful in technical writing. Figure 11-9 is a divided bar. Figure 11-10 is a circle graph. In both figures, note the orderly arrangement of the parts. In the case of the circle it is good practice to begin the largest segment at twelve o'clock and then proceed clockwise in descending order of magnitude.

Figure 11-9   Divided Bar Graph

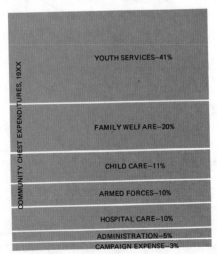

Figure 11-10   Circle Graph

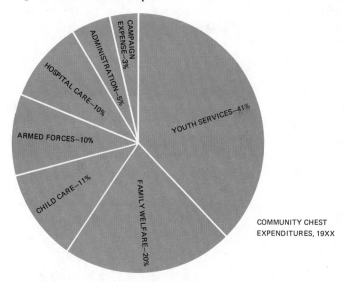

COMMUNITY CHEST
EXPENDITURES, 19XX

Pictographs

In drawing a pictograph the problem is to make sure that the reader interprets the drawing as a pictograph; for example, in Figure 11-11, as a pile of coins rather than as the volume or area covered. If the design is drawn so that it is seen in terms of the desired units, it

Figure 11-11   Pictograph

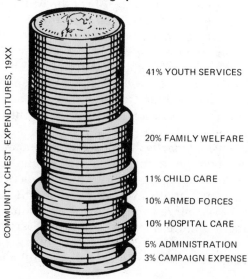

41% YOUTH SERVICES

20% FAMILY WELFARE

11% CHILD CARE

10% ARMED FORCES

10% HOSPITAL CARE

5% ADMINISTRATION
3% CAMPAIGN EXPENSE

can be read quite accurately. A pictograph often has the attention-getting value of novelty. Figure 11-11 presents the same information as Figures 11-9 and 11-10 but in pictograph form.

Table-Graph Relationship

Obviously, often you could present the same data in a table, in a layman-oriented graph such as a bar graph, or in a more technical line graph. Which method or combination of methods you choose will depend upon your intention and your audience.

In the table below, you see the number of hours and minutes that an automobile water pump survived testing at various revolutions per minute:

TABLE IV
TESTING TIME BEFORE FAILURE

Test No.	RPM	Hours-Minutes
1	4000	4:08
2	5000	4:02
3	6000	3:55
4	7000	3:45
5	8000	3:26
6	9000	2:45
7	10,000	1:15

This table shows clearly how the life of the water pump decreases as its speed of revolution is increased. But the table does not show very clearly the *shape* of the relationship of rpm to time before failure. If you were writing this paper for a predominantly lay audience you might choose to show this shape with a simple bar graph. Figure 11-12 shows clearly the relationship between rpm and the time before failure. Because it is customary to show time on the horizontal axis, the graph ignores the convention that the dependent variable is plotted on the ordinate.

But an audience primarily composed of technically minded people would probably prefer a line graph such as Figure 11-13 to a bar graph. They would recognize, as nontechnical readers might not, that the increases in rpm actually constitute a continuous variable. For them, the slope of the curve would show the continuous relationship of rpm and time to failure. Figure 11-13 also reverts to the conventional and plots the dependent variable along the vertical. The zero is suppressed on the abscissa, but the numbers on the scale make this fact quite clear.

Figure 11-12   Bar Graph: Testing Time Before Failure

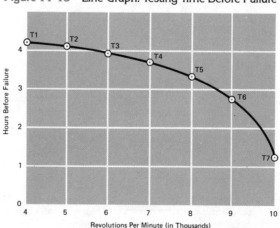

RPM (in Thousands)		
Test 7	1:15	
Test 6	2:45	
Test 5	3:26	
Test 4	3:45	
Test 3	3:55	
Test 2	4:02	
Test 1	4:08	

Hours Before Failure

Drawings, Diagrams, and Photographs

In this section we can touch only briefly upon the use of drawings, diagrams, and photographs. The production of these graphic aids generally calls for the services of a specialist in graphic aids. However, students can use these aids to a limited extent. And engineers, scientists, and technical writers should at least have some idea of how to use them.

Figure 11-13   Line Graph: Testing Time Before Failure

Drawings and Diagrams

Drawings and diagrams have one huge advantage over photographs. The artist can include as much or as little detail as needed. For example, Figure 11-14 reproduces a page from a Honeywell Technical Manual, *Instructions for Guarded Null Detector, Model 3990.*[1] The artist's only intention is to portray the dimensions of the

Figure 11-14   Diagram Showing Instrument Dimensions

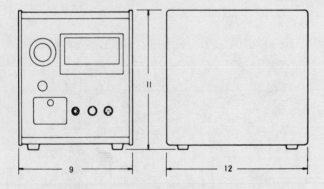

SECTION 2

INSTALLATION

2-1.   GENERAL

The Model 3990 null detector is shipped completely assembled and is ready for operation following unpacking, inspection, and checkout. Exercise normal care in unpacking the instrument from the shipping carton. After unpacking, carefully inspect the case, controls, and connectors for visible evidence of shipping damage. IN CASE DAMAGE HAS BEEN INCURRED, IMMEDIATELY NOTIFY THE RESPONSIBLE CARRIER AND INITIATE CLAIM PROCEDURE.

2-2.   INSTALLATION

For the bench model null detector, the requisites for placing the instrument in operation are a source of line (117 VAC $\pm 10\%$ or 230 VAC $\pm 10\%$, 50-60Hz) or battery (11 VDC or 14 VDC) power and an operating area relatively free of electromagnetic field. Refer to Figure 2-1 and 2-2 for outline dimensions of the bench and rack models.

Figure 2-1.   Outline Dimensions of the Null Detector, Bench Model

[1] Figures 11-14 through 11-16 and Figure 11-19 from Honeywell Test Instruments Division, *Honeywell Technical Manual: Instructions for Guarded Null Detector, Model 3990,* 1967. Reproduced by permission of Penn Airborne Products, Co.

bench model of the 3990. A simple drawing, free from a photograph's realistic clutter, is best for this.

Diagrams also can be as simple or as complicated as the writer wants them. Figure 11-15 shows a simplified block diagram of the

Figure 11-15   Block Diagram

is also made for safeguarding against accidentally introduced voltage over-loads through the use of diodes in the input filter. Stability is provided by a mechanical chopper modulator and a photoconductive demodulator. The modu-lator and demodulator are phase-synchronized by a common voltage source that operates at 94 Hz to minimize line frequency pickup effects.

B.   Figure 4-1 shows that the null detector consists of an input filter, a mechanical chopper feeding into an input transformer, an AC amplifier section feeding into an output transformer, a demodulator, an output filter, and a DC amplifier. Power for the modulator, demodulator, and amplifiers is furnished by an in-verter (DC to DC converter) driven by either a conventional 120/230-volt pow-er supply or a 12-volt lead acid battery. Three AC amplifiers are used in the 0.1 microvolt position, two amplifiers in the 1 microvolt and 10 microvolt po-sitions, and a single amplifier in the remaining ranges. Range switch S301 connects AC amplifiers to operating circuits as shown in the following diagram.

Switch Position	Amplifier(s) Affected
0.1 μv	All
1 μv	1st & 2nd
10 μv	1st & 3rd
Above 10 μv	1st only

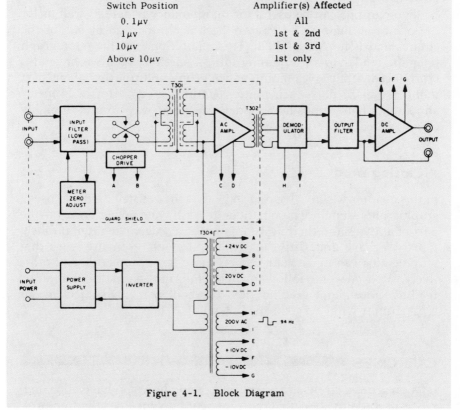

Figure 4-1. Block Diagram

Model 3990. In this particular case the author's intention is to give the technician who will *operate* the instrument a functional description of it. We have reproduced the entire page so that you can see how the writer introduces the diagram. Compare the simplification evident in Figure 11-15 with the complication in Figure 11-16. Figure 11-16 reproduces merely a portion of a complete schematic diagram of the Model 3990, intended for the technician who must *maintain* the instrument. Maintenance people obviously need complete detail whereas operators do not. As we have stressed many times, analyze your audience and give them what they need.

Photographs

Companion Figures 11-17 of U.S. Navy Sealab III and 11-18 of U.S. Navy Sealab II clearly depict the differences between a drawing and a photograph. The drawing provides information for the reader's mind. The photograph appeals to the mind's *eye*. Thus they are different not only in appearance but in their function and purpose. Both have their place in many reports.

Photographs can be useful for maintenance people as well as for the operator. Figure 11-19 reproduces a photograph of one of the circuit boards for Model 3990. The annotations on this photograph refer the reader to an adjoining table—part of which we have also reproduced—that gives maintenance people information about part numbers, spare parts, schematic locations, and so forth. Properly annotated photographs help readers find their way around in highly complex pieces of equipment.

A Closing Word

In this chapter we have tried to show that the construction and use of graphic aids demand much thought, skill, and experience on the part of authors and their assistants in the art department of publications. We have done little more than hint at the many ways that graphic aids can make your reports more understandable and more lively. We urge you to follow up on the information in this chapter by reading more about graphic aids. The books on this subject in our bibliography will be a good starting point.

EXERCISES

1. Construct a circle or bar graph showing how you have allocated, or propose to allocate, the time spent on your investigation project.

Figure 11-16   Schematic Diagram for Technician

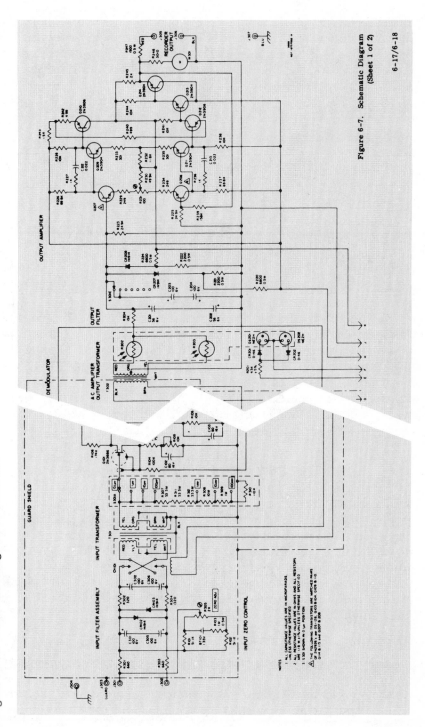

Figure 6-7.   Schematic Diagram
(Sheet 1 of 2)

6-17/6-18

Figure 11-17   Drawing of Sealab III

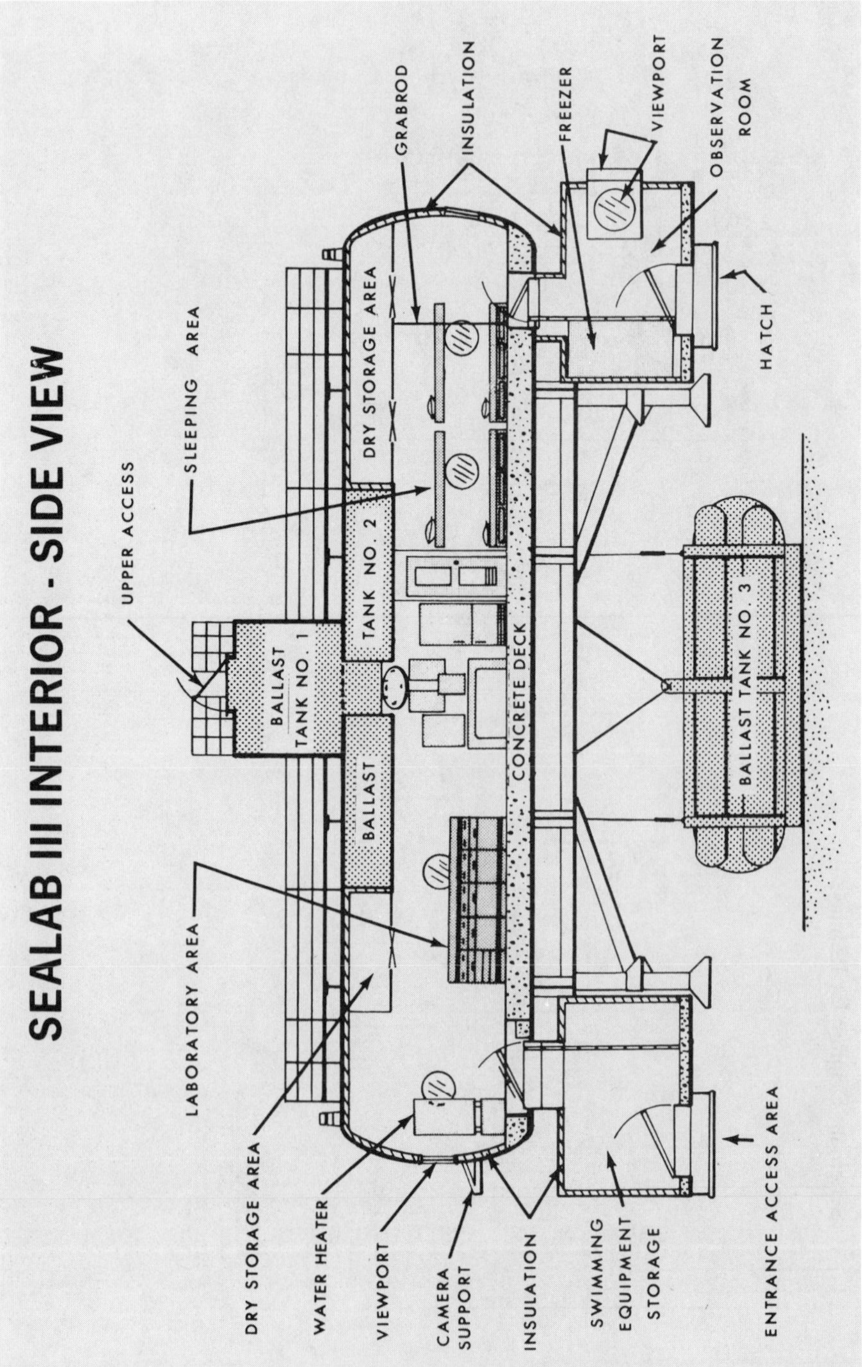

SEALAB III INTERIOR - SIDE VIEW

Figure 11-18 Photograph of Sealab II

Crane operator in staging vessel waits for Sealab II to be towed into position for lowering. Courtesy U.S. Naval Photographic Center.

Figure 11-19 Photograph and Corresponding Table

Figure 6-6. DC Amplifier Circuit Board

| References | | Description | Manufacturer's | | Honeywell | Qty/ | Spares |
Fig Index	Schem		Code	Part Number	Part No.	Assy	
10	CR206, CR207, CR208	Bridge, WO-2 Diode, 1N914	11711 08508	WO-2 1N914	16759755-002 16756845	1 2	
11	D8201, D8202						
12		Lamp, Glow	03508	NE-2H	16757777	2	2
13 14 15	QP01	Heatsink Plug, Housing Transistor	868n4	2N3053	16756892 16757774 16763169	2 4 1	
16	Q208, Q203, Q209, Q211, Q213, Q214	Transistor	04713	2N3904	16762172	6	
17	Q204, Q205, Q210, Q212, Q208	Transistor	04713	2N3904	16762173	4	
18	Q207, Q208	Transistor Assy			16758914	2	
19	Q206, Q215	Transistor	04713	2N3814	16761455-004	2	
20 21	R201, R203	Resistor, <3 K, 1/2 W, ±5% Photocell	44655 03911	LDDFA43000-5 CLK03CCL	16750076-589 16755808-004	1 2	2
22 23	R204, R205, R206	Resistor, 15 K, 1/8 W, ±1% Resistor, 750Ω, 1/8 W, ±1%	07716 07716	CEA-TO-1%-15 K CEA-TO-1%-750Ω	16757185-318 16757185-185	1 2	
24 25 26	R207, R208, R209, R233, R235	Resistor, 464Ω, 1/8 W, ±1% Resistor, 500Ω, 1/2 W, ±1% Resistor, 301Ω, 1/8 W, ±1%	07716 80294 07716	CEA-TO-1%-464Ω 308TP-1-501 CEA-TO-1%-301Ω	16757185-185 16750069-003 16757185-147	1 1 3	
27 28	R210, R211, R212	Resistor, 909Ω, 1/8 W, ±1% Resistor, 4.02 K, 1/8 W, ±1%	07716 07716	CEA-TO-1%-909Ω CEA-TO-1%-4.02 K	16757185-193 16757185-259	1 2	
29	R213	Resistor, 158Ω, 1/8 W, ±1%	07716	CEA-TO-1%-158Ω	16757185-130	1	
30	R214, R217	Resistor, 62Ω, 1/2 W, ±5%	44655	LDDFA82-5	16750076-520	2	
31	R215, R216	Resistor, 910Ω, 1/2 W, ±1%	44655	LDLD0910-10	16750076-568	2	
32	R218, R219	Resistor, 510Ω, 2 W, ±5%	44655	LDDXC310-4	16750076-044	2	
33	R220, R224	Resistor, 6.8 K, 1/2 W, ±10%	44655	LDLD48000-10	16750076-842	2	
34	R221, R222	Resistor, 2.2 K, 1/2 W, ±10%	44655	LDLD2200-10	16750076-836	2	
35	R223, R225	Resistor, 24.9 K, 1/8 W, ±1%	07716	CEA-TO-1%-24.9 K	16757185-339	2	
36	R226, R237	Resistor, 49.9 K, 1/8 W, ±1%	07716	CEA-TO-1%-49.9 K	16757185-382	2	
37	R227, R236	Resistor, 1 K, 1/8 W, ±1%	07716	CEA-TO-1%-1 K	16757185-201	1	

2. Assume that Bermuda has the following average monthly temperatures in degrees Fahrenheit:

January	63	July	79
February	63	August	80
March	63	September	78
April	65	October	74
May	70	November	69
June	75	December	65

Prepare a figure that plots these values as a continuous curve. Assume that Ottawa, Canada, has the following average monthly temperatures in degrees Fahrenheit:

January	12	July	69
February	13	August	67
March	26	September	59
April	41	October	46
May	54	November	33
June	64	December	18

Plot these values as a continuous curve and superimpose the curve on the graph you prepared for Bermuda.

Now, in two hundred words, point out the chief differences between the two curves and then discuss the causes and implications of the two temperature patterns.

3. Prepare a list of the illustrations (tables and figures) you intend to include in your project report.

CHAPTER 12

The Professional Report

In creating a professional report, you have to plan it, write it, and revise it. You must provide it with a proper mix of graphics and formal report elements such as a title page, a table of contents, and a proper ending. In this chapter, we discuss the entire process.

Planning

Nonfunctional, poorly planned reports do little for their readers or their authors. To understand the full import of that statement, you would have to spend some time in an organization watching reports make their way through the organization and observing how they are read and used. As consultants to business and governmental organizations, your authors frequently do exactly that. Let us share one of our experiences with you as an illustrative case study.

We recently spent an hour with the vice president of a large business. Directly working for him were three directors: one each for sales, investment analysis, and computer services. On his desk were four reports that he had to read that day, not an unusually large number. Other chores before him were three meetings, lunch with the president of another firm, and a stack of correspondence to read and answer. He was a busy person.

He picked up one of the reports, handed it to us, and asked us what we thought of it. The report—22 pages plus an appendix—was bare of any apparatus beyond a title and the author's name. It had no table of contents, no graphics or tables—nothing but 22 pages of closely packed typing and some computer company brochures. Its introduction gave no hint of the report's contents beyond what the title had already announced. The writer had plunged into a mass of statistics concerning input, output, power, remote consoles, ease of use, and cost for two computer systems. The paper was organized into two divisions, one for each system. Each division was further divided into a series of pros and cons. However, not all the criteria of input, output, and so forth had been applied consistently. Nor were there any captions to guide the reader through the report.

As we read the report, it seemed that one of the systems was better than the other, but we really couldn't be certain. Nowhere in the report were there any conclusions or recommendations. We looked quizzically at the vice president.

"Based on that report," he said," I have to decide which of those systems to buy for the company. I really don't know which is better. I don't even know who else has received the report. I'd like all three of my directors to have a copy, but I don't know whether they do or not. I'll have to have the writer and all the directors in and walk through the problem with them. It will take us the better part of a morning." He sighed, and we were glad we were not the author of the report.

This incident was quite revealing. The report was not useful precisely to the degree that it had ignored audience analysis, good organization, and report design. On this occasion, we worked with the writer to redo the report. We used the principles that are discussed in this book. We pointed out to the writer that the purpose of the report was to report research that led up to the *choice* of a computer system. Obviously, then, the report had to clearly state conclusions and recommendations. Next, we looked at the organization of the discussion. Because two systems were being compared, the use of the comparison-contrast pattern discussed in Chapter 5 (pages 107–110) was the obvious choice. The writer had already done that to a degree, but by using a loose pro-and-con division he had not applied his criteria consistently. We suggested that he organize his discussion around the six criteria—input, output, power, remote consoles, ease of use, and cost. Under each criterion, he would compare and contrast the two systems. In order that the reader could have some appreciation of how the two computer systems would function, we further suggested that he start his discussion with a unit describing how both systems would serve the users in sales and investment analysis. We therefore referred him to the techniques of process description (see pages 123–128). This organization would allow him to present the evidence fairly and evenly so that the reader could weigh it carefully. (Not to mention the fact that good organization may by its very nature help the writer-researcher reach a better decision.)

Looking through the material, we pointed out a few places where masses of statistics could be lifted out of paragraphs of densely packed prose and presented in simple tables and bar graphs. We cautioned the writer to refer to the tables and graphs at the appropriate times and also to use a few key facts from each in his text, thus tying his text and graphics together.

Next, we considered the needs of the audience—in this case, the vice president and the three directors. Only one of them, the director of computer services, was an expert in computers. She would be interested in some of the technicalities, particularly those concerning the criteria of input, output, and power. The vice president would be chiefly interested in both the general appreciation of how the systems would serve the users and cost. Because the directors of sales and investment analysis represented the users, they would be primarily interested in input, output, remote consoles, and ease of use. All four readers would want the criteria applied principally in terms of function—for example, how the output and remote consoles would benefit the users. And, of course, all four readers wanted to see conclusions and a recommendation.

To help the readers find the parts they were interested in, we reminded the writer that all divisions and subdivisions of the report should be well captioned. We decided also that a table of contents would help the readers find the parts they were most interested in. In other words, we considered the needs of a mixed audience as explained in the chapter on audience analysis (see pages 92–94).

In that same chapter, we cite evidence indicating that executives appreciate having key factual evidence summarized and are more likely to read conclusions and recommendations if they are either included in the summary or moved to the front of the report (see pages 80–83). In keeping with that, we recommended that the writer place an informative abstract near the front of the report that would include the conclusions and the recommendation. At the very front of the report we decided to place a title page that would include a concise but descriptive title, the author's name and title, a descriptive abstract, a distribution list naming all four readers, and a date.

We looked at the appendix material, which was largely performance and sales literature for the two computer systems. We urged the writer not to attach this material to the report but recommended that he instead mention the material in the report and offer to send it to anyone requesting it. Because the report was an internal report to a small group of readers, we also recommended a simple memorandum of transmittal. There were not enough graphics to warrant a list of figures. No references beyond the computer company literature were used, so a list of references did not seem justified. When done, we ended up with a report plan that contained the following in this order:

- A memorandum of transmittal
- A title page
- An informative abstract that contained
    —the chief facts of the report
    —conclusions that clearly, accurately, and succinctly pointed out the superiority of the better system
    —a recommendation to buy the better system
- A table of contents
- An introduction that announced the subject, purpose, scope, and plan of development of the report.
- A discussion organized as follows:
    —An appreciation of how the systems would function in the company
    —A division for each criterion that would explain the criterion and apply it to the two systems

This has been a true case study. Through it we have shown you that there is no *one* way to organize a report—no all-purpose format of title page, table of contents, and so forth. Rather, you now have at your disposal a wide range of organizational patterns and report elements. Choose carefully those that fit (need we say it?) your purpose, audience, and material.

As examples only, we show you a few of the possible patterns using the report elements explained in Chapters 9 and 10. See pages 185–186 for page references to the elements.

- *An internal report of 9 pages—informational, for executives:*
    —Memorandum of transmittal
    —Title page
    —Informative abstract
    —Introduction
    —Discussion
    —Complimentary close
- *An external report of 35 pages—decision-making, for executives and experts:*
    —Letter of transmittal
    —Title page
    —Table of contents
    —List of figures
    —Glossary

—Introduction
—Summary
—Conclusions
—Recommendations
—Discussion
—List of references
—Appendixes
- *Internal report of 12 pages—research report for scientists:*
  —Title page
  —Informative abstract
  —Table of contents
  —List of figures
  —Introduction
  —Discussion
  —Summary
  —Conclusions
  —Bibliography

In Part 3, "Applications," we discuss the variations on these formats that apply to specialized reports such as proposals and progress reports. For an example of a full-scale report, see Appendix A.

Writing the Rough Draft

After you have planned your report, the first thing you are likely to write is a draft of the longest section, the discussion. When do you write the introduction? Probably at the same time you write your ending elements and your abstracts—last. Successful introductions, like successful endings, require you to have an overview of the entire discussion. Some people, though, like to begin with the introduction as a means of setting their purpose and organization firmly in their minds. If you like to work that way, fine—but you may have to revise your introduction if your organization changes as you write the discussion.

Writing a draft is a very personal thing. Few people do it exactly alike. Some people write at a fever pitch; others do it slowly. All we can do is to describe in general the practices of most professional writers. Take our general hints and apply them to your own practices. Use the ones that make the job easier for you and revise or discard the rest.

Probably our most important hint is to begin writing as soon after the planning stage as possible. Writing is hard work. Most people, even professionals, procrastinate. Almost anything can serve as an excuse to put the job off: one more book to read, a movie that has to be seen, anything. The following column by Art Buchwald describes the problem of getting started in a manner that most writers would agree is only slightly exaggerated.

MARTHA'S VINEYARD—There are many great places where you can't write a book, but as far as I'm concerned none compares to Martha's Vineyard.

This is how I managed not to write a book and I pass it on to fledgling authors as well as old-timers who have vowed to produce a great work of art this summer.

The first thing you need is lots of paper, a solid typewriter, preferably electric, and a quiet spot in the house overlooking the water.

You get up at 6 in the morning and go for a dip in the sea, then you come back and make yourself a hearty breakfast.

By 7 a.m. you are ready to begin Page 1, Chapter 1. You insert a piece of paper in the typewriter and start to type "It was the best of times. . . ." Then you look out the window and you see a sea gull diving for a fish. This is not an ordinary sea gull. It seems to have a broken wing and you get up from the desk to observe it on the off chance that somewhere in the book you may want to insert a scene of a sea gull with a broken wing trying to dive for a fish. (It would make a great shot when the book is sold to the movies and the lovers are in bed.)

It is now 8 a.m. and the sounds of people getting up distract you. There is no sense trying to work with everyone crashing around the house. So you write a letter to your editor telling him how well the book is going and that you're even more optimistic about this one than the last

one which the publisher never advertised.

It is now 9 a.m. and you go into the kitchen and scream at your wife, "How am I going to get any work done around here if the kids are making all that racket? It doesn't mean anything in this family that I have to make a living."

Your wife kicks all the kids out of the house and you go back to your desk. It occurs to you that your agent may also want to see a copy of the book, so you tear out the paper and start over with an original and two carbons: "It was the best of times. . . ."

You look out the window again and you see a sailboat in trouble. You take your binoculars and study the situation carefully. If it gets worse you may have to call the Coast Guard. But after a half-hour of struggling they seem to have things under control.

Then you remember you were supposed to receive a check from the Saturday Review so you walk to the post office, pause at the drugstore for newspapers, and stop at the hardware store for rubber cement to repair your daughter's raft.

You're back at your desk at 1 p.m. when you remember you haven't had lunch. So you fix yourself a tuna fish sandwich and read the newspapers.

It is now 2:30 p.m. and you are about to hit the keys when Bill Styron calls. He announces they have just received a load of lobsters at

Menemsha and he's driving over to get some before they're all gone. Well, you say to yourself, you can always write a book on the Vineyard, but how often can you get fresh lobster?

So you agree to go with Styron for just an hour.

Two hours later with the thought of fresh lobster as inspiration, you sit down at the typewriter. The doorbell rings and Norma Brustein is standing there in her tennis togs looking for a fourth for doubles. **You don't want to hurt Norma's feelings so you get your racket and for the next hour play a fierce game of tennis, which is the only opportunity you have had all day of taking your mind off your book.**

It is now 6 p.m. and the kids are back in the house, so there is no sense trying to get work done any more for that day.

So you put the cover on the typewriter with a secure feeling that no matter how ambitious you are about working there will always be somebody on the Vineyard ready and eager to save you.

Reprinted by permission of Art Buchwald.

But you must begin, and the sooner the better. Find a quiet place to work, one with few distractions. Choose a time of day when you feel like working, and go to work. If you have followed our instructions on research and organization, much of your work is already done. You should have before you a complete plan of your discussion. Whether your plan is a formal outline or an informal one, you have your discussion divided into sections. Close at hand, you have your research material divided into matching sections. Your cards are distributed into well-organized sub-packs. Begin with the section that you think will be the easiest to write, and the whole task will seem less overwhelming. As you write one section, ideas for handling others will pop into your mind. When you finish an easy section, go on to a tougher one. In effect, you are writing a series of short, easily-handled discussions rather than one long one. Think of a 1500-word discussion as three short, connected, 500-word discussions. You will be amazed at how much easier this attitude makes the job.

Of course, when we advise you to write your discussion in sections, we assume that you plan to write a rough draft first, followed by a careful revision. Do not delude yourself that you can write a single draft of a long discussion. A rough draft followed by a revision, in all but the simplest discussions, is actually the most efficient procedure. While you are writing your rough draft leave plenty of space between the lines and at the margins for later revisions. Knowing that you can smooth out your draft later, you will be more relaxed. You will write faster and better.

How fast should you write? Again, this is a personal thing, but most professional writers write very rapidly, close to 1000 words an hour. Probably you should plan on at least 500 words an hour. Do not worry about phraseology or spelling in a rough draft. Proceed as swiftly as you can to get your ideas on paper. Later, you can smooth out your phrasing and look up the words with doubtful spelling.

Do not write for more than two hours at a stretch. This time span is one reason why you want to begin writing a long, important discussion at least a week before it is due. A discussion written in one long five- or six-hour stretch reflects the writer's exhaustion. Break at a point where you are sure of the next paragraph or two. When you come back to the writing, read over the previous few paragraphs to help you collect your thoughts and then begin at once.

Make your rough draft very full. You will find it easier to delete material later than to add it. Nonprofessional writers often write thin discussions because they think in terms of the writing time-span rather than the reading time-span. They have been writing on a subject for perhaps an hour and have grown a little bored with it. They feel that if they add details for another half-hour they will bore the reader. Remember this: at 250 words a minute, average readers can read an hour's writing output in about two minutes. Spending less time with the material than the writer must they will not get bored. Rather than wanting less detail, they may want more. This advice doesn't imply that you should pad your discussion. Brevity is a virtue in professional reports. But the discussion should include enough detail to demonstrate to the reader that you know what you're talking about. The path between conciseness on one hand and completeness on the other is often something of a tightrope.

As you write your rough draft, indicate where your references and visual aids will go. And don't be a slave to your outline. Often the writing process clarifies things in your mind—gives you insights into your work—in a way that organizing does not. Such clarification may call for a revised organization.

Revising the Paper

After you have finished your rough draft, let it cool for a while. If you go on to your revision immediately after writing your draft, you will miss many errors. Try to wait at least a day. Several days are better. When you are ready, sit down to it with a pencil, a pair of scissors, additional paper, a glue stick, and an open mind. A glue stick, which comes under many different trade names, is like a lipstick in which the stick of glue replaces the lip rouge. The solid glue is easy to roll out and apply. It does not wrinkle the paper as mucilage does, and, unlike rubber cement, it does not give off toxic fumes.

Read the draft first for organization and content. Try to put yourself in your reader's place. Does your discussion take too much for granted? Are questions left unanswered that the reader will want answered? Are links of thought missing? Have you provided smooth transitions from section to section, paragraph to paragraph? Do some paragraphs need to be split, others combined? Is some vital thought buried deep in the discussion when it should be put into an emphatic position at the beginning or end? Have you avoided irrelevant material or unwanted repetitions?

In checking content, be sure that you have been specific enough. Have you quantitified when necessary? Have you stated that, "In 1980, 52 percent of the workers reported taking at least 12 days of sick leave a year," rather than, "In a previous year, a majority of the workers took a large amount of sick leave each year"? Have you given enough examples, facts, and numbers to support your generalizations? And, conversely, have you generalized enough to unify your ideas and to put them into the sharpest possible focus?

Is your information accurate? Don't rely on even a good memory for facts and figures you are not totally sure of. Follow up any gut feeling you have that anything you have written seems inaccurate, even if it means a trip back to the library or laboratory. And check and double-check your math and equations. You can destroy an argument (or a piece of machinery) with a misplaced decimal point.

Be rigorous in your logic. Can you really claim that A caused B? Have you sufficiently taken into account other contributing factors? Examine your discussion for every conceivable weakness of organization and content and be ready to pull it apart. All writers find it difficult to be harshly critical of their own work, but you will find it a necessary attitude.

If you have done your job thoroughly in the research and planning stage, you will find few weaknesses in content and organization. But it is an unusual and lucky writer who finds none at all. Now is when the generous spacing and margins and the scissors and glue stick come into play.

Write small additions and corrections between the lines and in the margins. Use the proofreader's marks on the inside of the back cover of this book as needed. When you have major additions and reorganizations to do, use your scissors. Suppose you decide that you need a paragraph of transition between the two major sections. Write the needed paragraph. Cut your paper apart at the point where the paragraph is needed and glue the new paragraph into place between the sections. A professional writer always has scissors handy during revision. Is a heading botched with no room to write in a new one? Cut the botched heading out and insert a new one. Try to be neat. You or a typist will have to make sense of the revised draft. Figure 12-1 shows you what a revised manuscript page might look like. Compare it with our finished page 157–158.

Figure 12-1   Revised Manuscript Page

Observe the following conventions in ~~outlining:~~ *the body of your outline:*

● Make all entries grammatically parallel. Do not mix complete sentences with incomplete sentences. Do not mix ~~prepositional~~ *noun* phrases with ~~infinitive~~ *verb* phrases, and so forth. A formal outline with a hodgepodge of different grammatical forms will seem to lack/and, in fact, may lack/logic and consistency.

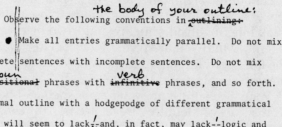

Incorrect	Correct
I. The overall view	I. The overall view
II. To understand the terminal phase	II. The terminal phase
III. About the constant-bearing concept	III. The constant-bearing concept

● Never have a single division. Things divide into two or more; so, obviously, if you have only one division you have done no dividing. If you have a "I" you must have a "II." If you have an "A" you must have a "B," and so forth.

Incorrect	Correct
I. Visual capabilities A. Acquisition	I. Visual capabilities A. Acquisition
II. Interception and closure rate	B. Interception and closure rate determination
III. Braking	II. Braking

● ~~Do not use periods or any other marks of punctuation after the entries in the body.~~

● ~~In the body~~ capitalize only the first letter of the entries and of proper nouns.

After you have revised your draft for organization and content, read it over for style. (We treat this as a separate step, which it is. But, of course, if you find a clumsy sentence while revising for organization and content, rewrite it immediately.) Use Chapter 8 to help you.

Rewrite unneeded passives. Cut out words that add nothing to your thought. Cross out the pretentious words and substitute simpler ones. If you find a cliché, try to express the same idea in different words. Simplify; cut out the artificiality and the jargon. Count the words in a few sentences here and there to see if you are close to the average you want. Break long involved sentences into two or even three shorter ones. Remember that you are trying to write effectively, not impressively. The final product should carry your ideas to the reader's brain by the shortest, simplest path.

Check your mechanics. Are you a poor speller? Check every word that looks the least bit doubtful. Develop a healthy sense of doubt and use a good dictionary. Some particularly poor spellers read their draft backwards to be sure that they catch all misspelled words. Do you have trouble with subject-verb agreement? Be particularly alert for such errors. We have provided you with a handbook, pages 453–485, that covers some of the more common mechanical problems.

When you are through with checking mechanics, complete your documentation and finish your graphics.

Once you have revised the draft of your discussion, write the formal elements you need, such as the introduction, abstracts, and a set of conclusions. You can now do the title page. You can also start such things as your table of contents and list of illustrations, though you will have to wait for the final draft to insert the proper page numbers.

When all is ready, you type your report or send it to the typist. No matter who types your report, proofread the final draft. The author of a report is responsible for any errors, even if someone else has done the typing.

We'll close here with a word of caution. We have been talking about your revising your own work. Therefore, we could cheerfully urge you to be ruthless with yourself. If someday you find yourself in the position of editing other people's work, you will find a gentler attitude more helpful. When editing, you are dealing with more than the paper lying in front of you. You are dealing with the human being who wrote the paper. It's your obligation to both author and reader to help make the work before you as clear and correct as possible, but you must have reasonable regard for the author's rights. So, avoid arbitrary attitudes. Don't just gleefully slap labels like "confusing" on the page. Why is the offending passage confusing? Perhaps the comment might better be phrased as a specific question, such as, "Isn't A really the same as B?" Be particularly careful about changing a writer's sentences. Don't change them just because they are not written the way you would have written them. Be sure that you can back up any change you do recommend in sentence structure—or anything else, for that matter—by demonstrating that the change really does clarify the thought or increase the efficiency or impact of the writing. If you can't justify a change, perhaps you had better allow the original to stand.

Checklist

Some writers feel that, like airplane pilots, they need a checklist; that is, a brief list of all the vital steps they must perform. We provide such a checklist for you here. (We also reproduce it on the inside front cover.) Used conscientiously in revision, this list will help you avoid many embarrassing oversights in your work.

1. Do you have a good title? Is it short but informative?
2. Have you stated clearly and specifically the purpose of the investigation and the report?
3. Have you stated clearly and fully the outcome of the investigation—what was actually accomplished?
4. Have you clearly and fully described the methods, materials, and equipment you used in conducting the investigation?
5. Have you put into the report everything required?
6. Does the report contain anything that you would do well to cut out?
7. Does the report present the substance of your investigation in a way that will be clear and readily understandable to your intended readers?
8. Are your paragraphs clear, well organized, and of reasonable length?
9. Is your prose style (diction) clear and readable?
10. Have you included all the mechanical and prose elements that your report needs?
11. Are your headings and titles clear, properly worded, and parallel?
12. Have you inserted the documentary reference numbers in your text?
13. Have you keyed the tables and figures into your text and have you sufficiently discussed them?
14. Are all parts and pages of the manuscript in the correct order?
15. Will the format of the typed or printed report be functional, clear, and attractive?
16. Does your manuscript satisfy stylebook specifications governing it?
17. Have you included required notices, distribution lists, and identifying code numbers?
18. Do you have written permission to reproduce extended quotations or other matter under copyright?
19. Have you proofread your manuscript for matters both large and small?

20. While you were composing the manuscript did you have any doubts or misgivings that you should now check out?

21. What remains to be done, such as proofreading your typist's final copy?

EXERCISES

1. Revise an earlier report for organization, content, style, and mechanics. Use scissors and glue stick where needed. Provide all needed prose and mechanical elements. Add graphics or improve the ones you already have. Type the final draft or have it typed. Turn in to your instructor the revised draft and the final typed version.

2. This exercise is both a class and individual effort. Your teacher will provide you with a draft of a discussion and a working bibliography for it. The class will decide the following:

   • Audience and purpose for the report.
   • The prose and mechanical elements the report needs to make it a functional report.
   • Where some of the prose should be replaced or supplemented by graphics.

   After your class discussion, take your scissors and glue stick to the draft. Revise it for organization, style, and mechanics. Prepare the needed graphics and indicate where in the draft they should go. Write the needed prose and mechanical elements. The exercise is complete when the report is ready for the typist.

3. Plan your final report in the manner discussed in the case study on pages 254–258. Decide upon purpose and audience, an organization for the discussion, needed prose and mechanical elements, the order of the parts in the report, and graphics. Put your plan in writing and discuss it with your instructor.

PART 3

Applications

In this part we discuss how to apply the basic techniques of Parts 1 and 2 in various types of technical reports and in speaking. Here we move from basic matters into professional work, covering correspondence, oral reports, and five kinds of technical reports: proposals, physical research reports, progress reports, instructions, and feasibility reports.

Our audience is still basically the student, but now it is the student who has mastered Parts 1 and 2 of this book. The nature of the material also makes Part 3 suitable for the professional writer or for the practicing scientist, engineer, or technician.

If you are still a student, we ask you to project yourself into the future. Imagine, as you write the exercises that your instructor assigns, that you are already practicing your chosen profession. Care about your work now as you will in the future. Do not be satisfied with anything less than a professional piece of work.

CHAPTER 13

Correspondence

You will as a technical writer often be responsible for a share of your organization's correspondence. Remember that to the receiver any letter from an organization represents that organization. If you write letters that are misleading, dull, or rude, you present an image of an organization that is misleading, dull, or rude.

We do not attempt in this chapter to give you a course in business correspondence. There are many fine books on the subject, some of them listed in our bibliography. However, we do cover those aspects of correspondence that will be most useful to you. We talk briefly about the tone and format of business letters. Then we give you specific instructions in how to write letters of inquiry, replies to letters of inquiry, letters of complaint and adjustment, the correspondence of the job hunt, and memorandums.

Style and Tone

Everything we said about style in Chapter 8 goes doubly when you are writing letters. Letters must be clear, so clear that the reader cannot possibly misunderstand them. Use short paragraphs, lists, active verbs, and specific words. Above all, avoid pomposity and the passive voice. Passive voice gives a cold formality to a letter's tone. Compare, "It is requested that you send ..." with, "I would appreciate your sending ..." Fill your letters with *I* and *you*, particularly *you*.

Do not make yourself a repository for all the clichés that writers before you have used. *Tell* people, do not *advise* them. Do not say "This is to acknowledge" or "I would like to thank you." (Simply say, "Thank you.") Do not open letters with "Re the" or "Referring to." Do not tell people "We are in receipt of your letter of the 10th and have noted its contents." They *know* that, or will know it when you answer their questions, grant their request, or whatever. If you want to pinpoint the date that a letter was written to you, do it in a courteous or an unobtrusive way. Say, "Thank you for your letter of November 10" or, "We don't blame you for being angry with us; your letter of November 10 made our mistake very clear." Perhaps the best solution is a subject or reference line that includes such file data. (See Figures 13-1 and 13-2.)

Many people have stressed the need for brevity in letters, and certainly it is a good thing to be concise. But do not get carried away with the notion. Letters are not telegrams. Too brief a letter gives an impression of brusqueness, even rudeness. Often a longer letter gives a better impression. The reader feels that you took more time to consider the problem. Particularly avoid brevity when you must refuse someone something. People appreciate your taking the time to explain a refusal.

Remember that a letter substitutes for conversation. Be natural. Consider the human being reading your letter when you write. Develop what has been labeled the *you-attitude.*

To some extent, we suppose, the *you-attitude* means that your letter contains a higher percentage of *you*'s than *I*'s. But it goes beyond this mechanical usage of certain pronouns.

With the *you-attitude* you see things from your reader's point of view. You think about what the letter will mean to the reader, *not* what it means to you. We can illustrate simply. Suppose you have an interview scheduled with a prospective employer and, unavoidably, you must change dates. You could write as follows:

> Dear Ms. Moody:
>
> A change in final examination schedules here at school makes it impossible for me to keep our appointment on June 10.
>
> I am really disappointed. I was looking forward to coming to Los Angeles. I feel inconvenienced, but I hope we can work out an appointment. Please let me know on what date we can arrange a new meeting.
>
> Sincerely yours,

Now this letter is clear enough, and it may even get a new appointment for its writer. But it has the *I-*, not the *you-attitude*. The writer thinks only of himself, and the reader may be vaguely or even greatly annoyed. This next version will please a reader far more:

Dear Ms. Moody:

A change in final examination schedules here at school makes it impossible for me to keep our appointment on June 10. When I should have been talking to you, I'll be taking an exam in chemistry.

I hope this change has not seriously inconvenienced you, Ms. Moody. Please accept my apologies.

Will you be able to work me in at a new time? Final exam week ends on June 12. Please choose any date after that convenient for you.

Sincerely yours,

Mechanically, the second letter contains more "you's," but, more to the point, it considers the inconvenience caused the reader, not the sender. And it makes it easy for the reader to set up a new date.

With the *you-attitude*, you consider both your reader's feelings and needs. Here is the way a camera company politely corrected an amateur photographer's film loading procedure:

Dear Mr. Bayless:

We are sending you a roll of film to replace the one over-exposed when you used your new Film-X camera for the first time. And we'll certainly replace the camera if it proves to be defective.

Trouble such as you have experienced is often caused by the photographer inadvertently forgetting to set the proper film speed while loading the camera. The ASA dial on the front of the camera must be set to match the ASA film speed number of the film you are working with. Failure to do so will result in either under- or overexposed film such as you experienced.

If you'll check Instruction 5 on page 3 of your Film-X Manual, you'll find complete information on how to set the ASA dial.

We hope that your problem was nothing more than an improperly set ASA. If not, however, follow the warranty instructions that came with the camera, and we'll replace your Film-X.

Sincerely yours,

With this letter, the writer begins by offering a gift. Next, he puts the reader's mind at ease by telling him that the company will replace the camera if need be. Next, he attempts to correct the reader's film loading procedure, the real reason, in the writer's opinion, for the trouble.

Suppose the letter-writer had taken an attitude like the one expressed in this letter:

```
Dear Mr. Bayless:

The type of film failure you wrote to us about stems from
faulty loading of the Film-X. Your method and not the cam-
era is defective.

Review the instructions that came with the camera. Pay par-
ticular attention to Instruction 5.
```

The second letter will surely put the reader's back up. Do people really write foolish, rude letters like this one? Unfortunately, perhaps because they are pressed for time, or because of a sense of superiority, they do—and worse. When people make a mistake, correct the consequences without making them feel like fools. We all make mistakes, and we appreciate gentle correction, not the lash of sarcasm. Taking time for the *you-attitude* is the good and human way to act. And, luckily for us all, it's also very good business.

Format

Almost any organization you join will have rules about its letter formats. You will either have a secretary to type your correspondence, or you will have to learn the rules for yourself. In this section we give you only enough rules and illustrations so that you can turn out a good-looking, correct, and acceptable business letter on your own. If for no other reason, you will find this a necessary skill when you go job-hunting.

Figures 13-1 through 13-4 on pages 272–275 illustrate four widely accepted business styles: the block, the semi-block, an alternative block, and the simplified. Figure 13-5 on page 276 illustrates the continuation page of a block letter. We have indicated in these samples the spacing, margins, and punctuation you should use. In the text that follows we discuss briefly the different styles and then give you some of the basic rules you should know about the parts of a letter. Before continuing with the text, look at Figures 13-1 through 13-5.

Figure 13-1   Block Letter

```
325 Fuller Road                                           Heading
Iowa City, IA 52240
September 2, 1980                                          Date Line

Doublespace Twice

Mr. John Pratt                                            Inside Address
Personnel Manager
Blank Corporation
325 Billingsley Drive
Los Angeles, CA 90211

Doublespace
Subject: Your letter of August 18, 1980                  Subject Line

Doublespace
Dear Mr. Pratt:                                           Salutation

Doublespace

━━━━━━━━━━━━━━━━━━━━━━━━━━━━━━━━━━━━━━
━━━━━━━━━━━━━━━━━━━━━━━━━━━━━━━━━━━━━━

Doublespace
━━━━━━━━━━━━━━━━━━━━━━━━━━━━━━━━━━━━━━
━━━━━━━━━━━━━━━━━━━━━━━━━━━━━━━━━━━━━━
━━━━━━━━━━━━━━━━━━━━━━━━━━━━━━━━━━━━━━

Doublespace
━━━━━━━━━━━━━━━━━━━━━━━━━━━━━━━━━━━━━━
━━━━━━━━━━━━━━━━━━━━━━━━━━━━━━━━━━

Doublespace
Sincerely yours,                                         Complimentary Close

Doublespace Twice

(Mrs.) Mary E. Clark                                     Signature Block
Doublespace
MEC:wge                                                  End Notations
Doublespace
Enclosure
Doublespace
cc: Ms. Georgia Mills
```

For most business letters you may have to write, any of the four styles shown would be acceptable. In letters of inquiry or complaint, where you probably do not have anyone specific in a company to address, we suggest simplified style. For letters of application we suggest the block or semi-block style without the use of a subject line. Some people still find the simplified letter without the conventional

Figure 13-2   Semi-Block Letter

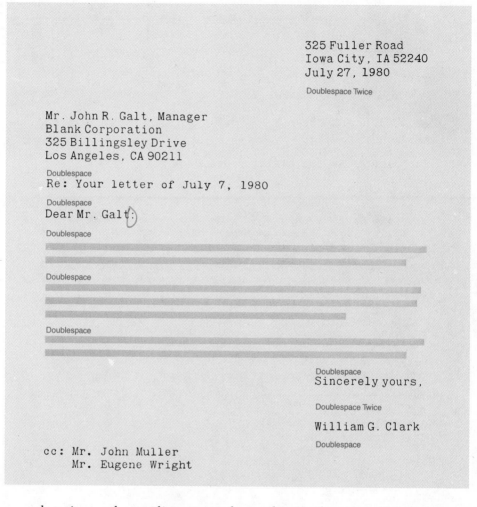

325 Fuller Road
Iowa City, IA 52240
July 27, 1980

Doublespace Twice

Mr. John R. Galt, Manager
Blank Corporation
325 Billingsley Drive
Los Angeles, CA 90211

Doublespace
Re: Your letter of July 7, 1980

Doublespace
Dear Mr. Galt:

Doublespace

Doublespace

Doublespace

Doublespace
Sincerely yours,

Doublespace Twice

William G. Clark

Doublespace

cc: Mr. John Muller
    Mr. Eugene Wright

salutation and complimentary close a bit too brusque. Unless you know for certain that the company you are applying to prefers the simplified form, do not take a chance with it.

If you are an amateur typist doing your own typing, we suggest the block style as the best all-around style. All the conventional parts of a letter are included, but everything is lined up along the left-hand margin. You do not have to bother with tab settings and other complications. Some people feel that a block letter looks a bit lop-sided, but it is a common style that no one will object to. The alternative block style is also an acceptable choice. No matter which style you choose, leave generous margins, at least an inch all around, and

Figure 13-3   Alternative Block Letter

December 18, 1980

Doublespace Twice

Mr. John R. Galt, Manager
Blank Corporation
325 Billingsley Drive
Los Angeles, CA 90211

Doublespace
Subject: Claim for elevator accident

Doublespace
Dear Mr. Galt:

Doublespace

Doublespace

Doublespace

Doublespace
Sincerely yours,

Doublespace Twice

William G. Clark
325 Fuller Road
Iowa City, IA 52240

Doublespace

Encl: Copy of accident report
      Filled-in claim

Doublespace

cc: Mrs. Nina Baldachi

balance the letter vertically on the page. Because letters should be good-looking with lots of white space, you should seldom allow paragraphs to run more than seven or eight lines. And don't be afraid of writing paragraphs as short as one or two lines, particularly at the end of a letter.

Figure 13-4   Simplified Letter

```
325 Fuller Road
Iowa City, IA 52240
13 November 1980
```

Doublespace Twice

```
Mr. John Galt, Manager
Blank Corporation
325 Billingsley Drive
Los Angeles, CA 90211
```

Doublespace

```
Subject: Your letter of 28 October 1980
```

Doublespace

Doublespace

Doublespace

Doublespace Twice

```
William G. Clark
```

Doublespace

```
WGC:mec
```

Doublespace

```
Enclosures (2)
```

Heading

We have assumed in our models that you do not have letterhead stationery of your own and have shown you how to type your heading. In the semi-block style, the heading is approximately flush right. In the other formats shown, the heading is flush left. Do not abbreviate words like *street* or *road*. Write them out in full. You may abbreviate your state. Figure 13-6 gives you the state abbreviations preferred by the Postal Service. If you do have business letterheads made up, have them printed on good quality white bond in a simple style.

Figure 13-5   Continuation Page of Block Letter

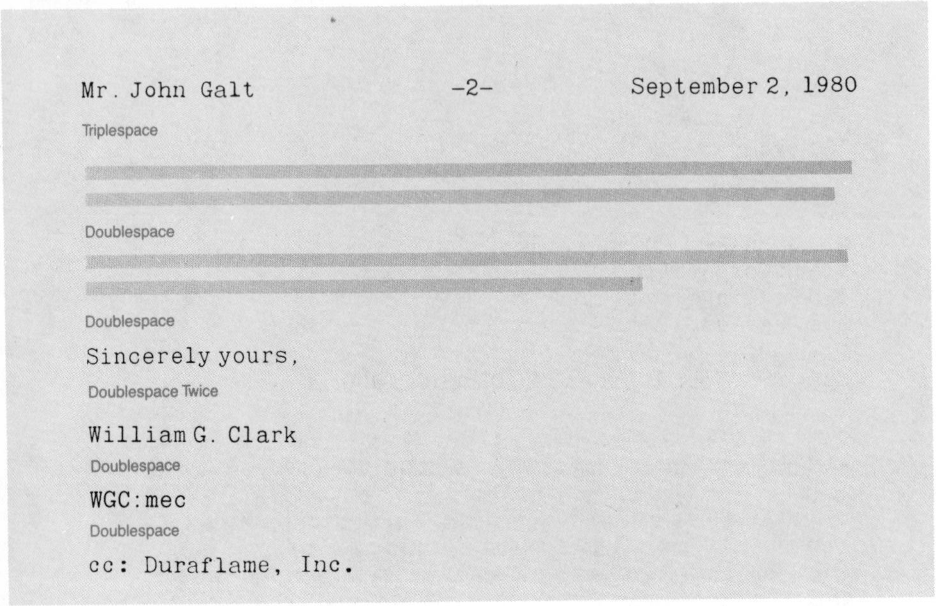

```
Mr. John Galt              -2-           September 2, 1980
Triplespace



Doublespace



Doublespace
Sincerely yours,
Doublespace Twice

William G. Clark
Doublespace
WGC:mec
Doublespace
cc: Duraflame, Inc.
```

Date Line

In a letter without a printed letterhead, the date line is part of
your heading in the block, semi-block, and simplified styles. In the
alternative block style, it's flush left and four spaces above the inside
address. When you do have a printed letterhead, the date line is flush
left in the block and simplified styles and approximately flush right in
semi-block. Place it two spaces below a printed letterhead. Write the
date out fully either as *June 3, 19XX* or *3 June 19XX*. Do not abbreviate
the month or add *st, nd*, etc. (e.g., 1st, 2nd), to the number of the day.

Inside Address

The inside address is placed flush left in all the formats shown.
Make sure the inside address is complete. Follow exactly the form
used by the person or company you are writing to. If your correspon-
dent abbreviates *Company* as *Co.*, you should also. Use *T. Edward
Smith* rather than *Thomas E. Smith* if that is the way Smith wants it.
Do use courtesy titles such as *Mr., Dr.*, and *Colonel* before the name.
The usual abbreviations used are *Mr., Mrs., Ms.*, and *Dr.* Place one-
word titles, such as *Manager* or *Superintendent*, immediately after the

Figure 13-6   State Abbreviations Recommended by the U.S. Postal Service

Alabama	AL*	Montana	MT
Alaska	AK	Nebraska	NE
Arizona	AZ	Nevada	NV
Arkansas	AR	New Hampshire	NH
California	CA	New Jersey	NJ
Colorado	CO	New Mexico	NM
Connecticut	CT	New York	NY
Delaware	DE	North Carolina	NC
District of		North Dakota	ND
Columbia	DC	Ohio	OH
Florida	FL	Oklahoma	OK
Georgia	GA	Oregon	OR
Guam	GU	Pennsylvania	PA
Hawaii	HI	Puerto Rico	PR
Idaho	ID	Rhode Island	RI
Illinois	IL	South Carolina	SC
Indiana	IN	South Dakota	SD
Iowa	IA	Tennessee	TN
Kansas	KS	Texas	TX
Kentucky	KY	Utah	UT
Louisiana	LA	Vermont	VT
Maine	ME	Virginia	VA
Maryland	MD	Virgin Islands	VI
Massachusetts	MA	Washington	WA
Michigan	MI	West Virginia	WV
Minnesota	MN	Wisconsin	WI
Mississippi	MS	Wyoming	WY
Missouri	MO		

*Notice that both letters of the abbreviation are capitalized and that no period is used.

name. When a title is longer than one word, place it on the next line by itself. Do not put a title after the name that means the same thing as a courtesy title; for example, don't write *Dr. Thomas E. Smith, Ph.D.*

Subject Line

Place the subject line flush left in all the styles. It is usually preceded by the word *Subject* or *Re,* though sometimes you will see it with nothing before it. In this line you can give the date of the letter being answered or a substantive heading such as "complaint adjustment."

Salutation

Place the salutation flush left. Convention still calls for the use of *Dear*. Always use a name in the salutation when one is available to you. When you use a name, be sure it is in the inside address as well. Also, use the same courtesy title as in the inside address, such as *Dear Dr. Sibley* or *Dear Ms. McCarthy*. You may use a first name salutation, *Dear Sarah*, when you are on friendly terms with the recipient. Follow the salutation with a colon.

What do you do when you are writing a company blindly and have no specific name to address? One solution, perhaps the best, is to choose the simplified style where no salutation is used. In the other formats, convention has called for the use of *Dear Sir* or *Gentlemen*. But because many people now feel uncomfortable assuming the recipient is always male, this convention is losing ground. Some people now use a *Dear* followed by the name of the department being written to, such as *Dear Customer Relations Department*. Still others substitute a *Hello* or *Good Day* for the traditional salutation. In other words, people are experimenting a bit instead of automatically following tradition. Whatever you do, we strongly urge you not to begin letters with *Dear Person* or *Dear People*. These salutations are distasteful to many.

Body

In the average length letter, single-space between the lines and double-space between the paragraphs. In a particularly short letter, double-space between the lines and triple-space between the paragraphs. Avoid splitting words between lines. Never split a date or a person's name between two lines. If you are going to have a continuation page, the last paragraph on the preceding page should contain at least two lines.

Complimentary Close

In the block and alternative block style, place the complimentary close flush left. In the semi-block style, align the close with the heading (or with the date line in a letterhead letter). Settle for a simple close, such as *Sincerely yours* or *Very truly yours*. Capitalize only the first letter of the close and place a comma after the close.

Signature Block

Type your name four spaces below the complimentary close. If you have a title, type it below your name. Sign your name immediately above your typed name. Avoid fancy flourishes and unintelligible signatures until you are president of the company.

If you are a man, sign your letters with your first name, middle initial, and last name. If your first name could be mistaken for a woman's, such as *Marion*, you might want to use your first two initials or use your first initial and middle name, as in *M. David Smith*. Your signature and your typed name should agree.

If you are a woman, sign your letters with first name, middle initial, and last name. If married, use your own first name, not your husband's. Again, the typed name below should agree with the signature. Before the typed name, in parentheses, women without an honorific title such as *Doctor* or *Professor* may choose to include a preferred title: *Miss, Mrs.,* or *Ms.* (See Figure 13-1.) The form *Ms.* (pronounced *miz* or by saying each of its two letters, *m-s*) is gaining popularity as a replacement for *Miss* or *Mrs.* It is a polite way to address a woman when you do not know her marital status, and it is a practical form for women who feel that marital status need not be identified in a professional setting.

End Notations

Various end notations may be placed at the bottom of a letter, always flush left. The most common ones indicate identification, enclosure, and carbon copy.

Identification. The notation for identification is composed of the letter writer's initials in capital letters and the typist's initials in lower case:

DHC:lnh

Enclosure: The enclosure line indicates that additional material has been enclosed with the basic letter. You may use several forms:

Enclosure
Enclosures (2)
Encl: Employment application blank

Carbon copy. Both courtesy and ethics require that you inform the recipient of a letter when you have sent a copy of the letter to someone else. In form, the copy notation looks like this:

cc: Ms. Elaine Mills

See Figures 13-1 to 13-5 for proper spacing and sequence of these three notations.

Continuation Page

Do not use letterhead stationery for a continuation page. Use plain white bond of the same quality as the first page. Try to begin the continuation page with a new paragraph. If you cannot, be sure the continued paragraph runs at least two lines. As shown in Figure 13-5, the continuation page is headed by three items: name of addressee, page number, and the date.

Letters of Inquiry

As a student, business person, or simply a private person, you will often have occasion to write letters of inquiry. Students often overlook the rich sources of information for reports that they can tap with a few well-placed, courteous letters. Such letters can bring brochures, photographs, samples, and even very quotable answers from experts in the field the report deals with.

Sometimes companies solicit inquiries about their products through their advertisements and catalogs. In such cases your letter of inquiry can be short and to the point:

> Your advertisement on page 89 of the January 1980 <u>Scientific American</u> invited inquiries about your new Film-X developing process.
>
> I am a college student and president of a 20-member campus photography club. The members of the club and I would appreciate any information about this new process that you can send us.
>
> We are specifically interested in modernizing the film developing facilities of the club.

As in the above letter, you should include three important steps:

1. Identify the advertisement that solicited your inquiry.
2. Identify yourself and establish your need for the information.
3. Request the information. Specify the precise area in which you are interested. Obviously, in this step you also identify the area where the company may expect to make a sale to you. You thus, in a subtle way, point out to the company why it is in their best interest to answer your inquiry promptly and fully: a good example of the *you-attitude* in action.

An unsolicited letter of inquiry cannot be quite so short. After all, in an unsolicited letter you are asking a favor, and you must avoid the risk of appearing brusque or discourteous. In an unsolicited letter you include five steps:

1. Identify yourself.

2. State clearly and specifically the information or materials that you want.
3. Establish your need for the requested information.
4. Tell the recipient why you have chosen him or her as a source for this information.
5. Close graciously, but do not say "thank you in advance."

The first four steps may be presented in various combinations rather than as distinct steps, or in different order, but none of the steps should be overlooked.

Identification

In an unsolicited letter, identify yourself. We mean more here than merely using a title in your signature block. Rather, you should identify yourself in terms of the information sought. That is, you are not merely a student, you are a student in a dietetics class seeking information for a paper about iron enrichment of flour. Or you are a member of a committee investigating child abuse in your town. Or you are an engineer for a state department of transportation seeking information for a study of noise walls on city freeways. Certainly, the more prestigious your identification of yourself, the more likely you are to get the information you want. But do not misrepresent yourself in any way. Misrepresentation will usually lead to embarrassment.

Some years ago, one of the authors of this text naively wrote for information on bookshelves using his college letterhead instead of his private stationery. He was deluged with brochures that were followed up by a long-distance telephone call from a sales manager asking how many shelves the college was going to need: "Is the college expanding its library facilities?"

"No," the lame answer came, "I just wanted to put up a few shelves in my living room."

Do not misrepresent yourself. But do honestly represent yourself in the best light you can. Most companies are quite good about answering student inquiries. They recognize in students the buyers and the employees of the future and are eager to court their goodwill.

Inquiry

State clearly what you want. Be very specific about what you want and do not ask for too much. Avoid the shotgun approach of asking for "all available information" in some wide area.

Particularly, do not expect other people to do your work for you. Do not, for example, write to Bell Telephone asking for references to articles on wave guides. A little time spent by you in the indexing and abstracting publications will produce the same information. On the other hand, it would be quite appropriate to write to a Bell engineer or scientist asking for a clarification or amplification of some point in a recent article that he or she had written. Science thrives on the latter kind of correspondence.

If you do need involved information, put your questions in an easy-to-answer questionnaire form. If you have only a few questions, include them in your letter. Indent the questions and number them. Be sure to keep a copy of your letter. The reply you receive may answer your questions by number without restating them.

For a larger group of questions, or when you are sending the questionnaire to many people, make the questionnaire a separate attachment. Be sure in your letter to refer the reader to the attachment. Questionnaires are tricky. Unless they are presented properly and carefully made up they will probably be ignored. Do not ask too many questions. If possible, phrase the questions for "yes" or "no" answers or provide multiple-choice options. Sometimes, meaningful questions just cannot be asked in this objective way. In such cases, do ask questions that require the respondent to write an answer. But try to phrase your questions so that answers to them can be short. Provide sufficient space for the answer expected. If you are asking for confidential information, stress that you will keep it confidential. The questionnaire should be typed, printed, well-mimeographed, or photocopied. Perhaps the best solution for multiple copies of anything these days is to use one of the many inexpensive photocopying services available. Look under *Photocopying* in the yellow pages of your telephone book. For further information about putting together a questionnaire, see pages 53–58.

Need

Tell the recipient why you need the information. Perhaps you are writing a report, conducting a survey, buying a camera, or simply satisfying a healthy, scientific curiosity. Whatever your reason, do not be complicated or devious here. Simply state your need clearly and honestly. Often this step can best be combined with the identification step. If there is some deadline by which you must have an answer, say what it is.

Choice of Recipient

Tell the recipient of the letter why you have chosen him or her or the recipient's company as a source for the information. Perhaps you are writing a paper about hi-fi equipment and you consider this company to be one of the foremost manufacturers of FM tuners. Perhaps you have read the recipient's recent article on space medicine. You specifically want an amplification of some point.

Obviously, this section is a good place to pay the recipient a sincere compliment. But do not flatter or be sickeningly sweet. Phoniness will get you nowhere.

Point out any benefit the recipient may gain by answering your request. For example, if you are conducting a survey among many companies, promise the recipient a tabulation and analysis of results when the survey is completed.

Gracious Close

Close by expressing your appreciation for any help the receiver may give you. But do not use the tired, old formula, "Thanking you in advance, I remain . . ." Do write a thank-you note *even* if all you get is a refusal. Who knows? The second letter may cause a change of heart.

Sample Letters

A complete letter of inquiry might look like this one:

Dear Mr. Hanson:

I am a second-year student at Florida Technological Institute. For a course in technical writing that I am currently taking, I am writing a paper on the proper way to educate Americans in the use of the metric system. The paper is due on 3 March.

The specific question that concerns me is whether metric measurements should be taught in relation to present standard measurements, such as the foot and pound, or whether they should be taught independently of other measurements. I see both methods in use.

In the journals where I have been researching the subject, you are frequently mentioned as a major authority in the field. Would you be kind enough to give me your opinion about how metrics should be taught?

Any help you can give me will be greatly appreciated, and I will, of course, cite you in my paper.

Sincerely yours,

We created the letter above to illustrate in a somewhat mechanical way all the steps in a good letter of inquiry. Reproduced below is a real letter received by one of us. The writer of the letter covers all the required steps in a relaxed, nonmechanical way. The letter also illustrates the use of a questionnaire approach within a letter. Notice, too, how the writer makes legitimate use of a person—Mary Schaeffer, past president of the Society for Technical Communication—known to both him and the letter's recipient.

Dear Professor Pearsall:

At the joint meeting of the STC Board and the Phoenix Chapter, Mary Schaeffer mentioned that you were preparing a survey of technical communication programs. We are interested in developing a technical communication degree program at our university. But, before we prepare a proposal, we want to study the programs at other schools.

Briefly, we would like to answer two questions:
    (1) Would such a degree program be of real value to both students and industry?
    (2) Would an interdisciplinary approach be best? (We have excellent schools of printing, journalism, engineering, etc.)

Of course, we are not asking you to answer these questions for us, but we would like to have your personal opinion and any information on proposed and existing technical communication programs that you can send us.

This is a particularly critical moment for us because the next catalog will cover <u>two</u> years. As you can see, we must move rapidly and surely.

We will greatly appreciate any assistance you can give us.

Sincerely,

Replies to Letters of Inquiry

In replying to a letter of inquiry, be as complete as you can. Probably you will be trying to avoid a second exchange of letters of questions and answers. If you can, answer the questions in the order in which they appear in the inquiry. If the inquirer represents an organization and has written on letterhead stationery, you may safely assume that a file copy of the original letter has been retained. In that case, you don't have to repeat the original questions—just answer them. If the inquirer writes as a private person, you cannot assume that a copy of the letter has been kept. Therefore, you will need to repeat enough of the original question or questions to remind the inquirer of what has been asked.

    You might answer the earlier letter about metric education in the following fashion:

Dear Ms. Montez:

The question of how to educate people about the metric system concerns a great many educators. I suppose it's inevitable that those familiar with the present system will be tempted to convert metric measurements to ones they already know. For example, people will say, "A kilogram, that's about two pounds."

In my opinion, however, such conversion is not the best way to teach metrics. Rather, people should be taught to think in terms of what the metric measurement really measures, to associate it with familiar things. Here are some examples:

--A paper clip is about a centimeter wide.
--A dollar bill weighs about a gram.
--A comfortable room is 20 degrees C.
--Water freezes at 0 degrees C.
--At a normal walking pace, we can go about five kilometers an hour.

These are the kinds of associations we have made all our lives with the present system. We need to do the same for metric measurements. As in learning a foreign language, we really learn it only when we stop translating in our heads and begin to think in the new language.

I hope this answers your question. You can get a valuable little book about metrics by writing to the Office of Public Information, Federal Reserve Bank of Minneapolis, Minneapolis, Minnesota 55480. Ask for The United States and the Metric System.

Sincerely yours,

Because the writer of the first letter wrote as a private person, she is, in this answer, reminded of the original question. The tone of the letter is friendly. The question is answered succinctly but completely. An additional source of information is mentioned, an excellent idea which should be done when possible.

When you can, include as enclosures to your letter previously prepared materials that provide adequate answers to the questions asked. If you do so, provide whatever explanation of these materials is needed. Reproduced below is the answer to the letter of inquiry on page 283 that asks for information about technical communication programs:

Dear Professor Garden:

The answer to both your questions is yes. Rather than take time to support my answer in this letter, I've enclosed several documents that should do the job for me.

The first is a proposal I submitted for a technical communication program at this university. My personal views on the value and nature of a program in TC are in the proposal. The program, by the way, was approved and is well under way.

The second enclosure is an article by John Walter that expresses his views on these subjects.

I've also enclosed for you a brochure explaining our program, a catalog in which our program is described (p. 44), and a list of other schools that have TC programs.

I hope all this helps you in your efforts to begin a TC program at your school. Write again if I can help in any way.

Sincerely yours,

Letters of Complaint and Adjustment

Mistakes and failures happen in business as they do everywhere else. Promised deliveries don't arrive on time or at all. Expensive equipment fails at a critical moment. If you're on the receiving end of such problems, you'll probably want to register a complaint with someone. Chances are good that you'll want an adjustment—some compensation—for your loss or inconvenience. To seek such adjustment, you'll write a letter of complaint or, as it is also called, a claim letter.

Your attitude in a letter of complaint should be firm but fair. There is no reason for you to be discourteous. You'll do better in most instances if you write the letter with the attitude that the offending company will want to make a proper adjustment once they know the problem. Don't bluster and threaten to withdraw your trade on a first offense. Of course, after repeated offenses you will seek another firm to deal with, and that should be made clear at the appropriate time.

Be very specific about what is wrong and detail any inconvenience caused you. Be sure to give any necessary product identification such as serial numbers. At the end of your letter, motivate the receiver to make a fair adjustment. If you know exactly what adjustment you want, spell it out. If not, allow the suggested adjustment to come from the company you are dealing with. Figure 13-7 shows you

Figure 13-7   Letter of Complaint

26 Shady Woods Road
White Bear Lake, MN 55101
28 October 1980

Customer Relations Department
Chapman Products, Inc.
1925 Jerome Street
Brooklyn, NY 11205

Subject: Broken sanding belts

This past September, I purchased two Chapman sanding belts,
medium grade, #85610, at the Fitler Lumber Company in St.
Paul, Minnesota.   I paid a premium price for the Chapman
belts because of your reputation.   However, the belts have
proved to be unsatisfactory, and I am returning them to you in
a separate package (at a cost to me of $1.36).

The belt I have labeled #1 was used only 10 minutes before it
broke. The belt labeled #2 was used only 5 minutes when the
glue failed and it broke.

I attempted to return the belts to Fitler's for a refund.
The manager refused me a refund and said I'd have to write to
you.

I am disturbed on two counts.   First, I paid a premium price
for your belts.   I did not expect an inferior product.   Second, does the retailer have no responsibility for the Chapman products he sells?   Do I have to write to you every time I
have a problem with one of your products?

I'm sure that Chapman is proud of its reputation and will want
to adjust this matter fairly.

John Griffin

John Griffin

what a complete letter of complaint looks like. Note that it is in
simplified format and addressed to the Customer Relations Department. You can safely assume that most companies have an office or
an employee specifically responsible for complaints; in any case, if
the letter is addressed in this manner it will reach the appropriate
person much more rapidly.

What happens at the other end of the line? You have received a
letter of complaint and must write the adjustment letter. What
should be your attitude? Oddly enough, most organizations welcome

letters of complaint. At least they prefer customers who complain rather than customers who think "never again" when a mistake occurs and take their trade elsewhere. Most organizations will go out of their way to satisfy complaints that seem at all fair. And a skillful writer will attempt to use the adjustment letter as a means of promoting future business.

Letters of adjustment fall into two categories—a granting of the adjustment requested or a refusal. The first is easy to write, the second more difficult.

When you are granting an adjustment, be cheerful about it. Remember, your main goal is to build goodwill and future business. Begin by expressing regret about the problem or stating that you are pleased to hear from the customer—or both. Our earlier comments about the *you-attitude* (pages 269–271) should be much in your mind while writing an adjustment letter. Explain the circumstances that caused the problem. State specifically what the adjustment will be. Handle any special problems that may have accompanied the complaint and close the letter. Figure 13-8 shows you such a letter.

A letter refusing an adjustment is obviously more difficult to write. You want, if possible, to keep the customer's goodwill. You want at the very least to forestall future complaints. In stating your refusal, you must exercise great tact.

Begin with a friendly opener. Try to find some common ground with the complainant. Express regret about the situation. Even though you may think the complaint is totally unfair, don't be discourteous. Incidentally, not everyone writes a letter as courteous as the one in Figure 13-7. Sometimes, in fact, people are downright abusive. If so, attempt to shrug it off. Just as you would not pour gasoline on a fire, don't answer abuse with abuse. Pour on some cooling words instead.

Second, explain the reason for the refusal. Be very specific here and answer at some length. The very length of your reply will help to convince the reader that you have considered the problem seriously. Third, at the end of your explanation, state your refusal. Sometimes, however, your explanation can be so complete that your refusal is obvious and does not need to be stated. If so, the best strategy is probably not to state the refusal directly. If, however, the complaint is serious enough that legal action may result, you must state your refusal quite clearly to avoid future misunderstandings. In any case, do not state the refusal until you have thoroughly presented the reasons for it.

Fourth, if you can, offer a partial or substitute adjustment. Finally, close your letter in a friendly way.

Companies selling products are not the only organizations that

Figure 13-8   Letter of Adjustment

CHAPMAN PRODUCTS, INC.
1925 Jerome Street
Brooklyn, NY 11205

November 11, 1980

Mr. John Griffin
26 Shady Woods Road
White Bear Lake, MN 55101

Dear Mr. Griffin:

Thank you for your letter of October 28. We're sorry that you had a problem with a Chapman product. But we are happy that you wrote to us about your dissatisfaction. We need to hear from our customers if we are to provide them satisfactory products.

The numbers on the belts you returned indicate that they were manufactured in 1970. Sanding belts, like many other products, have a "shelf life," and the belts you purchased had exceeded theirs. Age, heat, and humidity had weakened them.

Mr. Griffin, we stand behind our products. Although we are sure that the belts you purchased were not defective when we shipped them, we wish to replace them for you. You are being shipped a box of 10 belts, medium grade. We're sure that these belts will live up to the Chapman name.

We also suggest that you look in the yellow pages of your telephone directory under "Hardware-Retail" for authorized Chapman dealers. We can only suggest to independent dealers how they should shelve and sell our products. We can exert more quality control with our own dealers. We know that you can find a Chapman dealer who will give you excellent service.

Sincerely,

Theresa R. Brummer

Theresa R. Brummer
Customer Service

TB/ay

cc: Fitler Lumber Company

receive letters of complaint. Public service organizations do also. The letter that follows illustrates a refusal from such an organization—a state department of transportation. In this case, a citizen had written stating that a curved section of highway near her home was dangerous. She requested that the curve be rebuilt and straightened. The reply uses the strategy that we have outlined:

1. It begins in a friendly way.
2. It details why the highway department cannot rebuild the road immediately.
3. The refusal is so obvious that it is not stated explicitly.
4. A substitute is offered.
5. It closes in a friendly way.

In the case of this letter, the goal is not to keep a paying customer but to keep a taxpayer friendly. In either case, the strategy is to offer an honest, detailed, factual explanation in a cheerful way.

Dear Mrs. Ferguson:

Thank you for your letter concerning the section of Trunk Highway (TH) 50 near your home. The Department of Transportation shares your concern about the safety of TH 50, particularly the section between Prestonburg and Pikeville, near which your home is located.

TH 50, as you mentioned, has a goodly number of hills and curves and, because of the terrain, some steep embankments. However, its accident rate—3.58 accidents per million vehicle miles—is far from the worst in the state. In fact, there are 64 other highways with worse safety records.

We do have studies under way that will result ultimately in the relocation of TH 50 to terrain that will allow safer construction. These things take time, as I'm sure you know. We have to coordinate our plans with county and town authorities and the federal government. Money is assigned to these projects by priority, and, judged by its accident rate, TH 50 does not have top priority.

Accidents along TH 50 are not concentrated at any one curve. They are spread out over the entire highway. Reconstruction at any one location would cause little change in the overall accident record.

However, we are currently evaluating the need for guardrails along the entire length of TH 50. It is likely that within a year we will construct guardrails at a number of lo-

cations. Most certainly, a guardrail will be placed at the
curve that concerns you. This should correct the situation
to some extent.

We appreciate your concern. Please write to us again if we
can be of further help.

Sincerely yours,

The Correspondence of the Job Hunt

In many cases, the first knowledge prospective employers will have of
you is the letter of application and resume that you send to them. A
good letter of application and resume will not guarantee that you get
a job, but bad ones will probably guarantee that you do not. In this
section we tell you about the letter of application, the resume, and
several follow-up letters needed during the job hunt.

The Letter of Application

The good letter of application begins before you ever sit down to
your typewriter. Find out all you can about the company you are
applying to. Talk to friends who may know something about the
company. Read the company's advertisements. If they have a company magazine, get it and read it. Send away for copies of their
brochures and annual reports. (See Letters of Inquiry, page 280.)
Investigate the company in sources such as the *College Placement
Annual, Standard Statistics, Standard Corporation Reports*, and Dun
and Bradstreet's *Middle Market Directory, Million Dollar Directory*,
and *Reference Book of Manufacturers*. For information about the occupation you plan to enter look into the *Occupational Outlook Handbook*, published by the U.S. Bureau of Labor Statistics, or the *Career
Index*. Consult magazines and newspapers that regularly carry business news such as the *Harvard Business Review*, the *Wall Street Journal*, and *Business Week*. To see what has recently appeared in the
business press about the company, see the *F & S Index of Corporations
and Industries* and the *Business Periodicals Index*. For general coverage see *The New York Times Index* and the *Reader's Guide to Periodical
Literature*. Analyze the company to see what past achievements it is
most proud of and to determine its past and future goals.

For information about jobs with the federal government and how
to get them, look into *Working for the USA* and the *Federal Career
Directory*. The latter is published yearly. For jobs at the state level,
ask the state department of personnel for information.

We have listed only some of the most basic sources where you can find information to prepare for the job hunt. Your librarian can help you find many more, some perhaps quite specifically oriented to the occupation you are most interested in. We cannot emphasize enough how important this preparation is. Listen to what one interviewer, Michael Shaughnessy of General Motors, has to say on the subject:

> It's really impressive to a recruiter when a job candidate knows about the company. If you're a national recruiter, and you've been on the road for days and days, you have no idea how pleasant it is to have a student say, "I know your company is doing such and such, and has plants here and here, and I'd like to work on this particular project." Otherwise I have to go into my standard spiel, and God knows I've certainly heard myself give that often enough.[1]

Analyze yourself as well as potential employers. What are your strengths? What are your weaknesses? How well have you performed in past jobs? Have you shown initiative? Have you improved procedures? Have you accepted responsibility? Have you been promoted or given a merit raise? How can you present yourself most attractively? What skills do you possess that relate directly to what the employer seems to need? Are you really qualified for the job you want? Do not apply for a job if you lack the necessary qualifications. You may fake your way to the interview but not beyond.

Plan the mechanics of your letter of application carefully. Buy the best quality white bond paper. This is no time to skimp. Plan to type your letter, of course, or have it typed. Use a standard typeface such as pica or elite. Do not use italic. With pica you get ten letters to the inch, and with elite you get twelve. This can be useful information when you need either to expand or contract the space your information occupies in a letter or resume. Make sure your letter is mechanically perfect, free from erasures and grammatical errors. Be brief but not telegraphic. Keep the letter down to one page. Don't send a carbon or a letter duplicated in any way. Accompany each letter, however, with a duplicated resume. We have more to say about this feature later.

Pay attention to the style of the letter and the resume that accompanies it. The tone you want in your letter is one of self-confidence. You must avoid both arrogance and humility. You must sound interested and somewhat eager, but not fawning. Do not give the im-

---

[1] Avery Comarow, "Tracking the Elusive Job," *The Graduate* (Knoxville: Approach 13-30 Corporation, 1977), p. 42.

pression that you *must* have the job. Nor do you want to seem uncaring about whether you *do* get the job.

When describing your accomplishments in the letter and resume, use action verbs. They help to give your writing brevity, specificity, and force. For example, don't just say that you worked as a sales clerk. Rather, tell how you *maintained* inventories, *sold* merchandise, *prepared* displays, *implemented* new procedures, and *supervised* and *trained* new clerks. Here's a sampling of such words:

administer	edit	oversee
analyze	evaluate	plan
conduct	exhibit	produce
create	expand	reduce costs
cut	improve	reorganize
design	manage	support
develop	operate	was promoted
direct	organize	wrote

You cannot avoid the use of *I* in a letter of application. But take the *you-attitude* as much as you can. Think about what you can do for the prospective employer. The letter of application is not the place to be worried about salary and pension plans. Above all, be mature and dignified. Forget about tricky and flashy approaches. Write a well-organized, informative letter that highlights those skills your analysis of the company shows it desires most. We will discuss the application letter in terms of a beginning, a body, and an ending.

The beginning. Beginnings are tough. Do not begin aggressively or overly clever. Beginnings such as *"WANTED:* an alert, aggressive employer who will recognize an alert, aggressive young forester" will usually send your letter wastebasket-bound. A good beginning, if it is available to you, is a bit of legitimate name dropping. Use this beginning only if you have permission and if the name dropped will mean something to the prospective employer. If you qualify on both counts begin with an opener like this:

```
Dear Ms. Marchand:

Professor John J. Jones of State University's Food Science
faculty has suggested that I apply for the post of food
supervisor that you have open. In June I will receive my
Bachelor of Science in Food Science from State University.
Also, I have spent the last two summers working in food
preparation for Memorial Hospital in Melbourne.
```

Remember that you are trying to arouse immediate interest about yourself in the potential employer. Another way to do this is to refer to something about the company that interests you. Such a reference establishes that you have done your homework. Then try to show how some preparation on your part ties you into this special interest. See Figure 13-9 for an example of such an opener.

Sometimes the best approach is a simple statement about the job you seek accompanied by something in your experience that fits you for the job, as in this example:

> Your opening for a food supervisor has come to my attention. In June of this year, I will graduate from State University with a Bachelor of Science in Food Science. I have spent the last two summers working in food preparation for Memorial Hospital in Melbourne. I believe that both my education and work experience qualify me to be a food supervisor on your staff.

Be specific about the job you want. As the vice-president of one firm told us, "We have all the people we need who can do *anything*. We want people who can do *something*." Quite often, if the job you want is not open, the employer may offer you an alternative one. But employers are not impressed with vague statements such as, "I'm willing and able to work at any job you may have open in research, production, or sales."

The body. In the body of your letter you select items from your education and experience that show your qualifications for the job you seek. Remember always, you are trying to show the employer how well you will fit into the job and the organization.

In selecting your items, it pays to know what things employers value the most. One thorough piece of research[2] shows that for recent college graduates employers give priority as follows:

First priority   Major field of study
                  Academic performance
                  Work experience
                  Plant or home-office interview
                  Campus interview

---

[2] Jane L. Anton, Michael L. Russell, and the Research Committee of the Western College Placement Association, *Employer Attitudes and Opinions Regarding Potential College Graduate Employees* (Hayward, Calif.: Western College Placement Association, 1974), p. 10.

Figure 13-9   Letter of Application

635 Shuflin Road
Watertown, CA 90233
March 23, 1980

Mr. Morrell R. Solem
Director of Research
Price Industries, Inc.
2163 Airport Drive
St. Louis, MO 63136

Dear Mr. Solem:

I read in the January issue of <u>Metal Age</u> that Dr. Charles E.
Gore of your company is conducting extensive research into
the application of X-ray diffraction to the solutions of
problems in physical metallurgy. I have conducted
experiments at Watertown College in the same area under the
guidance of Professor John J. O'Brien. I would like to
become a research assistant with your firm and, if possible,
work for Dr. Gore.

In June, I will graduate from Watertown with a Bachelor of
Science degree in Metallurgical Engineering. At present, I
am in the upper 25 percent of my class. In addition to my
work with Professor O'Brien, I have taken as many courses
relating to metal inspection problems as I could.

For the past two summers, I have worked for Watertown
Concrete Test Services where I have qualified as a
laboratory technician for hardened concrete testing. I know
how to find and apply the specifications of the American
Society for Testing and Materials. This experience has
taught me a good deal about modern inspection techniques.
Because this practical experience supplements the theory
learned at school, I could fit into a research laboratory
with a minimum of training.

You will find more detailed information about my education
and work experience in the resume enclosed with this letter.
I can supply job descriptions concerning past employment
and my report of my X-ray diffraction research.

In April, I will attend the annual meeting of the American
Institute of Metallurgical Engineers in Detroit. Would it
be possible for me to talk with some member of Price
Industries at that time?

Sincerely yours,

Jane E. Lucas

Jane E. Lucas

Enclosure

Second priority	Extracurricular activities
	Recommendations of former employer
	Academic activities and awards
Third priority	Type of college or university attended
	Recommendations from faculty or school official
Fourth priority	Standardized test scores
	In-house test scores
	Military rank

Try to include information from the areas that employers seem to value the most, but put emphasis on those areas in which you come off best. If your grades are good, mention them prominently. If you stand too low in your class—in the lowest quarter, perhaps—maintain a discreet silence. Speak to the employer's interests and at the same time highlight your own accomplishments. Show how it would be to the employer's advantage to hire you. The following paragraph, an excellent example of the *you-attitude* in action, does all these things:

> I understand that the research team of which I might be a part works as a single unit in the measurement and collection of some data. Because of this, team members need a general knowledge and ability in fishery techniques as well as a specific job skill. Therefore, I would like to point out that last summer I worked for the Department of Natural Resources on a fish population study. On that job I gained electro-fishing and seining experience and also learned how to collect and identify aquatic invertebrates.

By being specific about your accomplishments, you avoid the appearance of bragging. It is much better to say, "I was president of my senior class," than to say, "I am a natural leader."

One tip about job experience: Mention job experience even if it does not relate to the job you seek. Employers feel a student who has worked is more apt to be mature than one who has not.

Do not forget hobbies that relate to the job. You are trying to establish that you are interested in, as well as qualified for, the job.

Do not mention salary unless you are answering an advertisement that specifically requests you to. Keep the *you-attitude*. Do not worry about pension plans, vacations, and coffee breaks at this stage of the game. Keep the prospective employer's interests in the foreground. Your self-interest is taken for granted.

If you are already working and not a student, you construct the body in much the same fashion. The significant difference is that you will put more emphasis on work experience than on college experience. Do not complain about your present employer. Such complaints will lead the prospective employer to mistrust you.

In the last paragraph of the body refer the employer to your enclosed resume. Mention your willingness to supply additional information such as letters concerning your work, research reports, and college transcripts.

Ending. The goal of the letter of application is an interview with the prospective employer. In your ending you request this interview. Your request should be neither humble nor overaggressive. Simply indicate that you are available for an interview at the employer's convenience and give any special instructions needed for reaching you. If the prospective employer is in a distant city, indicate (if you can) some convenient time and place where you might meet a representative of the company, such as the convention of a professional society. If the employer is really interested, you may be invited to visit the company at their expense. But you cannot ask for this expense-paid interview. If you really want the job, be prepared to pay your own way.

The complete letter. Figure 13-9 shows the complete letter of application. Take a minute to read it now. The beginning of the letter shows that the writer has been interested enough in the company to investigate it. The desired job is specifically mentioned. The middle portion highlights the course work and work experience that relate directly to the job sought. The close makes an interview convenient for the employer to arrange.

The Resume

With your letter of application you enclose a resume that provides your prospective employer with a convenient summary of your education and experience. To whom should you send letters and resumes? When answering an advertisement, you should follow whatever instructions are given there. When operating on your own, send them, if at all possible, to the person within the organization for whom you would be working; that is, the person who directly supervises the position. This person normally has the power to hire for the position. Your research into the company may turn up the name you need. If not, don't be afraid of calling the company switchboard and asking directly for the name and title you need. Write to personnel directors only as a last resort. Whatever you do, write to *someone* by name. Don't send "To Whom It May Concern" letters on your job hunt. It's wasted effort. Sometimes, of course, you may gain an interview without having sent a letter of application—for example, when recruiters come to your campus. When you do, bring a resume with you and give it to the interviewer at the start of the interview. He or she will appreciate this help tremendously.

Recent research indicates that, as in the letter of application, neatness and brevity—ideally only one page—are of major importance in your resume.[3] A tip here: If you have wide experience to report you will find that the smaller elite type will help you fit it on one page. With really extensive experience, you might want to go to a printing process that reduces the print in size. In any case, you must invest a little money and have the resume printed or photocopied. Never send carbons or mimeographed sheets. Make the resume good-looking. Leave generous margins and white space. Use distinctive headings and subheadings. The sample in Figure 13-10 provides you with a good example of what a resume should look like. Take a look at it now before you read the following comments.

Address. Use the address to which you want your mail sent. Give your phone number, and don't forget the area code.

Education. List the colleges or universities you have attended in reverse chronological order; in other words, list the school you attended most recently first, the one before second, and so on. Do not list your high school. Give your major and date or expected date of graduation. Do not list courses, but list anything that is out of the ordinary, such as honors, special projects, and emphases in addition to the major. Extracurricular activities also go here. For most students, educational information should be placed first in the resume. People with extensive work experience, however, may choose to put that first.

Business experience. As you did with your educational experience, put your business experience in reverse chronological order. To save space and to avoid the repetition of *I* throughout the resume, use phrases rather than complete sentences. The style of the sample resume makes this technique clear. As we advised you to do in the letter of application, emphasize the experiences that show you in the best light for the kinds of jobs you seek. Use active verbs in your descriptions. Do not neglect less important jobs of the sort you may have had in high school, but use even more of a summary approach for them. You would probably put college internships and work-study programs here, though you might rather choose to put them under education. If you have military experience, put it here. Give highest rank

---

[3] Rosemary Ullrich, "The Power of a Positive Job Search," *Business World Women* 11 (Fall 1977): 11.

Figure 13-10   Resume

```
                        RESUME OF JANE E. LUCAS

                          635 Shuflin Road
                      Watertown, California 90233
                        Phone: (213) 596-4236

Education       WATERTOWN COLLEGE            WATERTOWN, CALIFORNIA
1978-1980
                Candidate for Bachelor of Science degree in
                Metallurgical Engineering in June 1980. In
                upper 25% of class with GPA of 3.2 on 4.0 scale.
                Have been yearbook photographer for two years.
                Member of Outing Club, elected president in
                senior year. Elected to Student Intermediary
                Board in senior year. Oversaw promoting and
                allocating funds for student activities. Wrote
                a report on peer advising that resulted in a
                change in college policy. Earned 75% of
                college expenses.

1976-1978       SAN DIEGO COMMUNITY COLLEGE
                                        SAN DIEGO, CALIFORNIA
                Received  Associate of Arts degree in General
                Studies in June 1978. Made Dean's List three of
                four semesters. Member of debate team.
                Participated in dramatics and intramural
                athletics.

Business        WATERTOWN CONCRETE TEST SERVICES
Experience                          WATERTOWN, CALIFORNIA
1978, 1979      Qualified as laboratory technician for hardened
summers         concrete testing under specification E329 of
                American Society for Testing and Materials
                (ASTM). Conducted following ASTM tests: Load
                Test in Core Samples (ASTM C39), Penetration
                Probe (ASTM C803-75T), and the Transverse
                Resonant Frequency Determination (ASTM C666-73).
                Implemented new reporting system for laboratory
                results.

1978-1980       WATERTOWN ICE SKATING ARENA
                                    WATERTOWN, CALIFORNIA
                During academic year, work 15 hours a week as
                ice monitor. Supervise skating and and
                administer first aid.

1974-1978       Summer and part-time jobs included newspaper
                carrier, supermarket stock clerk, and sales-
                person for large department store.

Personal        Grew up in San Diego, California. Travels
Background      include Mexico and the Eastern United States.
                Can converse in Spanish. Interests include
                reading, backpacking, photography, and sports
                (tennis, skiing, and running). Willing to
                relocate.

References       Personal references available upon request.
                                                February 1980
```

This resume is modeled after the one taught students at the Harvard University Graduate School of Business Administration.

held, list service schools attended, and describe duties. Make a special effort to show how military experience relates to the civilian work you seek.

Personal background. Provide some personal information about yourself, so that the company can see you as a whole human being. Also, personal information can in a subtle way point out desirable qualities you possess. Recent travels indicate a broadening of knowledge and probably a willingness to travel. Hobbies listed may relate to the work sought. Participation in sports, drama, or community activities indicates a liking for working with people. Cultural activities indicate you are not a person of narrow interests.

If you indicate you are married, you might want to say that you are willing to relocate. Don't say anything about health unless you can describe it as excellent. Because of state and federal laws concerning fair employment practices, certain information should not appear in your resume. Do not include age or date of birth, sex, race, national origin, or religion. Do not mention handicaps and do not include a photograph. You may, if you wish, choose to omit this entire section.

References. Do not use up precious space listing references. Simply indicate that they are available. You should have at least three. Be sure to get permission from people before you send their names to anyone. It's a smart idea to send a copy of your resume to people you use as references. This will help them to discuss you in a specific— and positive—way.

Date line. At the bottom of the resume, place the date—month and year—in which you completed the resume.

Job objective. We have not shown a job objective entry on our sample resume. Some research indicates that you should include one. But we rather agree with the Harvard Graduate School of Business Administration, which says,

> We recommend the use of a job objective in a resume only if you really feel sure of what you want to do. Sometimes students who do not feel such a sense of certainty attempt a vague catchall job objective. It usually shows. Even if successful, such a bluff is unwise. A job objective is a two-edged instrument. It will heighten interest in some employers, but will diminish it in others.[4]

If you omit the job objective in your resume, you can, of course, mention it in your letter of application. If you do decide to include this entry, it should look something like this:

---

[4] Office of Career Development, *Writing Resumes* (Boston: Harvard University Graduate School of Business Administration, n.d.), p. 3.

```
Job Objective
  Desire work in food service management.
```

Place the job objective entry immediately after the address and align it with the rest of the entries.

Follow-Up Letters

Write follow-up letters (1) when you have had no answer to your letter of application in two weeks; (2) after an interview; (3) when a company refuses you a job; and (4) to accept or refuse a job.

When a company has not answered your original letter of application, write again. Be gracious, not complaining. Say something like this:

```
Dear Mr. Souther:

On 12 April I applied for a post with your company. I have
not heard from you, so perhaps my original letter has been mis-
placed. I enclose a copy of it.

If you have already reached some decision concerning my ap-
plication, I would appreciate your letting me know.

I look forward to hearing from you.

Sincerely yours,
```

After an interview, be smart and within a week's time follow up your interview with a letter. Such a letter draws favorable attention to yourself as someone who understands business courtesy and good communication practice. Express appreciation for the interview. Draw attention to any of your qualifications that seemed to be important to the interviewer. Express your willingness to live with any special conditions of employment such as relocation. Make clear your hope for the next stage in the process, perhaps a further interview. If you include a specific question in your letter, it may hasten a reply. Your letter might look like this one:

```
Dear Ms. Marchand:

Thank you for speaking with me last Tuesday about the food
supervisor position you have open.
```

Working in a hospital food service relates well to my
experience and interests. A feasibility study I am currently
writing as a senior project deals with a hospital food
service's ability to provide more varied diets to people with
restricted dietary requirements. Would you like me to send you
a copy?

I understand that my work with you would include alternating
weekly night shifts with weekly day shifts. This requirement
presents no difficulty for me.

Tuesdays and Thursdays are best for me for any future
interviews you may wish. But I can arrange a time at your
convenience.

Sincerely yours,

When a company refuses you a job, good tactics dictate that you
acknowledge the refusal. Express regret that no opening exists at the
present time and express the hope that they may consider you in the
future. You never know; they might.

Writing an acceptance letter presents few problems. Be brief.
Thank the employer for the job offer and accept the job. Settle when
you will report for work and express pleasure at the prospect of working for the organization. A good letter of acceptance might read as
follows:

Dear Mr. Solem:

Thank you for offering me a job as research assistant with your
firm. I happily accept. I can easily be at work by 1 July as you
have requested.

I look forward to working with Price Industries and particularly to the opportunity of doing research with Dr. Gore.

Sincerely yours,

Writing a letter of refusal can be difficult. Be as gracious as possible. Be brief but not so brief as to suggest rudeness or indifference.
Make it clear that you appreciate the offer. If you can, give a reason
for your refusal. The employer who has spent time and money in
interviewing you and corresponding with you deserves these cour-

tesies. And, of course, your own self-interest is involved. Some day you may wish to reapply to an organization that for the moment you must turn down. A good letter of refusal might look like this one:

Dear Ms. White:

I enjoyed my visit to the research department of your company. I would very much have liked to work with the people I met there. I thank you for offering me the opportunity to do so.

However, after much serious thought, I have decided that the research opportunities offered me in another job are closer to the interests I developed at the university. Therefore, I have accepted the other job and regret that I cannot accept yours.

I appreciate the courtesy and thoughtfulness that you and your associates have extended me.

Sincerely yours,

Letter and Memorandum Reports

Often a short report will be written as a letter or memorandum (usually called *memo*). Such reports seldom go over several pages. Consider memos as letters that stay within the organization, with a few differences. In a memo, more often than in a letter, you will be writing to someone you know quite well. The recipient may also be quite familiar with most of the background information. Therefore, more so than in letters, you can and should come directly to the subject and purpose of your message. Avoid irrelevant chit-chat, but do remember that memos go from one human being to another. Avoid being coldly formal. Do not be concise to the point of brusqueness. Use a relaxed, conversational style. Use *I* and *you*, particularly *you*.

Perhaps the chief difference between memos and letters is format. Figure 13-11 shows a typical memo format. Frequently, organizations have printed memo forms. The printed form saves secretarial time and reminds the writer of all the parts of a memo. Figure 13-12 illustrates such a printed memo form. (Incidentally, the memo in Figure 13-12 is a real one. It illustrates both a well-written memo and the interest that organizations have in good communication.) When continuation pages are needed in memos, use the same format shown at the top of the continuation page illustrated in Figure 13-5, on page 276.

Figure 13-11   Memorandum Format

MEMORANDUM

Doublespace twice

Date: 21 January 1980   cc: William Martin
Doublespace   Herbert Johnson
To: Keith Wharton   Gail McClure
Doublespace
From: James Tammen
Doublespace
Subject: Reallocation procedures
Doublespace

Doublespace

With these differences in mind, then, we can treat memos and letter reports the same. A common plan for both is as follows:

1. Begin by telling the reader subject and purpose. Perhaps you are report-ing on an inspection tour or summarizing the agreements reached in a consultation between you and the recipient. You may be reporting the results of a research project, or the beginning of one. Or you may be writing a progress report on a project that is underway but not com-pleted. Whatever the subject and purpose may be, state them clearly. If someone has requested the report, name the requester.

2. What Step Two should be depends to some extent on company preference and upon the situation. Many companies and organizations prefer that any conclusions, decisions, and recommendations reached should be given at this point in the report. In a neutral situation, where the reader is not likely to feel hostile or upset because of the conclusions, we rec-ommend this plan. But when you anticipate a hostile reader reaction, you would be safer if you gave your facts first and then your conclusions.

3. Develop your subject. Use the same techniques and rhetorical principles that you would use in a longer report. Quite often, however, because of the brevity desired in a letter or memo, your style may approach that of the informative abstract. (See pages 188–190.) You may also find listing a

Figure 13-12   Company Memorandum

OFFICE MEMO

Date   15 December 1978

To   All Departments
              DEPARTMENT OR OFFICE

Attention   Department Heads

From   Robert Lauer, Employee Relations
              NAME OF PERSON, OFFICE OR AGENCY (INCLUDE CITY AND STATE)

Subject   Report Writing Class Update

The following is a list of objectives for a Report Writing Class pilot program. This list is drawn from your comments on the needs analysis done December 5, 1978. Each objective is a statement of what the trainees will be able to do if training is successful. These needs will be communicated to the consultant selected to run our pilot.

Please review the list to determine relevance to your department's needs. Call me by December 29, 1978 if you have any suggestions for improving the list.

Report Writing Trainees should be able to

1. Identify the target audience(s) of each of their reports by functional area and job level.

2. Write a one sentence statement detailing the purpose of the report, that is, what the information is to be used for.

3. List the significant findings of the report in descending order of importance to the target audience.

4. Construct a framework (outline) for a report that is appropriate to both the target audience and stated purpose.

5. Compose (from the outline) a report that is

    A. Informative & concise

    B. Logically organized

    C. Free of departmental jargon (foreign to target audience)

    D. Free of mechanical errors

    E. Readable

    F. An accurate depiction, where appropriate, of statistical data.

WHEN A REPLY IS REQUIRED, PLEASE RETURN THIS MEMO WITH YOUR HANDWRITTEN COMMENTS IF THAT IS CONVENIENT.

6696 Rev. 3-72 Printed in U.S.A.

Courtesy of Robert Lauer, St. Paul Fire and Marine Insurance Company.

useful technique in letters and memos. (See page 165.) Remember to consider your audience precisely as you do in longer reports. If you are writing to an executive, do not fill your letter or memo with jargon and technical terms. If you must report a mass of statistics, try to round them off. If absolute accuracy is neccessary, perhaps you can give the figures in an informal table. (See pages 233–234.) You may use captions in a letter or memo just as you would in a longer report. Many a business communication would profit by a few well-placed captions to provide better transition between points. (See pages 215–216.)

4. Frequently, you will conclude by telling the reader you will be happy to follow up with additional information, personal consultations, or other appropriate action.

All of the reports that we discuss in Chapters 14 through 18 can be and often are written as memos or letters. All of the rhetorical modes discussed in Chapters 5 and 6, used singly or in combination, are found in memos and letters.

However, when a report gets beyond four or five pages long, it's probably too long to be presented as a memo or letter. At this point, you must consider changing to a format that allows you to use such report elements as descriptive abstracts, title pages, and prefaces. See Chapter 12, particularly pages 254–258, for more on this matter.

How to Write a Business Letter

Recently in *Business Week* magazine we came upon an advertisement, "How to Write a Business Letter," written for the International Paper Company by Malcolm Forbes, publisher of *Forbes* magazine. Mr. Forbes' thoughts about successful business letter writing agree so well with ours that, by way of a conclusion to this chapter, we reproduce the ad for you in Figure 13-13. Like Mr. Lauer's memo in Figure 13-12, it illustrates that business people take good communication seriously.

EXERCISES

1. Write an unsolicited letter of inquiry to some company asking for sample materials or information. If you really need the information or material, mail the letter, but do not mail it as an exercise.

2. Write a letter to some organization applying for full- or part-time work. Accompany your letter with a resume. It may well be that

you are seeking work and can write your letter with a specific organization in mind.

3. Imagine that you are working for a firm that provides a service or manufactures a product you know something about. Someone has written the firm a letter of inquiry asking about the service or product. Your task is to answer the letter.

4. Think about some service or product that has recently caused you dissatisfaction. Find out the appropriate person or organization to write to and write that person or organization a letter of complaint.

5. Swap your letter of complaint written for Exercise 4 for the letter of complaint written by another member of the class. Your assignment is to answer the other class member's letter. You may have to do a little research to get the data you will need for your answer.

6. Write a memo to some college official or to an executive at your place of work. Many of the papers you have written earlier in your writing course are probably suitable for a memorandum format. Or you could choose some procedure, such as college registration, and suggest a new and better procedure. Perhaps your memo could be to a teacher suggesting course changes. A look at Figure 4-3 on page 81 could also suggest a wide range of topics and approaches that could be used in a memo.

Figure 13-13

How to write a business letter

Some thoughts from Malcolm Forbes
President and Editor-in-Chief of Forbes Magazine

International Paper asked Malcolm Forbes to share some things he's learned about writing a good business letter. One rule, "Be crystal clear."

A good business letter can get you a job interview.

Get you off the hook.

Or get you money.

It's totally asinine to blow your chances of getting *whatever* you want—with a business letter that turns people off instead of turning them on.

The best place to learn to write is in school. If you're still there, pick your teachers' brains.

If not, big deal. I learned to ride a motorcycle at 50 and fly balloons at 52. It's never too late to learn.

Over 10,000 business letters come across my desk every year. They seem to fall into three categories: stultifying if not stupid, mundane (most of them), and first rate (rare). Here's the approach

I've found that separates the winners from the losers (most of it's just good common sense)—it starts *before* you write your letter:

Know what you want
If you don't, write it down—in one sentence. "I want to get an interview within the next two weeks." That simple.

List the major points you want to get across—it'll keep you on course.

If you're *answering* a letter, check the points that need answering and keep the letter in front of you while you write. This way you won't forget anything—*that* would cause another round of letters.

And for goodness' sake, answer promptly if you're going to answer at all. Don't sit on a letter—*that* invites the person on the other end to sit on whatever you want from *him*.

Plunge right in
Call him by name—not "Dear Sir, Madam, or Ms." "Dear Mr. Chrisanthopoulos"—and be sure to spell it right. That'll get him (thus, you) off to a good start.

(Usually, you can get his name just by phoning his company—or from a business directory in your nearest library.)

Tell what your letter is about in the first paragraph. One or two sentences. Don't keep your reader guessing or he might file your letter away—even before he finishes it.

In the round file.

If you're answering a letter, refer to the date

"Be natural. Imagine him sitting in front of you—what would you say to him?"

it was written. So the reader won't waste time hunting for it.

People who read business letters are as human as thee and me. Reading a letter shouldn't be a chore—*reward* the reader for the time he gives you.

Write so he'll enjoy it
Write the entire letter from his point of view—what's in it for *him*? Beat him to the draw—surprise him by answering the questions and objections he might have.

Be positive—he'll be more receptive to what you have to say.

Be nice. Contrary to the cliché, genuinely nice guys most often finish first or very near it. I admit it's not easy when you've got a gripe. To be agreeable while disagreeing—that's an art.

Be natural—write the way you talk. Imagine him sitting in front of you—what would you *say* to him?

Business jargon too often is cold, stiff, unnatural.

Suppose I came up to you and said, "I acknowledge receipt of your letter and I beg to thank you." You'd think, "Huh? You're putting me on."

The acid test—read your letter *out loud* when you're done. You might get a shock—but you'll know for sure if it sounds natural.

Don't be cute or flippant. The reader won't take you seriously. This doesn't mean you've got to be dull. You prefer your letter to knock 'em dead rather than bore 'em to death.

Three points to remember:

Have a sense of humor. That's refreshing anywhere—a nice surprise in a business letter.

Be specific. If I tell you there's a new fuel that could save gasoline, you might not believe me. But suppose I tell you this:

"Gasohol"–10% alcohol, 90% gasoline–works as well as straight gasoline. Since you can make alcohol from grain or corn stalks, wood or wood waste, coal–even garbage, it's worth some real follow-through.

Now you've got something to sink your teeth into.

Lean heavier on nouns and verbs, lighter on adjectives. Use the active voice instead of the passive. Your writing will have more guts.

Which of these is stronger? Active voice: "I kicked out my money manager." Or, passive voice: "My money manager was kicked out by me." (By the way, neither is true. My son, Malcolm Jr., manages most Forbes money–he's a brilliant moneyman.)

"I learned to ride a motorcycle at 50 and fly balloons at 52. It's never too late to learn anything."

Give it the best you've got

When you don't want something enough to make *the* effort, making *an* effort is a waste.

Make your letter look appetizing –or you'll strike out before you even get to bat. Type it–on good-quality 8½" x 11" stationery. Keep it neat. And use paragraphing that makes it easier to read.

Keep your letter short–to one page, if possible. Keep your paragraphs short. After all, who's going to benefit if your letter is quick and easy to read?

You.

For emphasis, underline impor-

tant words. And sometimes indent sentences as well as paragraphs.

Like this. See how well it works? (But save it for something special.)

Make it perfect. No typos, no misspellings, no factual errors. If you're sloppy and let mistakes slip by, the person reading your letter will think you don't know better or don't care. Do you?

Be crystal clear. You won't get what you're after if your reader doesn't get the message.

Use good English. If you're still in school, take all the English and writing courses you can. The way you write and speak can really help –or *hurt*.

If you're not in school (even if you are), get the little 71-page gem by Strunk & White, *Elements of Style*. It's in paperback. It's fun to read and loaded with tips on good English and good writing.

Don't put on airs. Pretense invariably impresses only the pretender.

Don't exaggerate. Even once. Your reader will suspect everything else you write.

Distinguish opinions from facts. Your opinions may be the best in the world. But they're not gospel. You owe it to your reader to let him know which is which. He'll appreciate it and he'll admire you. The dumbest people I know are those who Know It All.

Be honest. It'll get you further in the long run. If you're not, you won't rest easy until you're

found out. (The latter, not speaking from experience.)

Edit ruthlessly. Somebody ~~has~~ said that words are ~~a lot~~ like inflated money–the more ~~of them that~~ you use, the less each one ~~of them~~ is worth. ~~Right on.~~ Go through your entire letter ~~just~~ as many times as it takes. ~~Search out and~~ **A**nnihilate all unnecessary words, ~~and~~ sentences–even ~~entire~~ *paragraphs*.

"Don't exaggerate. Even once. Your reader will suspect everything else you write."

Sum it up and get out

The last paragraph should tell the reader exactly what you want *him* to do–or what *you're* going to do. Short and sweet. "May I have an appointment? Next Monday, the 16th, I'll call your secretary to see when it'll be most convenient for you."

Close with something simple like, "Sincerely." And for heaven's sake sign legibly. The biggest ego trip I know is a completely illegible signature.

Good luck.

I hope you get what you're after.

Sincerely,

Malcolm S. Forbes

CHAPTER 14

Instructions for Performing a Process

You will find, once on the job, that instructing others to follow some procedure is a common task. Sometimes the instructions are given orally. But when the procedure is done by many people or is done repeatedly, written instructions are a better choice. Instructions may be quite simple—perhaps comprising half a typed page—or exceedingly complex—a bookshelf full of manuals. They may be highly technical, dealing with operating machinery or programming computers, for example. Or they may be executive or business oriented—how to create a certain kind of brochure or how to route memorandums through a company. The task is not to be taken lightly. A Shakespearean scholar who had also served in the British Army wrote the following:

> The most effective elementary training [in writing] I ever received was not from masters at school but in composing daily orders and instructions as staff captain in charge of the administration of seventy-two miscellaneous military units. It is far easier to discuss Hamlet's complexes than to write orders which ensure that five working parties from five different units arrive at the right place at the right time equipped with the proper tools for the job. One soon learns that the most seemingly simple statement can bear two meanings and that when instructions are misunderstood the fault usually lies with the original order.[1]

---

[1] G. B. Harrison, *Profession of English* (New York: Harcourt, Brace and World, 1962), p. 149.

Sets of instructions may contain six possible sections:

Introduction
Theory or Principles of Operation
List of Equipment and Materials Needed
Description of the Mechanism
Performance Instructions
Troubleshooting Procedures

We do not present the above list as a rigid format. For example, you may find that you do not need a theory section, or you may include it as part of your introduction. You may want to vary the order of the sections. You may want to describe or list equipment as it is needed while performing the process rather than in a separate section. Often nothing more is provided than the performance instructions. We describe all six sections so that you'll know how to write all of them. Take the information we give you and adjust it to suit your needs.

The Introduction

Introductions to sets of instructions are usually short and to the point. Normally, they tell what the instructions cover and for whom they are intended. They may be as short as this introduction for a brochure on tree planting:

> Most home owners are interested in having healthy, attractive trees on their property. If you pay little attention to planting, your trees may grow slowly, continually lack vigor, or even die. However, your trees are more likely to thrive if they are properly planted and cared for. This bulletin describes the proper methods of handling and planting to give your landscape trees the best chance of success.[2]

After reading this introduction you know the audience—home owners. You know the subject matter is tree planting and care. You know the purpose is to show the reader the procedures to follow for healthy trees. Included in the scope of the bulletin are the proper handling and planting of landscape trees.

Look at Figure 14-1. There the introduction for a government consumer bulletin on paint and painting clearly lays out audience,

---

[2] Richard Rideout, *Planting Landscape Trees* (St. Paul, Minn.: University of Minnesota, Agricultural Extension Service, 1978), p. 2.

Figure 14-1   Sample Introduction

Introduction

A good paint job not only adds beauty to your home, both interior and exterior; it also protects your investment. The right paint, properly applied to a surface carefully prepared, is an excellent barrier to weathering and decay.

But therein lies a problem. Surface preparation is an exacting task. The paint must be of good quality, and of a type designed for the particular job you want done. And it must be applied as the manufacturer intended or it may not cover well or dry properly to provide a satisfactory appearance.

PAINT AND PAINTING has been written for the nonprofessional painter. Its simple, clearcut instructions and illustrations will help you to do a good job of painting the interior and exterior of your home with a minimum of trouble and expense.

Because there are so many types of modern paints, however, it is essential that you READ LABELS CAREFULLY AND FOLLOW INSTRUCTIONS EXACTLY AS RECOMMENDED BY THE MANUFACTURER. The guidelines in this booklet are general in nature; the instructions from the manufacturer are specific to the particular product you are using. For the best results, use both of them.

Contents

From U.S., General Services Administration, *Paint and Painting* (Washington, D.C.: U.S. Government Printing Office, 1977).

subject, and purpose. The scope and plan of organization are shown by printing the table of contents on the same page as the introduction—an imaginative breakaway from the conventional method of dealing with these two items separately.

Both the planting and painting bulletins attempt to persuade the reader of the desirability of attending carefully to the instructions.

Such motivation is common. The use of motivation is even more obvious in the following introduction to a short course on "capabilities books":

> Whether new company or old, large or small, industrial or consumer, almost every firm whose management expects to stay in business needs an updated "capabilities" book.
>
> Type of product sold, or kind of service rendered, makes no difference: Capability books are designed to convince prospects that *this* company must be given serious consideration.
>
> There are many reasons why the need today is almost universal. Most companies have a completely different profile from that of just a few years ago.
>
> ☐ *Diversification* is one reason.
> ☐ *Change in ownership* is another.
> ☐ *Space-age technologies and materials* have created new products for new markets.
> ☐ *Computerization* has resulted in other changes.
>
> Unless the company takes the trouble to tell its story of these changes, and to provide a broad explanation of what it can do *now*, important new users of its services and its products will be lost.
>
> In essence, the successful capability book is a door opener into new markets. It also provides significant side benefits.
>
> This fourth volume in the Mead Short Course examines the new capabilities book, and reproduces current examples from a number of different kinds of companies.[3]

Good introductions to instructions, then, are not much different from the introductions we have already described for you in our chapter, "Prose Elements." (See pages 192–196.) They usually state subject, purpose, scope, and plan of organization. Frequently, they also attempt to motivate the reader.

When introductions are longer than the ones we have shown you, it is usually because the writers have chosen to include theory or principles of operation in the introduction. This is an accepted practice. But we'll tell you how to give such information in the next section.

Theory or Principles of Operation

Many sets of instructions contain a section that deals with the theory or principles of operation that underlie the procedures explained.

---

[3] *The Mead Short Course in the Graphic Arts* (Dayton, Ohio: Mead Paper, 1974), vol. 4, *Capabilities Booklets*, p. 2. Copyright 1974 by Mead Paper. All rights reserved.

Sometimes historical background is also included. These sections may be called "Theory" or "Principles of Operation," or they may have substantive titles like "Color Do's and Don'ts," "Purpose and Use of Conditioners," and "Basic Forage Blower Operation." Information about theory is presented for several reasons. People have a natural curiosity about the principles behind a procedure. Also, they want to know the purpose and use of the procedure. The good TV repair technician wants to know why turning the "vertical control" steadies the picture. Understanding the purposes behind simple adjustments enables the technician to investigate complex problems. What if nothing happens when the "vertical control" is turned? With a background in theory, the technician will know more readily where to look in the TV set to find a malfunction.

Such sections can be quite simple. In the following excerpt labeled "Color Do's and Don'ts," some basic color design theory is presented in easy-to-understand language:

DO - Use light colors in a small room to make it seem larger.

DO - Aim for a continuing color flow through your home—from room to room—using harmonious colors in adjoining areas.

DO - Paint the ceiling of a room in a deeper color than walls, if you want it to appear lower; paint it in a lighter shade for the opposite effect.

DO - Study color swatches in both daylight and nightlight. Colors often change under artificial lighting.

DON'T - Paint woodwork and trim of a small room in a color which is different from the background color, or the room will appear cluttered and smaller.

DON'T - Paint radiators, pipes and similar projections in a color which contrasts with walls or they will be emphasized.

DON'T - Choose neutral or negative colors just because they are safe, or the result will be dull and uninteresting.

DON'T - Use glossy paints on walls or ceilings of living areas since the shiny surface creates glare.[4]

As simple as this excerpt is, it presents color principles. Readers are not only told to use light colors in a small room, but why—"to make it seem larger."

Theory sections can also be lengthy and complex. Here is an excerpt about hay baling from a John Deere manual:

Purpose and Use

Baling is essentially a "packaging" operation. The range of materials that can be packaged with a baler is wide—from high-quality hays to crop residues—especially small-grain straw.

---

[4]U.S., General Services Administration, *Paint and Painting* (Washington, D.C.: U.S. Government Printing Office, 1977), p. 15.

The popularity of baling is greater than any other hay-packaging method. Such wide acceptance has come because farmers like the size, shape, and density of bales. Bales are small enough to be man-handled for stacking and feeding, but dense enough for efficient inside storage. Loose hay may take twice as much storage space as an equal weight of baled hay. This high density (10 to 14 pounds per cubic foot) is extremely important to commercial hay growers because it makes long-distance transportation of baled hay economically feasible.

Baling used to be primarily a custom operation. Now, most hay farmers own balers. Small, lower-priced PTO balers match the needs of many smaller operations. Owning a baler lets a farmer time his own haying operations, which is essential for high-quality hay. Some very-small-acreage hay growers cannot economically justify ownership of a baler, but custom baling is readily available in most areas.[5]

In baling hay, a farmer must pay a good deal of attention to having the hay at the proper moisture content. Why is this so? John Deere explains:

Moisture Content

Moisture content of the crop at baling time also influences the effectiveness of a baler. Hay baled at the proper moisture content yields more nutrition and forms well-shaped, solid bales. Such bales stack and store efficiently. Also, hay baled at the proper moisture content maintains high density throughout the storage period, and the bales are easier to handle and feed than poorly-shaped, loose bales.

When hay is baled too dry, leaves shatter easily. Excessive leaf loss may result in a drastic reduction in quality. Hay baled too dry will not compact tightly; result: loose, poorly-formed bales. In addition, the windrow pickup is not efficient in lifting dry hay.

Wet hay can be baled, but the baler operates under excessive strain. Wet hay is difficult to push through the bale chamber. This increases the load on the baler plunger and other baler parts. Also, wet bales are normally heavy, even with the bale tension bars under reduced tension. Wet bales may dry and shrink in storage and become loose and sloppy. Or, bales of wet hay may heat and mold. Bales that have heated and molded in storage have lower quality and may even cause livestock to become sick.

Proper hay moisture content for baling depends on type and maturity of crop, weather conditions, and desired storage life. Effects of each factor must be carefully evaluated. Good managers and operators learn to estimate the best baling moisture content for their local conditions.[6]

---

[5] *Fundamentals of Machine Operation: Hay and Forage Harvesting* (Moline, Ill.: Deere and Company, 1976), pp. 88–89. Copyright © 1976 by Deere and Company. All rights reserved.
[6] Ibid., p. 100.

Theory sections are not found only in instructions that deal with technical subjects such as hay baling. They can be—and should be—found in any instructional document. In Mead Paper's short course on capabilities booklets, the writer tells the reader the objectives of the booklets:

> Today's successful capabilities booklet is a well designed, carefully planned, broad-brushed presentation of a company's various abilities.
>
> ☐ *It is not* a parade of products and prices.
> ☐ *It is not* a financial report.
> ☐ *It is not* a bare-bones listing of equipment.
> ☐ *It is not* a history or anniversary recital of corporate milestones.
>
> The capabilities book educates and informs. It tells the reader in general terms how the company can help him, and why. It presents convincing photography with factual deductive writing. It is prepared specifically as a marketing tool—but its advantages range far beyond that goal. *It is a long-term sales builder.*
>
> As previously mentioned, the successful capabilities book is prepared with a primary objective of expanding customers by generating sales in new markets.
>
> Important *secondary uses* often can be anticipated. Here are a few:
>
> *Salesmen* have a powerful reference tool to present.
>
> *Inquiries* to the company can be answered with a short covering note accompanying the new book.
>
> *Employee recruiting* is strengthened—at college and other levels.
>
> *Internal distribution* helps educate employees who rarely are in a position to comprehend the corporate scope.
>
> *Financial people* get a broad overview to supplement the company's financial reporting.
>
> *Community leaders* and the local press get a new, accurate picture of the company's importance.[7]

As our excerpts illustrate, many diverse items of information can be placed in a theory or principles section. Remember, however, that the major purpose of the section is to emphasize the principles that underlie the actions later described in the performance instructions. The emphasis is on objectives, purpose, and use. In this section you're telling your readers *why*. Later you'll tell them *how*. You tell your readers what they need to know. Often, you should also tell them what, out of curiosity, they want to know. And you use language suitable to their level of knowledge in the subject matter. Obviously, good audience analysis is a must for any instruction writer.

---

[7]*Capabilities Booklets*, pp. 4–7.

List of Equipment and Materials Needed

In a list of equipment and materials, you tell your readers what they will need to accomplish the process. A simple example would be the list of cooking utensils and ingredients that precedes a recipe. Sometimes in straightforward processes, or with knowledgeable audiences, the list of equipment is not used. Instead the instructions tell the readers what equipment they need as they need it: "Take a rubber mallet and tap the hubcap to be sure it's secure."

When a list is used, frequently each item is simply listed by name, as in the following list of painter's equipment. In this list the writer felt that the audience needed little further explanation and provided it in only three places:

Check List for Painters[8]

Before starting to paint, make sure you have all the tools needed to do the job right. Following are some of the items usually required:

Paint brushes	Paint scraper
Dust brush	Paint strainer, wire mesh, or cheesecloth
Stiff bristle brush	***Paint tray
Wire brush	Patching plaster
Caulking gun	Putty or glazing compound
*Cans	Putty knife
Drop cloths	Rags
**Emery cloth	Rollers
Hammer	Roller extension poles
Ladder	Sandpaper or production paper
Masking tape	Spackling compound
Mixing paddle	Steel wool
Paint	Turpentine or other solvents
Paint bucket	Wire combs

*For cleaning brushes with solvents
**For cleaning and polishing metal surfaces.
***For painting with rollers.

Sometimes, however, your audience analysis may indicate that more information is needed. You may want, for example, to explain the properties and uses of the items used—as in this partially reproduced annotated list:

---

[8] General Services Administration, *Paint and Painting*, p. 24.

Paint Types, Properties, and Uses

Types	Properties	Typical Uses
Oil base primers	Good adhesion and sealing; resistant to cracking and flaking when applied to unprimed wood; good brushing and leveling; controlled penetration; and low sheen. Unsuitable as a top coat and should be covered with finish paint within a week or two after application.	As primer on unpainted woodwork or surfaces previously coated with house paint.
Anti-rust primers	Prevent corrosion on iron and steel surfaces. Slow-drying type provides protection through good penetration into cracks and crevices. Fast-drying types are used only on smooth, clean surfaces, and those which are water resistant are effective where surfaces are subject to severe humidity conditions or fresh water immersion.	Priming of steel and other ferrous metal surfaces when good resistance to corrosion is required.
Galvanizing primers	High percentage of zinc dust provides good anti-rust protection and adhesion. Galvanizing/zinc dust primers give excellent coverage, one coat usually being sufficient on new surfaces. Two coats are ample for surfaces exposed to high humidity.	Priming of new or old galvanized metal and steel surfaces. Satisfactory as finish coat if color (metallic gray) is not objectionable.
House paints (oil or oil alkyd base)	Made with drying oils or drying oil combined with alkyd resin. Excellent brushing and penetrating properties. Provides good adhesion, elasticity, durability, and resistance to blistering on wood and other porous surfaces. Often modified with alkyd resins to speed drying time. Apply with brush to obtain strong bond, especially on old painted surfaces.	General exterior use on properly primed or previously painted wood or metal surfaces.

Types	Properties	Typical Uses
House paints (latex type)	Exterior latex paints have durability comparable to oil base paints. Resistant to weathering and yellowing, and so quick-drying that they can be recoated in one hour. Can be applied in damp weather over a damp surface. Easy to apply and brush or roller can be cleaned quickly with water. Free from fire hazard.	Covers properly primed or previously painted concrete, stucco, and other masonry and wood surfaces.[9]

Often a table is useful in a materials or equipment list. Figure 14-2 is an example of a well-made table. It brings together paint products with their uses. Notice, also, the classification scheme in the table, based upon the substance being painted.

If the equipment or materials needed cannot be easily obtained, you'll do your readers a service by telling them where they can find the hard-to-get items.

Description of the Mechanism

Instructions devoted to the operation and maintenance of a specific mechanism include a section describing the mechanism. Also, when it is central in some process, the mechanism is frequently described. In such sections, follow the principles for technical description given on pages 128–135. Break the mechanism into its component parts and describe how they function. The following excerpt is typical. (We have not reproduced all the figures referred to below. For the ones we have reproduced, see accompanying Figure 14-3.)

The primary components of a baler are as follows (see Figs. 12 and 13):

- Pickup
- Auger and/or feed tines
- PTO shaft
- Feeder teeth
- Plunger
- Hay dogs

- Bale chamber
- Bale-measuring wheel
- Tying mechanism
- Needles
- Wire twister or twine knotter
- Bale chute

---

[9] Ibid., p. 19.

Figure 14-2   Sample Table

	House Paint (Oil or Oil/Alkyd)	Cement Powder Paint	Exterior Clear Finish	Aluminum Paint	Wood Stain	Roof Coating	Trim Paint	Porch and Deck Paint	Primer or Undercoater	Metal Primer	House Paint (Latex)	Water Repellent Preservative
MASONRY												
Asbestos Cement	X•								X		X	
Brick	X•	X		X					X		X	X
Cement & Cinder Block	X•	X		X					X		X	
Concrete/Masonry Porches And Floors								X			X	
Coal Tar Felt Roof						X						
Stucco	X•	X		X					X		X	
METAL												
Aluminum Windows	X•			X			X•			X	X•	
Steel Windows	X•			X•			X•			X	X•	
Metal Roof	X•									X	X•	
Metal Siding	X•			X•			X•			X	X•	
Copper Surfaces			X									
Galvanized Surfaces	X•			X•			X•			X	X•	
Iron Surfaces	X•			X•			X•			X	X•	
WOOD												
Clapboard	X•						X		X		X•	
Natural Wood Siding & Trim			X		X							
Shutters & Other Trim	X•						X•		X		X•	
Wood Frame Windows	X•			X			X•		X		X•	
Wood Porch Floor								X				
Wood Shingle Roof					X							X

X• = Black dot indicates that a primer, sealer, or fill coat may be necessary before the finishing coat (unless the finishing surface has been previously finished).

What To Use and Where

From U.S., General Services Administration, *Paint and Painting.* (Washington, D.C.: U.S. Government Printing Office, 1977), p. 18.

Let's consider the functions of these components.

Component Functions

The *pickup* lifts hay from the windrow and conveys it to the auger or feed rake (Fig. 14). *Hay compressors* on the pickup hold hay down for uniform feeding, and keep strong winds from blowing hay out of the pickup.

Depending on the baler design, an *auger* or *feed rake* or *tines* deliver hay to the edge of the *bale chamber. Feeder teeth* then deliver hay into the baling chamber. The feeder teeth are timed to enter the baling chamber when the plunger is retracted (in the forward position) (Fig. 15).

The *plunger* is driven by a crank arm and pitman, and moves back and forth in the baling chamber about 80 strokes a minute. As the plunger moves into the baling chamber, the plunger knife (Fig. 16) moves past a stationary knife to slice off hay not totally within the chamber. The plunger then compresses the new charge of hay into the bale chamber (Fig. 17).

To keep the partly formed bale compressed in the bale chamber as the plunger retracts, *hay dogs* engage the bale (Fig. 18).

The process of feeding hay into the bale chamber and compressing it with the plunger is repeated until the bale is formed. The density of the

Figure 14-3   Illustrations to Accompany a Description of a Mechanism

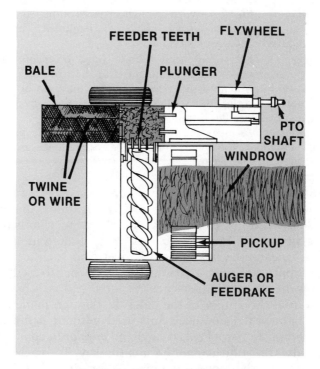

Fig. 15—Hay Movement into the Baling Chamber

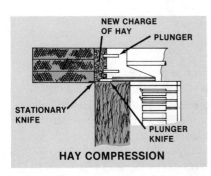

Fig. 17—Hay Is Compressed by the Plunger

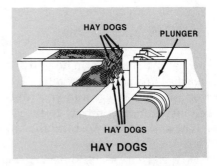

Fig. 18—Hay Dogs Hold Compressed Hay

bale is determined by adjusting the spring-loaded upper and lower tension bars on the bale chamber (Fig. 19). On some models, tension bars are controlled hydraulically.

The *bale-measuring wheel* rotates as the bale progresses through the bale chamber (Fig. 20). When the wheel has completed a preselected cycle, the *tying mechanism* is tripped. Bale length may be changed by adjusting the bale-measuring cycle.

Operation of the tying mechanism is timed to synchronize with the plunger movement. When the plunger is at the rearwardmost position and hay is fully compressed, *needles* deliver the wire or twine to the tying mechanism (Fig. 21). As the wire or twine is grasped by the tying mechanism, the needles retract, and the bale is tied.

The entire process is repeated as each completed bale passes through the bale chamber. The bale is finally forced out onto the ground (Fig. 22) or loaded as described earlier.[10]

As in the above example, the description is generally accompanied by numerous photographs and illustrations, many of them annotated. Sometimes exploded views of the mechanism are provided, as in Figure 14-4. We hasten to add that in this context "exploded" means the mechanism is drawn in such a way that its component parts are separated and are thus easier to identify. Figure 14-4 makes the concept quite clear.

Notice that the author of the baler description made certain assumptions about the audience and used without definition terms such as *windrow* (a row of raked hay) and *pitman* (a connecting rod). Again, as always, audience analysis is the key. To reach the knowledgeable farmers the manual is intended for, the writer had only to define those terms that would be new to the audience or used in an unusual way.

Performance Instructions

The actual instructions on how to perform the process obviously lie at the heart of any set of instructions. For a sample, let's look at a small portion of John Deere's instructions concerning balers—in this instance instructions on how to thread the needles of a baler with the wire that will eventually bind the hay bales:

For typical wire needle threading (Fig. 40) [See our Figure 14-5]:

- Make sure all wire pulleys turn freely.
- Thread the wire from the right-hand coil through the guide (1).

---

[10]*Fundamentals of Machine Operation*, pp. 92–94.

Figure 14-4 Exploded View of Equipment Assembly

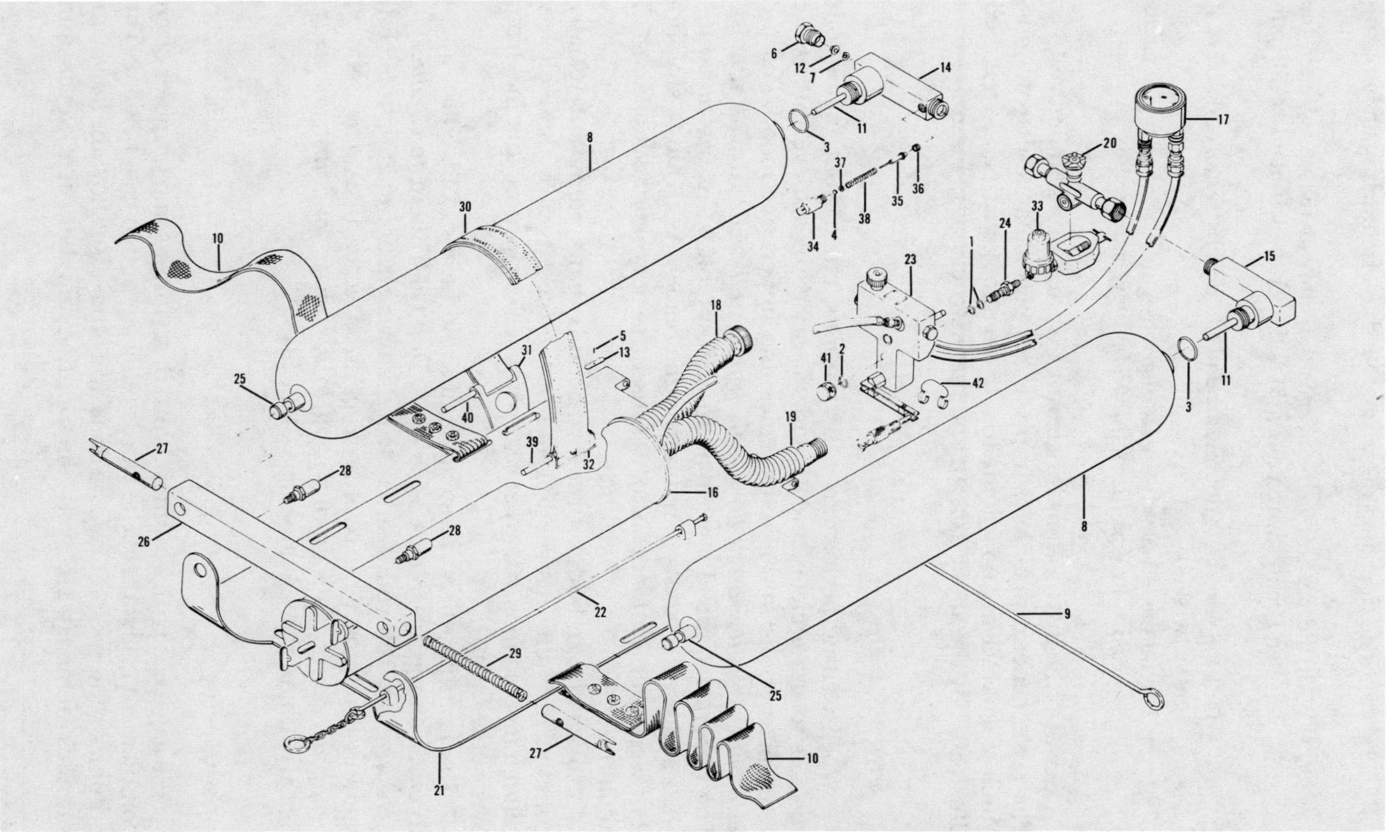

- Continue threading the wire around the front left-hand wire pulley (2).
- With the needles in home position, thread the wire under the left-hand center wire pulley (3) and over the left-hand needle pulley.
- Pull the wire back, loop it around the needle frame, and secure it with a twist (4).
- Thread the left-hand wire through the guide (5), then repeat Steps 2, 3, and 4 through the right-hand pulleys and needle.[11]

See also the instructions for cleaning a paint brush in Figure 14-6. The same general principles apply to all performance instructions. Much of what we are now going to tell you about writing them is well illustrated by the above excerpt and the instructions in Figure 14-6.

Style

When writing performance instructions, one of your major concerns is to use a clear, understandable style. To achieve this, write your instructions in the active voice and imperative mood: *Pull the wire back, loop it around the needle frame, and secure it with a twist.* The imperative mood is normal and acceptable in instructions. It's clear and precise and will not offend the reader. As an aid to simplicity, keep your sentences short. Ten- to twenty-word sentences are about right.

Notice also in both sets of instructions that a paragraph usually contains only one instruction and at the most two or three closely related instructions. Keep each step in a series clear and distinct from every other step.

Use familiar, direct language and avoid jargon. Tell your readers to *check* things or to *look them over.* Don't tell them to *conduct an investigation.* Tell your readers to *use* a wrench, not to *utilize* one. Fill your instructions with readily recognized verbs like *adjust, attach, bend, cap, center, close, drain, install, lock, replace, spin, turn, and wrap.* See also Chapter 8, "Achieving a Clear Style."

Graphics

Be generous with graphics. Word descriptions and graphics often complement each other. The words tell *what* action is to be done. The graphics show *where* it is to be done and often also show the *how.* Our samples demonstrate well the relationship between words and

---

[11] Ibid., p. 102.

Figure 14-5   Threading Needles of a Typical Wire-Tie Baler

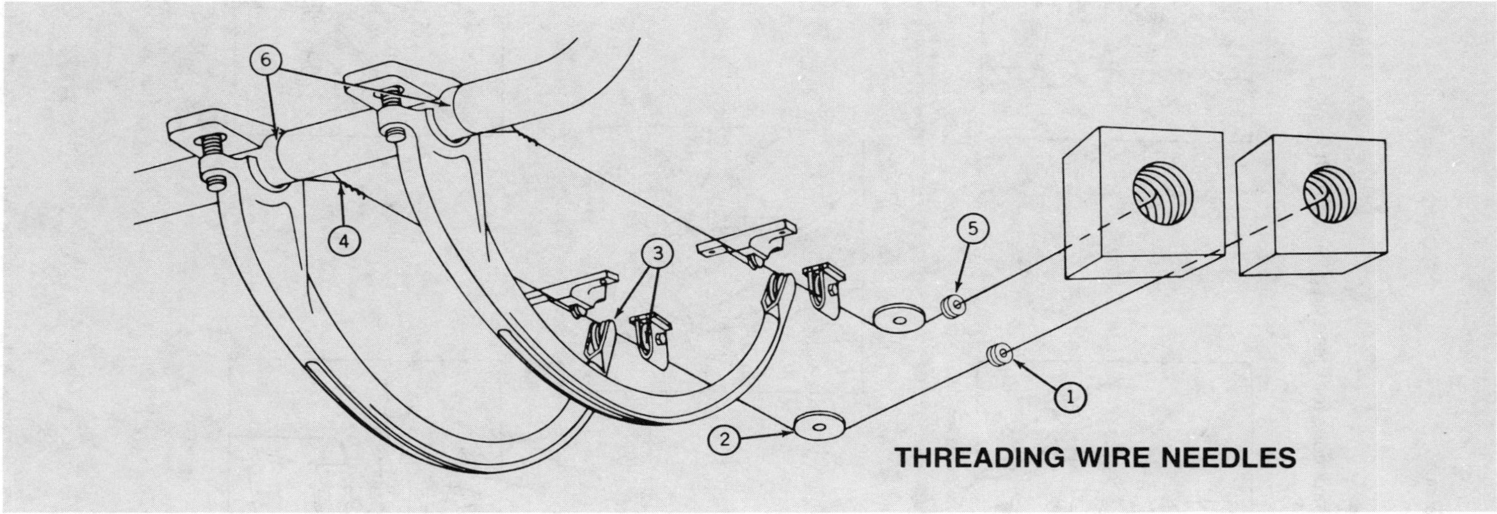

THREADING WIRE NEEDLES

From *Fundamentals of Machine Operation: Hay and Forage Harvesting* (Moline, Ill.: Deere and Company, 1976), p. 102. Copyright © by Deere and Company. All rights reserved.

Figure 14-6   Instructions for Cleaning a Paint Brush

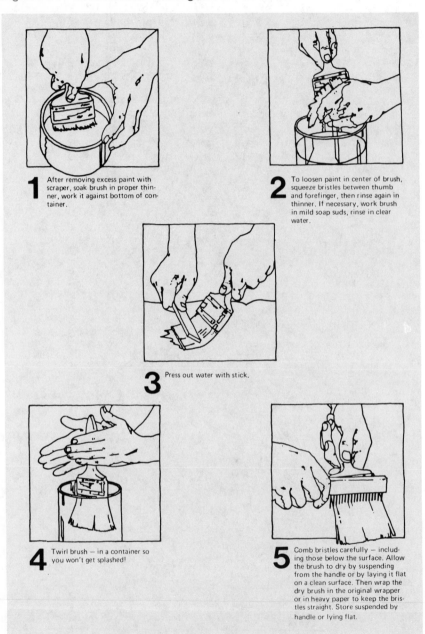

1 After removing excess paint with scraper, soak brush in proper thinner, work it against bottom of container.

2 To loosen paint in center of brush, squeeze bristles between thumb and forefinger, then rinse again in thinner. If necessary, work brush in mild soap suds, rinse in clear water.

3 Press out water with stick.

4 Twirl brush — in a container so you won't get splashed!

5 Comb bristles carefully — including those below the surface. Allow the brush to dry by suspending from the handle or by laying it flat on a clean surface. Then wrap the dry brush in the original wrapper or in heavy paper to keep the bristles straight. Store suspended by handle or lying flat.

From U.S., General Services Administration, *Paint and Painting* (Washington, D.C.: U.S. Government Printing Office, 1977), p. 3.

graphics. Note that graphics are usually annotated with numbers or words to allow for easy reference to them. As Figure 14-6 demonstrates, graphics can be used to show the worker, or at least the worker's hands, actually performing the job.

Organization

When writing performance instructions, organize the process being described into as many major routines and subroutines as needed. For example, a set of instructions for the overhaul and repair of a piece of machinery might be broken down as follows:

- Disassembly of major components
- Disassembly of components
- Cleaning
- Inspection
- Lubrication
- Repair
- Reassembly of components
- Testing of components
- Reassembly of major components

Notice that in this case the steps are in chronological order. Both our samples in Figures 14-5 and 14-6 also demonstrate chronological order. This matter of chronological order can be important in the extreme. Don't tell the reader,

> Place the mouthpiece in the mouth, squeeze the inhalation hose closed and attempt to inhale through the mouthpiece. If it is possible to inhale with the inhalation hose closed off, the check valve is missing or defective.

And then add,

> Place mouthpiece shut-off valve in the DIVING position before the test.

With the shut-off valve in the wrong position, the test will not come out right. Remember that your reader may be following you step by step.

If not following a step in sequence might damage the equipment or injure the operator, word the step in the form of a warning:

WARNING

Avoid all contact with oil and grease. Oil coming in contact with high pressure connections may result in an explosion. USE NO OIL.

Note that in our baler excerpt on page 324 the writer saved space and repetition by telling the reader to "repeat Steps 2, 3, and 4 through the right-hand pulleys and needle." This practice is legitimate, but use it with great care. Visualize your reader. Maybe he or she will be perched atop a shaky ladder, your instructions in one hand, a tool in the other. Under such circumstances the reader will not want to be flipping pages around to find the instructions that need to be repeated. You will be wiser and kinder to print once again all the instructions of the sequence. But if the reader will be working in a comfortable place, you would probably be safe enough to say, "Repeat steps. . . ."

This reader and situation analysis can help you make many similar decisions. Suppose, for example, that your readers are not expert technicians, and the process calls for them to use simple test equipment. In such a situation, you should include the instructions for operating the test equipment as part of the routine you are describing. On the other hand, suppose your readers are experienced technicians following your instructions at a comfortable workbench with a well-stocked library of manuals nearby. Then you can assume that they know how to operate any needed test equipment, or you can refer them to another manual that describes how to operate the test equipment.

Must your performance instructions always be in chronological order? When sequence is important in the proper performance of the process, then the answer is yes. But sometimes you will find that sequence is not crucial or perhaps can't even really be determined. In that case some sort of logical topical arrangement is called for (see pages 97–98). Look at the following instructions for driving a tractor on the road:

 CAUTION: Observe the following precautions when operating tractor on a road.

1. Be sure brakes are evenly adjusted, and couple pedals together before driving on a road. Avoid hard applications of brakes. A towed load of more than twice the weight of the tractor should have brakes. If not, drive slowly and avoid hills.

2. Be sure SMV emblem and warning lights are clean and visible. If towed or rear-mounted equipment obstructs these safety devices,

install SMV emblem and warning lights on equipment. See your John Deere dealer.

3. Turn light switch to "H" position. Never use flood lights or any lights which could blind or confuse other drivers. Always dim lights when meeting another vehicle.

4. If equipped use turn signals when turning. Be sure to return lever to center position after turning.

5. Drive slowly enough to maintain safe control at all times. Slow down for hillsides, rough ground, and sharp turns, especially when transporting heavy, rear-mounted equipment.

6. Before descending a hill, shift to a gear low enough to control speed without using brakes.[12]

These instructions are really a series of cautionary notes and sequence is not vital; therefore, the writer chose a topical arrangement.

Often a combination of chronological and topical instructions is used. For example, in Mead Paper's publication on capabilities booklets, the major routines of the performance instructions are in chronological order: Preparation, Production, and Distribution. But look at the organization of the Distribution section:

> Key to the ultimate success of the capabilities book is selecting the right people to receive it. In this case, building the list is far more difficult than it is for a product catalog or an annual report.
>
> Great care is required because in many cases it will be new prospects in new markets who will prove the most fruitful of all recipients. You won't want to miss important names, nor should you select poor prospects. Here are a few of the groups to be considered—either by mail or personal distribution:
>
> *New prospects*—executives and purchasing agents in companies in every industry where you should now be doing business but are not.
>
> *Existing customers*—especially executives in other divisions or areas not using your services and who possibly are not even aware of your new profile.
>
> *Financial community*—possibly the same list used for the annual report: financial analysts and business editors of newspapers and periodicals.
>
> *Shareholders*—depending on the nature of your company and the number of its shareholders. If the booklet will help strengthen stockholder relations—or help sell them as "customers" for company products—then it probably should be distributed to them.
>
> *Prospective employees*—for recruiting at campuses and other levels.

---

[12]*Operator's Manual: 850 and 950 Tractors* (Moline, Ill.: Deere and Company, n.d.), p. 43.

Suppliers—sources of many indirect sales, suppliers are often neglected in the distribution of corporate literature. They can and should be loyal supporters, but are often in woeful need of "education."

Answering inquiries—of a general nature directed to the company. In the case of big corporations, this can save a great deal of expensive office time.

Coupon returns—companies which regularly advertise in print frequently include a coupon return for requesting new literature such as the capabilities book.

Salesmen—certainly almost every sales representative should carry an adequate supply of the capability book, and hand it out wherever feasible.

Employees—probably not all of them, but certainly all those in key positions who should be aware of the company's new capabilities but for one reason or another are not.[13]

No particular sequence is possible, so a topical organization is used.

Ordinarily, then, you will use a chronological order in performance instructions. But sometimes you will find that a topical plan or a combination chronological-topical plan is the proper approach.

Format

Modern practice in format tends toward simplicity in numbering systems and generous spacing. Instruction writers do not want their readers making mistakes because they read cramped and crowded lines incorrectly. Figure 14-7 is a good example of today's style. It also illustrates rather well how to handle the computer-human being interaction. In side-by-side columns the instructions tell you what the computer does or displays and what the operator's proper response is.

Sometimes no numbering is used at all. The writer depends upon white space alone to keep the steps distinct or perhaps uses bullets as in the excerpt on page 327. However, in a long set of instructions you may wish to use a numbering system to keep the sequence straight or to allow for cross reference, such as *See Section 4.2c*. Often the conventional outline system of Roman numerals, Arabic numerals, capital letters, and lowercase letters will suffice (see page 217). Or you may want to use the decimal system (see page 218).

---

[13] pp. 26–27.

Figure 14-7   Sample Format

ZEDIT for customer summary file

- Edits the customer summary file
- Prints a listing of those items in the customer summary file which have errors
- Activates customer summary file for sales analysis

What you need to run this procedure . . .

SALP30

Before you begin . . .

- Any errors in the customer summary file must be corrected before the file can be activated for sales analysis.
- You will be prompted to activate the files for sales analysis. Once you have activated the customer summary file, you will not be prompted to activate it again.

	What the system does or displays	What you do
1		Ready the system.
2	←READY	If the last procedure you ran loaded SALP30, go to Step 3. If not, insert diskette SALP30, then type in: ZLIBRARY SALP30
	Loads diskette.	
	←READY	After ←READY is displayed, remove diskette SALP30, and go to Step 3.
3	←READY	Type in: ZEDIT CUSTSUM
4	ACTIVATE FILE FOR SALES ANALYSIS? (YES/NO)	Type in: YES to activate sales analysis. Otherwise, type in: NO
5	Prints an edit listing. ←READY	File the listing for audit.

End of ZEDIT for customer summary file

From *System/32 Manufacturing Management Accounting System Sales Analysis Run Book* (Atlanta, Ga.: IBM Corporation, 1975). Reprinted by permission.

Troubleshooting

Many sets of instructions include a troubleshooting section designed to aid the technician. They usually take the form of a three-column chart with headings such as *Problem, Possible Cause,* and *Possible Remedy.* Figure 14-8 shows two typical examples. Notice that the top chart in the figure uses graphics to illustrate the problem, an excellent technique which should be used where possible. Notice also that the remedies are given as instructions in the active voice, imperative mood.

The second example at the bottom of Figure 14-8 gives page references when appropriate to guide the reader to additional information.

Putting It All Together

In the preceding sections we have discussed the possible parts of a set of instructions. To put it all together for you, we now present and discuss a short instructional document—a fact sheet called *Driving a Wellpoint.*[14] In content and style, it represents a paper well within the reach of a student.

Driving a Wellpoint

ROGER E. MACHMEIER   Comments

A wellpoint or drive point is a pipe with wall openings large enough to allow water to enter and small enough to keep the water-bearing formation in place. Wellpoints suitable for hand driving are available in sizes from 1¼ to 2 inches in diameter and from 18 inches to 3 feet long. The size of openings in the wellpoint is determined by the relative grain size of the material in the water-bearing formation. Some of the finest grains adjacent to the wellpoint should be removed by pumping to make the well more productive (see figure 1). However, a wellpoint should not be ex-

The first two paragraphs combine an introduction and a theory section. Key terms are defined, and the use of wellpoints and shallow wells are explained. The third paragraph contains a warning. The author has wisely placed the warning early in the paper. Basically, the introduction is good. It could be improved, however. The first and second paragraphs would actually be better reversed. Also, some mention of the intended audience would have been helpful.

---

[14] Roger E. Machmeier, *Driving a Wellpoint* (St. Paul, Minn.: University of Minnesota, Agricultural Extension Service, 1972).

Figure 14-8   Sample Troubleshooting Charts

TROUBLE-SHOOTING

Most baler operating problems are caused by improper adjustment or delayed service. This chart is designed to help you when a problem develops, by suggesting a probable cause and the recommended solution.

Apply these suggested remedies carefully. Make certain the source of the trouble is not some place other than where the problem exists. A thorough understanding of the baler is a must if operating problems are to be corrected satisfactorily. Refer to the operator's manual for detailed repair procedures.

TROUBLE-SHOOTING CHART

PROBLEM	POSSIBLE CAUSE	POSSIBLE REMEDY
Knotter Difficulties — Twine Baler		
	Tucker fingers did not pick up needle twine or move it into tying position properly.	Adjust tucker fingers. Adjust needles or twine disk. Check twine disk and twine-box-tension. Install plungerhead extensions.
	Hay dogs do not hold end of bale.	Free frozen hay dogs. Replace broken hay-dog springs. Reduce feeding rate. Install plungerhead extensions.
	Extreme tension on twine around billhook during tying cycle causes twine to shear or pull apart.	Loosen twine-disk-holder spring. Smooth off all rough surfaces and edges on billhook.

ELECTRICAL SYSTEM

PROBLEM	POSSIBLE CAUSE	POSSIBLE REMEDY	PAGE REFERENCE
Battery will not charge	Loose or corroded connections.	Clean and tighten connections.	66
	Sulfated or worn-out battery.	Check electrolyte level and specific gravity.	67
	Loose or defective alternator belt.	Adjust belt tension or replace belt.	52
"CHG" indicator glows with engine running	Low engine speed.	Increase speed.	
	Defective battery.	Check electrolyte level and specific gravity.	67
	Defective alternator.	Have your John Deere dealer check alternator.	

Top sample from *Fundamentals of Machine Operation: Hay and Forage Harvesting* (Moline, Ill.: Deere and Company, 1976), p. 111. Copyright © by Deere and Company. All rights reserved. Bottom sample from *Operator's Manual: 850 and 950 Tractors* (Moline, Ill.: Deere and Company, n.d.), p. 76.

Figure 1. Installed wellpoint.

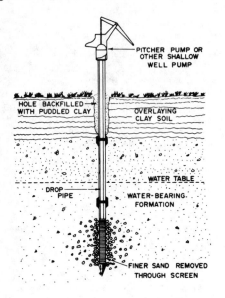

The fact sheet contains three excellent graphics. Figures 1 and 3 clarify the concept of a shallow well and the procedure for driving a wellpoint. Figure 2 supplements and amplifies the written definition of a wellpoint. Notice that the text refers to the figures whenever such reference is useful. All three figures are well annotated.

pected to yield large quantities of water.

Why Wellpoints are Used
It may be desirable to develop a water supply for sprinkling lawns, gardens, etc. An economical supply can often be obtained from a shallow aquifer (water-bearing formation) through a wellpoint. The water table should be high, preferably within 10 feet and no farther than 15 feet underground. The wellpoint must be driven deep enough to penetrate a water-bearing formation below the water table, but it should not exceed 25 feet in depth.

Shallow water tables are susceptible to pollution. Drain fields, dry wells, animal wastes, heavy fertilizer applications, etc. can contaminate a shallow water table. Recharge is usually from rainfall, falling directly above and percolating downward to the water table. As the water moves downward it may carry contaminants. Thus, extreme care and periodic testing are necessary if the water is used for drinking.

Materials Needed

The following materials are needed: wellpoint, riser pipe (in 5- or 6-foot lengths, with 6-inch nipple), couplings, drive cap, and pipe thread compound.

Openings in the wellpoint should be large enough to permit the finer particles of the water-bearing formation to enter the wellpoint while keeping the coarser particles out. With proper sized openings, development of the well removes the finer particles and forms an envelope of porous and permeable material around the screen (see figure 1).

A local well contractor or hardware dealer may have valuable suggestions on the most suitable opening size for the wellpoint.

Several types of wellpoints are constructed of different materials. A few types are shown in figure 2. For mesh-covered wellpoints, the size of openings is designated by the mesh size. Common sizes, from larger to smaller openings, are 40-, 50-, 60-, 70-, and 80-mesh. For wellpoints with slot type openings, common slot sizes are 18, 12, 10, 8 and 7 slot. Slot sizes are the opening width in thousandths of an inch. No. 18 slot is 0.018-inch wide, No. 12 is 0.012-inch wide, etc.

The second major section, "Materials Needed," details the materials needed. (Note the use of headings to separate the sections.) The first paragraph under this heading lists the materials. The following paragraphs present details to help the reader select and use the materials wisely. The writer's audience analysis is evident in this and the next section, "Tools Needed." Because the writer goes into so much detail about the materials, we can deduce that he sees his audience as people inexperienced in the specific task, driving a wellpoint. But in the next section, "Tools Needed," he lists tools without comment, indicating that the writer sees his audience as being skilled tool users. The writer probably visualizes his audience as farmers or home owners used to doing their own routine maintenance.

Figure 2.   Types of wellpoints.

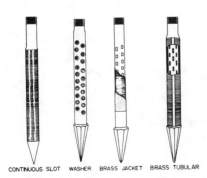

CONTINUOUS SLOT   WASHER   BRASS JACKET   BRASS TUBULAR

The riser pipe should be galvanized pipe in 5- or 6-foot lengths for convenient hand driving. A standard 21-foot length of pipe cut into four pieces normally is adequate for a driven well.

Special drive pipe couplings, which allow the pipe pieces to butt together, are desirable. The impact of driving is then transmitted through the pipe and not the pipe threads. Under severe driving conditions, standard pipe couplings may cause thread cracking or failure.

The drive cap is placed on the top of the pipe section being driven. The cap transmits the blow to the pipe and protects the threads.

All pipe joints should be screwed tightly after threads are carefully cleaned and oiled. White lead or pipe thread compound should be used to improve airtightness.

Tools Needed

A posthole digger or soil auger, sledge hammer, carpenter's level or plumb bob, and pipe wrenches are needed for hand-driving a wellpoint.

Installation Method

The well site should be at least 75 feet from sources of contamination such as a septic tank, cesspool, dry well, drain field, foundation drain, downspout, marshy area, or animal yard.

Dig or drill a vertical hole 1-2 inches larger than the wellpoint and as deep as possible with the posthole digger or auger (figure 3). The hole usually cannot be drilled more than a foot below the water table.

Rub a bar of soap over all the openings of the wellpoint to help prevent clay and sand from entering and to reduce friction during driving.

Attach a length of riser pipe to the open end of the wellpoint. Clean the threads, add the pipe thread compound to the outside threads, and make a tight connection with the pipe wrenches. In-

The "Installation Method" is the performance instructions. Notice that it begins with a wisely placed cautionary note about well location. The writer uses short sentences and paragraphs to keep his instructions clear and separate. He uses the active voice and imperative mood in most of this section. He assumes an experienced tool user who does not have to be told, for example, how to use a plumb bob or carpenter's level. But he does take the time to explain the use of the soap, a point even an experienced tool user might be curious about.

Figure 3.   Driving a wellpoint.

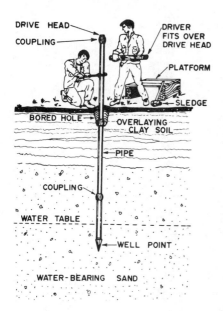

sert the wellpoint into the hole. Attach the nipple and drive cap to the top of the riser pipe. Do not use pipe thread compound on the drive cap.

Make sure that the pipe and wellpoint assembly are vertical by using the carpenter's level or the plumb bob. Use the sledge or driver to strike the drive cap with square, solid blows (figure 3).

When the drive cap is about 4 inches above the ground, unscrew the drive cap and nipple and place the unit on another section of pipe.

Clean the threads on the new section of pipe and on the pipe in the ground and add pipe thread compound to the outside threads. Tighten the joint using two pipe wrenches working in opposite directions to avoid twisting the assembly in the ground, and resume driving.

Continue the above procedure, adding sections of riser pipe as needed. To

The writer follows chronological order throughout the performance instructions. He frequently gives the reader helpful tips to make the job easier or to prevent complications.

keep the threaded joints tight during driving, give the riser pipe an occasional half turn with a wrench. Use the wrench only to take up slack in the threaded joints and take care not to twist the pipe severely.

Pour water into the top of the pipe at regular intervals. This makes driving easier and will determine when a water-bearing formation is reached. When a gallon of water disappears into the formation within 2 minutes after being poured into the well, it is unnecessary to drive any deeper.

Enough of the riser pipe should extend above ground so the desired pump can be attached conveniently. For example, a hand pitcher pump should be about 3 feet above the ground. For an electric pump, the top of the riser should be a foot or less above ground. The pipe may have to be cut and threaded to obtain the proper height, or it may be possible to drive the pipe to the proper height.

To prevent pumped water and other surface water from moving downward along the pipe to the water-bearing formation, backfill the drilled hole with clay soil (figure 1). The clay should be extremely wet so it will puddle or flow like heavy grease. Fill the hole somewhat higher than ground level. Cement can be poured in a base around the riser pipe. However, using cement makes it extremely difficult to pull the well at a later date.

There are several ways to clean a new well by surging and pumping. One simple method is to take a wooden rod or closed-end pipe and—simulating piston action—rapidly work it up and down for about 5 minutes just below the water level in the well. This surging effect will draw fine, loose sand and silt into the well, leaving the coarser and more permeable material outside the

The format is simple. No numbering system is used. The writer puts only single instructions or very closely related instructions into each separate paragraph.

wellpoint. Remove the fine sand from the well with a pitcher pump or other pump capable of handling sand.

Another means of cleaning out the sand is to jet water into the well with a garden hose inserted to the bottom. The sand and silt particles will wash out around the hose. Repeat until no more sand is obtained by pumping.

Attach the pump to be used following the manufacturer's instructions. To use the well water for domestic purposes, the pump should be connected to a drop pipe installed inside the well (figure 1). A drop pipe is used so that the underground pipe would be under suction if the pump were connected directly to the well.

The well should be disinfected before the well water is used for drinking. For a 1½- to 2-inch well, add 1 cup of 5.25 percent hypochlorite laundry bleach to 3 quarts of water and pour the solution into the well. After 12 hours, pump the well until the chlorine odor is no longer objectionable. For additional information on disinfection procedures, see Agricultural Extension Service M-156, *Chlorination of Private Water Supplies.*

As is sometimes done, this set of instructions refers the reader to other instructions—*Chlorination of Private Water Supplies*—for additional information. However, the writer includes enough information on disinfection for the reader to accomplish the task. Never refer readers to documents that are not easily obtainable.

Forty-eight hours after pumping, the water can be sampled for bacteriological examination. Write to the Minnesota Department of Health, Minneapolis 55440 for a sampling kit that includes instructions and a sterile container. The water will be tested at no charge for bacteria, nitrates, and detergents. If a shallow well is used for domestic purposes, the water should be tested annually or semi-annually.

It is usually a sound policy to have a reliable well driller develop a sanitary water supply for a home. Water purity and personal health should be the major consideration in developing a domestic water supply.

The concluding paragraph presents once again the warning that shallow wells are not the best source for drinking water. The writer obviously wants this point taken seriously.

The fact sheet you have just read is typical of the thousands of sets of instructions that are turned out yearly to help people carry out various procedures. It contains most of the sections we have described in this chapter. It does lack a section for "Description of the Mechanism." But the wellpoint, the major mechanism involved, is described in the introduction and again in the "Materials Needed" section. Only a troubleshooting section is completely missing. The paper illustrates well our advice to combine the sections as needed and to omit those sections not needed.

Notice that the paper has no particular conclusion, unless the final warning could be considered as such. For the most part, sets of instructions have no conclusions. They simply end with the last instruction. On occasion, particularly when writing for laypeople, you might wish to close with a summary of the chief steps of the process or, perhaps, a complimentary close (see page 200). However, such endings are not general practice.

A Final Word

Let us close this chapter on instructions with an important but often overlooked bit of advice. When you are writing instructions, check frequently with the people who are going to use them. Bring them a sample of your theory section. Discuss it with them. See if they understand it. Does it contain too much theory or too little? Submit your performance instructions to the acid test. Let members of the audience for whom the instructions are intended—but who are not familiar with the process—attempt to perform the process following your instructions. If they cannot, don't blame them. Examine your instructions to see where you have failed. Probably you will find you have left out some vital link in the process.

EXERCISES

1. Writing instructions offers a wide range of possible papers. Short papers might consist of nothing more than an introduction and a set of performance instructions. Examples—good and bad—of such short instructions can be found in hobby kits and accompanying such things as toys, tents, and furniture that must be put together. Textbook laboratory procedures frequently are examples of a short set of instructions. Write a short set of instructions. Here are some suggested ideas:

Developing film
Making corrections on a typewriter
Drawing a blood sample
Applying fertilizer
Using a lawn mower
Setting a bicycle gear
Building a dog house
Replacing a broken window
Cleaning a carpet
Balancing a checkbook

2. The fact sheet presented on pages 332–339 is a good example of a longer set of instructions that uses most of the possible sections described in this chapter. Write a similar set of instructions that includes an introduction, a theory section, a list of equipment and materials needed, and performance instructions. If needed, also provide a description of a mechanism and troubleshooting procedures. Here are some suggested topics:

Laying a concrete patio
Tuning a radio transmitter
Writing (or following) a computer program
Setting up an accounting procedure for a small business
Conducting an agronomy field test
Checking blood pressure
Painting an automobile
Baking bread

CHAPTER 15

Proposals

All projects have to begin somewhere and with someone. Initiative is required to attack the endless amount of work to be done in the world, and much of this work consists of research and experimentation. We can do some of it for ourselves. At other times we have others do it for us, and sometimes we do it for others.

Think of the projects and programs that have been started in your own community this past year. For example, the citizens of your community may have desired a public swimming pool for many years, but nothing happened until some individual took up the question and presented it to the town council as a proposal. The individual may have been a private citizen, a member of the council, or the editor of the local newspaper.

In business, industry, and government, proposals are an everyday necessity. As in community affairs, the original suggestion may come from the person, company, or agency that would like someone else to do a job that needs doing. On the other hand, a person, company, or agency that desires additional work may offer to render service to another. In either case, an oral or written *proposal* is required.

The flow chart (Figure 15-1) depicts a typical investigative project from its inception through award of contract.

Block #1 indicates the "unsolicited situation," in which one offers, unasked, to perform work for another. Block #1a indicates the opposite or "solicited situation," in which someone requests the services of someone else. Scanning across the flow chart, you see the various stages that ensue until either a work contract is awarded or the proposed project dies out before any productive work is done.

Figure 15-1   Flow Chart of Activities from Proposal to Award of Contract

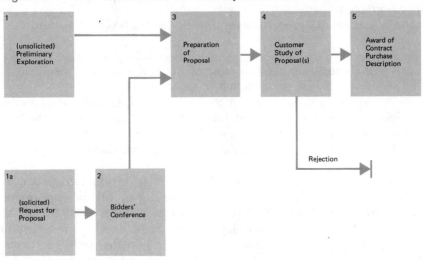

BASIC PLAN OF PROPOSALS

Referring to block #1a of Figure 15-1, you see that companies, institutions, and agencies solicit research to be done by outsiders. Informally, they spread the news by word of mouth. Officially and formally, they issue a Request for Proposal (RFP) and distribute it to likely prospects. The RFP may be a one-page letter or it may be a bound document of a hundred pages or more.

The recipients who are interested in the project outlined in the RFP respond with a proposal (that is, they bid on it) in the hope they will win out against the competition. The interim is a time of anxious waiting. During times of normal prosperity the qualified competitor stands, at best, one chance in five of winning out in the competition. Happy is the contractor who is awarded the contract. Happier still is the person who can live with the contract and gain a profit from it.

Refer again to Figure 15-1; specifically, to block #1, at which point the course of the unsolicited proposal begins.

The term "unsolicited" as applied to proposals is frequently misleading. There is a great deal of exchange of information within the research community. The word gets around that X-Company *might* be interested in having someone else conduct a research project for it. Those who are privileged to hear the word take it back to their own Y-Company. Heads get together, inquiries are discreetly made, and a "feeler" letter may go to X-Company. As a result, a proposal may be prepared on the basis of often slim information. At best there is one chance in eight that this kind of unsolicited proposal will be accepted, but the chance is worth taking if Y-Company has research personnel to spare.

The customary procedure for unsolicited proposals is as follows:

1. An individual or group desiring to do research for pay prepares an unsolicited proposal and sends it to an appropriate sponsor—a company, a private foundation, or a government agency.
2. The sponsor, who will pay for the research if approved, considers the merits of the proposal, the qualifications of the researcher, and the benefits that might come if the research has a productive outcome.
3. If the benefits seem worth the money, the proposal is funded.

When and if you reach the stage that an unsolicited proposal seems to be in order, go to the library to dig up all the supporting data you can find. Identify one or more sponsors who might be interested. Consult with your department head or vice president in charge of long-range research and other knowledgeable persons. Write a letter to the proposed sponsor to obtain its specifications for proposals. Then settle down for weeks of work to create an attractive and persuasive proposal. The chances—say one in twenty—may be worth your investment of time and energy.

Because the "unsolicited proposal" brings all proposal functions into play, we limit the rest of our discussion to it. With this understanding, we can safely assert that a proposal is designed to discharge two salesmanship functions:

1. To get a proposal accepted.
2. To get you (or your company) accepted to perform work.

Every item of content, every point of diction, every matter of display and format, must be designed to accomplish these dual objectives of all proposals.

Basic Plan of Proposals

Fortunately for us, proposals are so common and so important to the world's work that they have now settled down to a pattern that would seem to exhaust the possibilities.

Here we will limit our attention to the short-form letter proposal. Not only is this the most commonly used form, but it is most likely to satisfy your proposal requirements as a student, technician, or junior engineer.

1. *Introduction.* Here you set down everything needed to inform your reader

about the problem you are presenting. Identify the letter as a proposal. Explain how the idea came to you. Identify the subject matter of the proposed research. Comment upon the importance of the problem. Present your qualifications for the work. Possibly preview the plan of your letter.

2. *Body.* Here you develop the substance of your proposal in all necessary detail. This should be the longest section of the proposal. For topical contents, see Items for Inclusion.

3. *Conclusion.* Here you try to consolidate what you have accomplished in the body of the proposal letter. You say whatever might help to encourage acceptance and precipitate action. Again, you may stress the importance of the problem. You may also suggest an interview or assert your willingness to modify portions of the proposal.

4. *Attachments.* Here you insert display matter and back-up matter that would interrupt your main presentation in the body of the proposal letter. Possibilities for inclusion here are testimonial letters from previous customers, flow charts of your intended work program, and descriptions of past projects.

Understand that this outline applies to a small-scale proposal. This kind of proposal can be, and usually is, prepared by one person. You should understand, however, that the paper work for mammoth proposals (investigations in the millions or billions of dollars) may fill a five-foot shelf.

Items for Inclusion

Whenever you have to prepare a proposal or contribute to a proposal under preparation, be selective. Include only those items that are pertinent. Headings may often be combined to advantage. Each item has its own work to do in accomplishing the dual objective of any proposal. You should consider the following items for inclusion.

Reference to earlier association	Time and work schedule
Subject and purpose	Facilities available
Definition of the problem	Previous experience
Immediate background of the problem	Personnel and their qualifications
	References
Need for solution of the problem	Likelihood of success
Benefits that will come from solution	Products of the project
Feasibility of solution	Cost and method of payment
Scope	Descriptive and advertising
Methods to be used	literature
Task breakdown	Urge to action

Reference to Earlier Association

In drafting a proposal you should use every legitimate device that will work to your advantage. Therefore, in either the cover letter or the proposal itself, mention any work that you did for your intended client in the past. It is also helpful to introduce the names of any employees you may know, particularly if the client has a high regard for them. Even an earlier telephone conversation with your client, a chance meeting, or an exchange of letters can be adroitly introduced to your psychological advantage. Any of these techniques may give you the slight "edge" that will get you a hearing and a sympathetic reading.

Subject and Purpose

Do not be vague or coy about revealing the subject matter of the investigation you are proposing or about the fact that you are proposing your services to conduct the investigation. Early in the proposal, take care of these points with firmness and clarity. Make the "what" and the "why" clear beyond all possibility of misunderstanding.

The "what" and "why" may be stated under separate headings, as here:

Subject

I propose to investigate the suitability of various rug fiber types, conditional requirements, and availability of rug fiber choices, with respect to the Woodward Nursing Home.

Purpose

As the major outcome of the proposed investigation, I will recommend from among the options available the rug fiber most suitable for use in the Woodward Nursing Home.

Alternatively, subject and purpose may be presented in a combined statement, sometimes with a dual heading, *Subject and Purpose:*

Subject and Purpose

I propose that the Textile Consultant Division of Henley Rug Manufacturing Co. assess the conditional requirements of the Woodward Nursing Home and investigate the advantages of the options available, in order to determine which rug fiber type would be most suitable to the Nursing Home's needs.

When a proposal involves a complicated and highly technical problem, separate statements of subject and purpose are preferable. Each of these statements may run to a hundred words or more. Relatively simple problems and brief proposals, by contrast, tend to use combined statements. It is largely a matter of scale and individual emphasis.

Definition of the Problem

Persons and companies often suffer for years from a problem they have never defined and may not suspect they have. Exorbitantly high telephone bills, delays in invoicing and shipping, poor inventory control, poor customer relations, ineffective advertising, high maintenance costs—such "facts of life" may steadily grow worse through inattention and lack of recognition. An outsider, not having been blinded by day-to-day living with the problem, is usually in a far better position to detect its presence and to formulate a description of it.

Depending upon the scale of the proposal, you should spend from a paragraph to several pages in defining, locating, and describing the problem you propose to solve. By this means you may "shock" your intended client into sudden and full awareness that a problem really does exist. However, you should guard against overstatement and overdramatization, because these techniques can boomerang.

Immediate Background of the Problem

Trash collection and disposal is a generally recognized problem today. Populous cities along the Atlantic Coast transport their refuse hundreds of miles where it is dumped in landfills in lightly populated areas. Natives near these dumping grounds are increasingly raising their voices in protest. Who wants the natural countryside to be the garbage dump for the big cities?

How did we get into this sorry plight? We really do not know. It seems clear, however, that before we can devise solutions we must first discover what generated the problem in the first place. Discovering the origin and development of the problem may lead to means for its solution. Have Americans become an increasingly wasteful people? How much of the blame can be placed on marketing practices that emphasize bottles, cans, and cartons? We don't know, but we must find out.

A historical review can be quite enlightening and helpful in proving your grasp of the problem. However, your job is not to write history but to solve an existing problem. Therefore, keep the historical review brief and concentrate upon the recent past. Be selective. Make the main points stand out.

Need for Solution to the Problem

In the 1920s and 1930s automobile exhausts were discharging contaminants into our atmosphere just as they are today—and possibly at a higher rate per vehicle mile. But it was not until the 1950s that a few clear-sighted individuals pointed with alarm to the extent of air pollution from this source alone. Respiratory irritation and pulmonary congestion, smog that has killed the frail and aged, soiling of buildings, and grime that settles everywhere are now costing us hundreds of thousands of dollars every year. And the cost in health, comfort, and cleanliness cannot be stated in dollars.

Air pollution from automobile exhausts and other sources is now generally recognized as a problem by people in government at all levels, by automobile manufacturers, and by the ordinary citizens. But the pioneers in the fight against air pollution have had to fight— and are still fighting—a lusty campaign to convince America that the *need* to do something about the problem is urgent and critical.

For the most part, people are not really aware of local problems they have grown up with and become accustomed to. In a proposal dealing with such a problem you may have to shock them into acknowledging that the problem is a crucial one that will adversely affect their health or well-being. Spell out the consequences of ignoring the problem, as the former Secretary of Health, Education and Welfare has done with cigarette smoking.

Benefits that Will Come from Solution

This section is often combined with the preceding one because they overlap. However, "benefits" seems to be more positive and specific.

With respect to the trash disposal problem, you might stress that avoiding the long-distance shipping of trash would save hundreds of thousands of dollars a year for a city the size of Philadelphia. Freight cars and trucks would be released for more profitable work. One possible local solution might be to convert garbage into alcohol and then into gasohol. On the whole, such a solution might prove to be income producing rather than tax consuming.

Feasibility of Solution

Work plans and methods are fine and necessary for inclusion in proposals, but such plans are seldom taken at face value. Will they solve the problem? Arguments for their feasibility are to be found in the laws of physics and chemistry, the practices of engineering, mathematical models, and analyses. Such demonstrations tend to prove that the means to be used can be made to work; that is, are both practical and practicable in terms of present knowledge and attainments.

If you turn to your dictionary, you will find that *feasible* has two meanings. A feasible action or undertaking is one that:

1. Can be accomplished in a purely *practical* sense (i.e., is not precluded by physical, economic, or similar concerns); and/or
2. Is suitable or appropriate (i.e., generally conceded to be *desirable* as well as practical).

The first meaning is more commonly applied than the second. However, since the second meaning overlaps the first—in the sense that a totally inappropriate project also ceases to be practical—our example will consider both.

Suppose you wish to determine whether it would be feasible to allow students beyond the fifth grade to use electronic calculators in some or all of their classrooms. First, the project would have to be assessed in practical terms. Could the local school system equip its classrooms with calculators in view of purchase costs, available power outlets, maintenance costs, class size, room space and configuration, and so on?

Your research might establish that these practical obstacles are insurmountable, in which case there would be no reason to go any further. If the school can neither afford nor install the calculators, their desirability becomes irrelevant.

Suppose, however, that the practical problems appear to be manageable. At this point, you need to consider the more subtle issue of desirability. For instance, if the calculators were readily available to students, would they come to depend on them so much that they would forget the basic mathematical rules involved? Many parents and teachers might raise this question to illustrate the seeming "unsuitability" of calculators. That response is a natural one, based on the tradition that tedious mental calculations are essential to learning mathematics. But in fact, as your research would indicate, a math calculator is helpful only when the person using it *already* understands the processes involved. The calculator, then, would not permit students to bypass the customary route to higher mathematics—addition, subtraction, multiplication, division, decimals, fractions, percentages, and so on. Rather, it could enable students who *did* have a sound grasp of these fundamentals to progress much more rapidly toward advanced concepts.

Your study might conclude, then, that classroom calculators are not only practical but appropriate at the level in question. Their presence could serve as an incentive for younger students to master the techniques for using them, and could stimulate older students to attempt more challenging problems.

In this example, to clarify our meaning, we have stripped the problem down to less than its bare essentials. A proposal for this project might run to four or five pages, and the resulting feasibility study report might run to twenty or more pages. (For more detailed treatment of feasibility, see Chapter 18, "Feasibility Reports.")

Scope

The scope statement sets boundaries and states what is to be done within these boundaries. You may state that you will use only off-the-shelf commercial equipment or that you will design, develop, and test equipment that has not been devised to date. You may take economics into account or ignore all matters of cost. You may sample public opinion or disregard the question entirely. You may adopt existing techniques and modify them, or create entirely new techniques. You may limit study to estimates of technical feasibility, or you may agree to demonstrate its practicability. You may study all possible sites within your state, only one site, or a selected few. In other words, the scope statement establishes the depth, breadth, and means of your approach. Also, for everyone's protection, you may include in it negative observations covering what you will *not* do.

Methods to Be Used

It is always heartwarming to have someone say that he intends to do something we very much want to have done. But immediately, realists that we are, we ask *how?* A plumber without tools can diagnose the leak in our shower-faucet and sympathize with us about the drip. But without knowledge gained from experience and without wrenches, screwdrivers, and washers, he is powerless to stop the leak.

When you propose a problem for solution, you must demonstrate that you have the means required to solve it. A plan of attack, a method of operation, a systematic analysis of intended procedures, will attest to your practical good sense and know-how.

Task Breakdown

All but the simplest projects require a breakdown into tasks. One task may consist of initial exploration and planning, another of a search of the literature, another of correspondence and interviews, and so on. (See Chapter 3, "Gathering Information.") To try to do everything at one time produces confusion and dispersion of effort. Furthermore, one or several tasks may have to be completed before others can be started. You cannot, for example, assemble equipment before you have collected parts, and you cannot analyze data before you have acquired them. Breaking a project into its component tasks also gives some assurance that the total job will be done in an orderly manner and be completed on schedule.

Time and Work Schedule

With few exceptions, work proposals and agreements specify the calendar period within which projects are to be completed. The period may extend for days, months, or years. Further, the schedule may stipulate that portions of the work are to be done in a stated order and are to be completed by a given date. For complicated projects time and effort must be carefully allocated. It is common practice to prepare a time-based flow chart showing activities and their intended duration and to display this chart on a wall. Even limited projects, such as student research problems, require planning for efficient performance to guarantee that work will be completed by target or deadline dates.

Facilities Available

For the simplest of projects you may need only a pad of writing paper and a ball-point pen. This problem you can solve by spending a few cents at the community five-and-ten. But many projects, apparently simple on first consideration, may shortly prove to require more than you bargained for. These means or "facilities" may include a typewriter or typing service, a desk calculator, a technical library, use of a car, surveying instruments, testing apparatus, materials for constructing models, photographic equipment—and so on. On a still larger scale, you might need a truck, the use of a light aircraft, and electronic computers.

Before you decide to prepare a proposal, you must realistically determine the means and facilities the proposed work will require. If you do not already possess the means, then all or some of the cost of acquiring the means must be included in the cost and financing of the project. If you already have the means, some or all of their original cost should be paid off, amortized, during the life of the project proposed.

Most proposals should include a section on facilities. With this section you prove two points: first, that you appreciate the depth and ramifications of the project you are proposing, and second, that you are prepared to do the work because you have the necessary means or are fully prepared to acquire the means. Your intended customer will find this demonstration reassuring.

Previous Experience

As with the person who is seeking a first job, lack of experience in someone offering a proposal makes prospective employers pause, whereas successful accomplishment in the past promises success in the future. Whenever possible, therefore, you should cite earlier successes with similar problems when you prepare the new proposal. It is not essential that the past and present problems be identical—only that they overlap and have major points in common. You should include dates, contract numbers, names, and addresses so that your prospective client can verify your statements.

Personnel and Their Qualifications

Many persons in research, industry, and government enjoy national and international reputations for their knowledge and performance. Institutions seek to attract celebrated "names." If you qualify, or if people in your organization qualify, you would do well to include their names and biographical sketches in a proposal. Known names tend to be regarded as a known quantity. However, if you cannot guarantee the availability of such persons for the project, do not use wording that can be construed as a firm promise. Otherwise, your contract may be broken on the grounds of misrepresentation.

References

To support your performance record, you would do well to include the names and addresses of individuals, companies, and agencies that are able and willing to testify in your behalf. Satisfied customers can write letters that spell the difference between a hit and a near-miss with proposals. Be sure, however, that they hold you in favor and will respond promptly and generously to inquiries concerning you. Obtain their permission well in advance of using them as references.

Likelihood of Success

At first thought, you may feel that you would be foolhardy to write a proposal concerning any problem that you could not guarantee to solve. But understand that every problem exists simply because it has not been solved. If the solution was self-evident, then the problem would very probably have been solved before your arrival on the scene. Furthermore, some problems—cancer prevention and cure, the common cold, racial tensions, and war—must be attacked, and are being attacked, even though the leads are slim and the prospects for early and full solution are dim. The universality and severity of a problem warrant the expenditure of effort even though the solution may not be found in our lifetime. Polio would still be killing and maiming our children and young adults if researchers had not had the courage to gamble against the odds.

Given infinite time, money, and means, we can reasonably hope to solve all problems that now afflict humanity. But when you prepare a proposal, you must set bounds on time, money, and means. Often you simply guarantee a limited solution. Therefore, in part to protect yourself and your professional reputation against charges of misrepresentation and nonfulfillment, you should consider including an estimate of success in every proposal, perhaps in percentage form.

Products of the Project

Stating what you are to produce throughout a project and at its termination does commit you to definite performance and productiveness. Sometimes you may find it burdensome and costly to satisfy all points of your agreement. However, if you succeed in dodging this issue, and do not specify what you will produce and supply, you may be placing yourself in a still worse situation. Your customer may "construe" your proposal as promising delivery of a working model, or prototype equipment, or as promising follow-up maintenance and consulting services free of extra cost to him. Therefore, a clear and specific statement of products to be delivered, while it commits you, also sets bounds on what you are to deliver. For the moral here, read the guarantee on the next roll of photographic film you buy. You will see that the manufacturer limits its obligation to replacing the original roll of film—and only if the original was defective in manufacture or processing. If you leave the lens cap on your camera, you, and not the film manufacturer, will have to stand the loss.

Cost and Method of Payment

Practices differ in handling money matters in connection with proposals. Sometimes the "cost accounting" is included in the proposal proper. Sometimes it accompanies the proposal as an addendum or rider. And sometimes it is handled in a separate financial and legal contract. In any case, you must never commit yourself to working for others without a binding agreement covering dollar amounts, hours of labor, fees and profits, timing, and method of payment. Penalty clauses for nonperformance or late performance, limits on fees and profits as a percentage of the whole, and delayed payment practices should be scrutinized closely. If you are not adept in these matters, go to a qualified financial expert or attorney.

Descriptive and Advertising Literature

Most companies, as well as experienced consultants to industry, keep a supply of "exhibit" literature on hand. This literature may include some or all of the following items:

Biographical sketches	Security clearances held
Descriptions of earlier projects	Statements of financial condition
Company organization chart	Employment practices
Statements of company policy	Physical location and setting

Although these write-ups have to be revised periodically, a supply of them is kept in stock so that they can be selectively included in proposals.

Urge to Action

As we commented at the start of this chapter, every proposal involves sales functions. A good product, of course, sells itself, but it does no harm to advertise its merits also. Therefore, at one or more places in a proposal, particularly at the end of the proposal or in the transmittal letter, you may wish to include several sentences whose purpose is to trigger acceptance of the proposed project. These statements should be given an objective basis, however. Horn blowing, banner waving, and self-praise may work against you rather than for you. If you have proposed the construction of a community swimming pool, you may wish to state that adoption of the proposal this fall would make it possible to have the pool in operation by June 1. You may stress the need and the benefits. You may give evidence of your sincere intentions by requesting an interview or by stating your willingness to modify the proposal. Whatever techniques you use to precipitate acceptance of the proposal should reflect your appreciation of the reader's situation and should express your own tact and good taste. Perhaps you also have had the experience of an overzealous salesperson's killing a sale by talking long and loud after you were willing to buy.

Planning the Proposal

Please understand that only very long and comprehensive proposals include all of the topics described on the preceding pages. A brief proposal, one extending to four or five pages, may include only seven or eight of the topics. Also, there are various ways of combining the topics so that more than one can be treated under a single heading.

A properly drafted proposal contains neither too much nor too little. The proposal (actually a suggestion) that the office try out a new brand of pencil would not warrant forty pages of text, for the labor cost of preparing the proposal would be greater than any conceivable savings from using a better brand of pencil. But the federal government will not sponsor and underwrite the development of a new nuclear-powered aircraft carrier on the shaky basis of an anonymous note dropped into the suggestion box at the Bureau of Ships.

Writing the Proposal

Do not try to write more than one or two pages at a sitting. It will take you a few minutes to warm up to the job. Then you will reach a peak, after which the quality of the content and diction will taper off. We suggest that you use the "panel method"; that is, write up one topic per sitting. Then come back after an intermission and write up another topic. In this way you should produce panels of roughly equal and good quality, for you will have concentrated on them one by one while you were feeling fresh and alert.

Assembling the Proposal

When you have written up all the topics you intend to include in the proposal, spread the write-ups on a table or desk. If one topic seems too brief or scrappy, you may decide to rewrite and expand it. Or you may be able to tuck it in with some other topic write-up under a combined heading. Or you may decide to chuck it into the wastebasket. Similarly, extended or mixed write-ups may be pruned or split into two panels with more specific headings.

Now you should scan through the write-ups with the intention of rearranging them. The individual compositions should read progressively, so that those coming early pave the way for those that follow. Shift the panels from one position to another until you have them arranged in an order that makes the best and clearest sense.

Additional Considerations

Overzealousness

The very fact that you initiate a proposal is proof of your vigor and aggressiveness. But these instincts, which are basically good, have to be curbed. You should therefore screen your ideas and weed out those that seem impractical. Ask whether you have a good fighting chance against the opposition. Ask whether you have the time, manpower, and capital backing. Ask whether you have the needed facilities and technical competence. Getting a proposal accepted is one matter, but will you be able to fulfill the contract? Unhappy are the individuals who bite off more than they—or their company—can chew.

Timing

Too much too soon, too little and too late, the right thing at the wrong time—these labels may be applied to basically sound proposals that are rejected because of bad timing. You should follow the daily news, read business magazines, watch the trends in industry to find the time that is ripe for a particular proposal. What are today's problems and issues and needs? What will be tomorrow's? It is too late for yesterday and too soon for the day after tomorrow. A few hours with your newspaper and magazines like *Time* or *Newsweek* will prove to you that today's problems include racial tensions, riot control, rehabilitation of our cities, air and water pollution, care of the aged, spiraling costs—to name a very few. Many agencies will be receptive to proposals bearing on these problems.

Appearance

A proposal reaches your prospective customer's desk as a "package." Once the wrappings are removed, the contents will trigger a reaction. The initial reaction will be visual and tactile. A firm binding, an attractive cover, clean printing, and standard format will place the reader in a receptive mood. You must avoid "brochuremanship"—conspicuous overdressing, such as using hand-laid paper, half-calf covers, and deckled edges. But there is a happy medium between too much and too little. Neatness, good lay-out, commercial-quality printing, and careful workmanship may tip the scale of judgment in your favor.

Overstatement

Do not promise more than you can do with the time and means available to you. In the flush of excitement while preparing a proposal, your generous impulses may lead you to promise your prospective customer things verging on the miraculous—the moon on a silver platter—all done in record time for mere pennies. "Blue sky" and panacea proposals, fortunately, are rejected by ethical and knowledgeable potential customers, but you are obligated to safeguard your own livelihood and professional reputation. Customer acceptance of unwise proposals has brought financial hardship to many research companies. Don't let it happen to you.

Specifications

You should be aware that some companies and agencies have very strict specifications governing the proposals they will accept for consideration. These specifications may set a maximum on the number of pages, or require certain headings to appear, or stipulate a certain internal arrangement of contents. They may stipulate five copies, one of which must be "reproducible." If the proposal is submitted in response to a general or specific invitation, it may have to arrive by a strict deadline. For their own good—and for yours, too—companies and agencies with strict proposal specifications often are glad and able to furnish you a copy of their proposal style manual. Study it and adhere to the very letter of the detailed specifications.

Sample Proposals

Commercial Problem

The student-written proposal starting on page 358 covers four single-spaced typewritten pages in its original format. It simulates the commercial or business situation. That is, the student author, by projecting her imagination, did all those things she would have to do if working for hire as a researcher. Despite these fictional surroundings, the proposal is realistic and professional.

Figure 15-2   Sample proposal: Commercial Problem

Henley Rug Manufacturing Co.
Textile Consultant Division
1020 Old Meadowcrest Road
King of Prussia, PA 19117

Proposal Letter
117-F-70-11-P

17 October 1979

Dr. Ernest P. Snyder
Director
Woodward Nursing Home
200 South Providence Road
Woodward, PA

Dear Dr. Snyder:

An article in the Philadelphia Inquirer on Sunday, September
27, 1979, dealing with the conversion of the Woodward Orphan-
age into a nursing home caught my attention because of the
scope of the renovations you are contemplating.   I am familiar with the old building, a forbidding structure, and was
pleased to see your sensitivity to the patients' needs, both
functional and aesthetic.   I wondered if you had considered
the installation of carpeting as one way of enhancing comfort
and appearance, and a telephone conversation with your associate, Dr. Blakely, assured me that you had.   It seems,
however, that you are in a quandary as to just what type of rug
to choose.

Technology has complicated the picture with the development
of a variety of colors, textures, patterns, and fiber types.
The average rug retailer is only equipped to offer advice
as to what color, texture, and pattern would best suit your
needs, especially on the basis of appearance, and he knows
little of the advances in fiber development and the properties that each fiber possesses.   There is little incentive
for him to impart the knowledge he does have, since sometimes
the more expensive fibers he wishes to sell receive lower ratings in areas such as flammability.

My company has experience in rug fiber technology and has successfully matched the right fiber type with the individual

Courtesy of Anne Bellegia.

Figure 15-2 (continued)

Dr. Ernest P. Snyder   −2−   17 October 1979

needs of many businessmen and hospitals in the area at great
savings to them, both immediate and long−run.   Therefore, I
submit the following proposal for your consideration.

Proposal

I propose that the Textile Consultant Division of Henley Rug
Manufacturing Co. assess the conditional requirements of the
Woodward Nursing Home and investigate the advantages of the
options available in order to determine which rug fiber type
would be most suitable to your needs.

Background

Rug fiber types consist of natural fibers such as wool and
cotton, artificial fibers such as viscose and acetate rayon,
and synthetic fibers such as nylon, polyesters, and polyac−
rylonitriles, not to mention blends.

Each fiber type has its advantages as to strength; flammabil−
ity; resistance to mildew, inorganic acids, electricity;
and, of course, cost.   Every building, depending on its func−
tion, has its built−in requirements.   If a critical humidity
must be maintained for a certain plant operation, the ab−
sorptivity of the fiber is important.   A hallway receiving
heavy traffic will require a fiber less subject to abrasion.

A mistake in picking the rug fiber for the needs of the estab−
lishment may lead to unnecessary expenses for rug upkeep and
replacement, or, in the case of the tragic fire in the nursing
home in Marietta, Ohio, loss of lives.

Procedure

The method of determining the right fiber type for the Wood−
ward Nursing Home would involve two phases and would take ap−
proximately six weeks.

    Phase 1:   This would consist of consultations with a
              representative of the nursing home and examination
              of the premises to establish any adverse conditions
              to which carpeting would be subjected.

Figure 15-2   (continued)

Dr. Ernest P. Snyder   —3—   17 October 1979

   Phase 2:   This phase would entail investigation into
        the published literature comparing the various fi-
        bers through tests of their physical properties,
        and an analysis of these factors, together with
        cost, in order to arrive at a decision as to the most
        durable and economical fiber for your needs.

A progress report would be submitted halfway through the
study, and the final report would be presented on or before
December 7, 1979.

Cost

The cost of this investigation is minimal, determined solely
by the salaries of the personnel involved in the study.   We
are able to offer such a low cost because we have found that
satisfaction with carpeting has led to its extended usage,
thereby increasing our over-all sales.   We feel confident
that this amount would be far less than what you would save in
rug up-keep and replacement as a result of accepting our help
in choosing a durable carpet.   The cost break-down is outlined below.

                Salary of textile engineer $200
                Salary of textile researcher $150
                Secretarial fee $ 25
                                                    375

This amount would be due within one week following the receipt
of the final report.

Personnel and Qualifications

The team that would be assigned to this project would consist
of William Turner and David McCarthy, whose educational and
professional qualifications are listed below.

   William Turner: B.S. in textile engineering from North
        Carolina State University; three years of textile
        research for American Viscose Division of FMC Corp.;
        five years with Henley.

   David McCarthy: B.S. in textile technology from the
        Philadelphia School of Textiles; two years in the

Figure 15-2   (continued)

Dr. Ernest P. Snyder   —4—   17 October 1979

   physical testing laboratory of E. I. du Pont de
   Nemours & Company, Inc.; three years with Henley.

Feasibility

When considering this proposal you will naturally be concerned with the likelihood of finding a solution.   To date we
have helped seventeen companies, five hospitals, three nursing homes, and even one private home confronted with your di-
lemma.   Our most recent project was to choose replacement
carpeting for the Park View Nursing Home in New Haven, Connecticut.   The original carpet had lasted only two years.   We
were able to provide advice that led to the choice of an inexpensive, but remarkably durable, new synthetic fiber whose
properties indicate a lifetime of five to ten years.

— — — — — — — — — — — —

In order to insure that the rugs for the Woodward Nursing Home
are the most durable, attractive, and practical ones available, I urge you to accept this proposal.   Our investigation
could begin immediately and could be completed in time for
your January opening.   We are open to suggestions for ways to
modify the investigation in order to suit your needs.   In future correspondence please refer to this project by its designated number, 117-F-70-11.

              Respectfully yours,

              Anne Bellegia

              Anne Bellegia
              Director
              Textile Consultant Division

AHB/dls

Figure 15-3   Sample Proposal: Classroom Problem

STATE UNIVERSITY
INTER-OFFICE CORRESPONDENCE

Date: April 8, 1979
From: C. A. Yartz
To: K. W. Houp

This is a proposal for the study and development of a problem
of current interest and relevance in the field of education.
As an associate teacher during the winter term at State University in 1979, I became increasingly interested in and concerned with the current blossoming of advanced placement
programs in public high schools.   This experience, corre-
lated with similar experiences of fellow associate and professional teachers, has led me to believe that there is a need
for an in-depth study of the problem of developing and maintaining an effective and relevant advanced placement program for the public schools.   Because my field of study is
English, I propose to concentrate on the field of English.

Because of your involvement in the field of education, and
English in particular, I felt that you would be interested in
this topic.   With a revolution under way in the educational
network of our nation and the ever-present cry for relevance,
advanced placement courses are snowballing in high schools
across the country.   And because of the complaints heard on
the campus about college freshmen being ill-prepared in
English, steps must be taken to insure that these "advanced"
students get from their courses what they would get from any
basic college course:

In my study I plan to review and analyze existing data pertinent to the subject, sample public opinion of the endeavor,
explore advanced placement programs for their strengths and
weaknesses, consult experts in the field and interview involved teachers and students.   The final report will
include a pilot model of what appears to be the ideal advanced
placement program, and it should take about eight weeks to
complete.   My experience and familiarity with the demands
of college English and with the "typical" high school program
convince me that a study of the problem is urgently needed,
before the advanced placement programs develop to the point

Figure 15-3 (continued)

K. W. Houp   -2-   April 8, 1979

where change will be looked upon with traditional suspicion
and apprehension.

The time for a comprehensive and effective advanced placement
program in the high schools is now, before it is too late.
Without the proper perspective and orientation of the program, it would soon become just another lackadaisical school
effort which will lead to even more poorly prepared college
applicants. My time, experience and efforts are at your disposal and I eagerly await your approval of my proposed project.

Classroom Problem

The brief proposal starting on page 362, which was submitted by
a student, is sufficient for setting up a short-term, limited problem of
several weeks' duration for classroom use, in which student and instructor will work in close cooperation.

By doubling its length and including more particulars, the resulting proposal might be sufficient for handling a nonclassroom problem when researcher and intended customer do not have the opportunity for daily association.

Whether the proposal you write is a one-paragraph memo for
your department head down the hall or a 120-page document for a
governmental agency, it has to be tailored to fit the particular situation. If you have met the person for whom the proposal is intended,
even briefly, this fact will subtly affect your diction (whether or not
you are consciously aware of it). The more known elements you have
to work with, the more confidence you will have in drafting the proposal. Past experience, if any, will also sharpen your instincts, but
there is a first time for everything. Simply do the best you can, in
accordance with your present capabilities.

EXERCISES ▐███████████████████████████

1. Assume that you have been asked by the Community Planning Commission to identify what you believe to be your town's most urgent problem. Here are some likely probabilities:

    Traffic and parking congestion   A larger and more effective police
    Need for new business and   force
    industry   A deteriorated business district
    Housing for low-income families

    Select one of these topics, or some comparable and realistic problem in your community, and then compose a Statement of the Problem in about 300 words.

2. List the means you would use to gather information for a follow-up proposal you would submit to the Community Planning Commission. Consider questionnaires, interviews, and the like. Be as specific as possible.

3. If you were hired to research the problem referred to in Exercises 1 and 2, how would you break down the whole project into tasks? How much calendar time do you estimate you would need to complete the investigation?
    Prepare answers to this double question as your instructor directs.

4. What section titles do you foresee for the final report? List these for submission to your instructor.

5. What sales functions enter into the drafting of proposals? Under what circumstances might the "hard sell" approach backfire?

6. What principles of human psychology govern the physical placement of the Cost or Cost and Method of Payment section in proposal letters? Prepare in writing a sentence or two on each of these principles.

7. In your local newspaper, do you find any examples of proposals, whether or not so labeled? Does the paper contain any news items that might inspire someone to generate one or more proposals in response? If your instructor directs, bring these examples to class.

8. Having studied this chapter, you should be aware of the essen-

tials of preparing a proposal. You may now be asked to prepare a brief proposal on the major course project you intend to undertake.

9. You may then be asked to prepare a full-scale proposal for the same project.

CHAPTER 16

Progress Reports

Time and distance, as well as differences in main interests, normally separate researchers from those who will make use of their findings. An accounting of progress, whether submitted as a bound report or as a letter, helps to keep the customer in touch with the work being done.

The main and obvious function of any progress report is to give the company, department, or individual an accounting of the work that has been done. It explains how you have spent your hours and the customer's money and what you have accomplished as a result of the investment. Though we admit that this purpose is dominant, we must not lose sight of four other purposes that are discharged by a progress letter or report:

1. It enables the customer to check on progress, direction of development, emphasis of the investigation, general conduct of the research. Thus the customer can alter the course of the work before too much time and money have been invested.
2. It enables the researchers to estimate work done and work remaining with respect to the total time and effort available.
3. It compels researchers to shape their material and focus their attention.
4. It provides a sample report that helps both the customer and the researchers to decide upon the tone, content, and plan of the final report.

Several different arrangements are used to report progress. A popular plan for a year-long project is shown in Figure 16-1. A progress letter is sent at the close of work each month, except at the end of

Figure 16-1   Two-Level Method of Reporting Progress

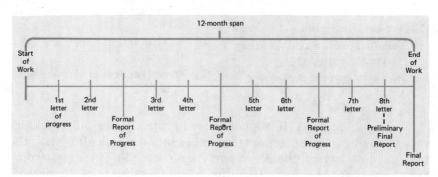

a month closing the first, second, and third quarters of the year. At the end of each quarter except the last, a formal bound report of progress is sent to the customer. This quarterly report recapitulates the main contents of the preceding monthly letters and adds the work done during the month since the last letter. In short, the bound quarterly report is sent instead of a third letter for each of the first three quarters.

Toward the end of a project (about eleven months into a year's project), when affairs often have reached a critical stage, a preliminary copy of the proposed final report may be sent either in addition to or instead of the eighth monthly letter. Usually the prose has yet to be edited and polished; some or all of the illustrations may be roughly sketched or omitted entirely. Because of its ephemeral nature, the preliminary version of the final report is usually reproduced by the least costly means that will give legible copy. The submission of this report enables the customer to react and criticize and thereby get a final report more to his or her liking (see Figure 16-1).

Work is done and progress is made with the passage of time. This fact gives us a useful clue to the basic nature of every progress report, whether or not the particular format makes time the dominant element. However, the progress-with-time philosophy is clearly evident in the following scheme of reporting progress:

1st Progress Report

(1) Introduction
(2) Project Description
(3) Work done in the period just closing
(4) Work planned for the next work period
(5) Work planned for periods thereafter
(6) Overall appraisal of progress to date

2nd Progress Report

(1) Introduction

(2) Project Description

(3) Summary of work done in the preceding period(s)

(4) Work done in the period just closing

(5) Work planned for the next work period

(6) Work planned for periods thereafter

(7) Overall appraisal of progress to date

The opening parts, Introduction and Project Description, seldom vary from one progress report to another in a series dealing with the same project. In fact, these two parts of a succeeding progress report are often created by correcting (updating) these parts of the preceding progress report.

If changes have been made in the contractual agreement, the Project Description (also called Work Statement or Contractual Requirements) has to be updated to accord with the most recent agreement. This description spells out what the researcher is required to do and produce. That is, it represents the total job the researcher is prepared to undertake.

The "work done" sections are of chief interest in most progress reports, for these sections describe, for better or worse, what has been accomplished during the work period (or periods) just closing.

Progress Report Plans

For our purposes here, we can regard every progress report as having three main ingredients in the body: (1) the consumption or passage of time and the order of events throughout that time; (2) the allocation of effort from the task point of view; and (3) the activity or problem point of view. Awareness of these three ingredients suggests to us what we call the "Three T's" of report planning: *time, tasks,* and *topics.* These three plans will now be discussed separately, and afterwards their combinations will be treated.

Time Plan

If the time scheme is used for the overall plan of a progress report, the reader understands what time period is being discussed at any point in the report. Let us look at four possible headings for proof of this point:

Work Previously Done
Work Done in the Period Closing

Work Scheduled for the Next Period
Work Scheduled for Periods Thereafter

Obviously, this is the plan foreshadowed on pages 367–368 of this chapter. This often-used plan has a dynamic character. It gives the recital a cumulative effect. And it offers few problems in obtaining coherent flow from section to section.

We have commented that the time plan tends to be dynamic—that is, it gives the impression that something is being accomplished. Having said this, we should also remark that the time plan may lead to windy generalization and too much emphasis on procedure. But, with all its possible faults, it is widely used and generally has proven effective.

Task Plan

The project description that should be included in every progress report often gives the task breakdown of the entire project. These tasks may be performed at different times, by separate individuals, working at different locations. If this, or something close to it, is the working arrangement, then it is convenient to organize the progress report on the task plan.

When you were younger your mother very likely told you to "straighten up your room." Whether you or she used the term "task," that is exactly what your mother was assigning: a specific action, chore, or undertaking. It is a task to prepare, administer, and tabulate a questionnaire. It is a task to take daily readings of temperature and humidity and to record them. It is a task to construct a model of balsa wood. Because the term is not at all self-defining, you have to take time out to define "task" whenever you use it.

Once you have clearly defined and described the tasks, then the main headings of the body may be simply stated:

Task One
Task Two
Task Three
Task Four

The task plan is realistic and objective. The customer and project supervisor can readily and firmly estimate the amount of work done and remaining on each task. But, by the same token, it throws the glaring spotlight of unfavorable publicity on tasks that are not going well. Also, if the tasks are to be done in sequence, with little or no overlapping, then a given progress report may have solid prose to devote to only one of the tasks, with the other tasks being essentially blank for the time being. Therefore, the task plan seems most appropriate when several or all tasks are being performed concurrently.

Topics Plan

The topical plan ordinarily uses substantive headings such as "Railroad Trackage," "Rolling Stock," "Public Demand." In other words, the topics denote subject areas for investigation.

Railroad Trackage
Rolling Stock
Public Demand
Expected Revenue

Combination Plans

For a well-organized report, the Three-T plans described above can and often must be used in combination. To illustrate, we offer these permutations:

I. Work Previously Done
    A. Task One
    B. Task Two
    C. Task Three
    D. Task Four
II. Work Done in the Period Just Closing
    A. Task One
        -etc.-

I. Task One
    A. Work Previously Done
    B. Work Done in the Period Just Closing
    C. Work Planned for the Next Period
II. Task Two
    A. Work Previously Done
        -etc.-

We feel it would be unwise to offer any one plan or combination of plans as being generally superior to others. We do urge that you collect the information you wish to include in the progress report and then select the plan that will best hold and present the information. And you must always be prepared to try a different plan if the first choice proves unwieldly.

Additional Considerations

The total impression a progress report conveys to the reader is, of course, the important thing. But this total impression is the integrated product of more particular things, including those discussed in the following pages.

Physical Appearance

To put the reader (the client or customer) in a receptive frame of mind, a progress report must be physically attractive. By this we do not mean expensiveness and glamor but rather neatness, appropriateness, and good design. The report should have a protective cover both front and back. The title page should be tasteful and uncluttered. The print (or typing) should be clean and legible. We do not expect diamonds to come in a soiled paper bag, nor do readers expect logic, sound content, and honest performance to come in shoddy wrappings. Nevertheless, all the progress letters and reports arriving from a project should not exceed a very small percentage (perhaps five percent) of the total funding of the project. After all, we are paid to *make* progress and not to linger lovingly over the *reporting* of progress.

Style and Tone

Progress reports are a project's emissaries. If these emissaries seem tired, confused, and unhappy, then what is the customer to think of the workers "back home"? Progress reports therefore should read with vigor, firmness, and authority—one might risk *optimism.* Yet their forcefulness must lie in more than artful writing. Glowing generalizations must be bolstered by recitations of detailed factual accomplishment. Snags, problems, and delays should be admitted, of course, but the accent should be on positive accomplishments. A neat balance between these two aspects of a project will prevent progress reports from reading like either a trail of disaster or an outpouring of giddy optimism. Excess in either direction will always have its day of reckoning.

Originality

The first progress report often leaves much to be desired. For one thing, the project just recently got under way, and whatever progress has been made cannot yet be crystallized. For another thing, the first progress report, lacking precedent, seems relatively tentative and experimental. The next two or three progress reports usually represent a substantial improvement over the first, for they have accomplishments to report and they profit from earlier experience. However, after the third or fourth progress reports of a series, the reports tend to hit a plateau or go downhill. A feeling of ennui and repetition may be disturbing to authors. This slacking-off must be prevented at any cost. Bringing in new blood to the writing staff may help. An oral project review in the auditorium may help. But a clear recognition of the need to maintain reading interest, verve, and originality is a necessity. Keep out of the rut. Do not simply warm over last month's progress report like the proverbial Sunday roast. Take a fresh look at the whole problem of reporting progress.

Accomplishment and Foresight

Your customer will find it heartening to learn all you have accomplished on his behalf. Past performance is probably the most reassuring promise of future performance. Yet investigators should not seem to be moving blindly into the future work periods, like an automobile driver about to run off the margin of his only road map. Therefore progress reports, while stressing what has been done, should give adequate attention to plans for the future. For the next work period, the plans should be firm and detailed; for work periods thereafter, the plans may understandably be less specific. Showing your plans reveals you to be workmanlike, and also makes it possible for your customer to suggest modifications. Getting your customer into the act usually works to everyone's benefit.

Exceeding Expectations

A progress report should give your customer a pleasant but mild surprise. Notification that a new task has been started a few days before its scheduled beginning, three or four graphic aids, a technical appendix, some noticeable improvements in format are useful ways of cheering your customer without undue labor or expense on your part. On the other hand, avoid sudden and excessive novelty, such as using multiple overlays and color printing that have no precedent in previous progress reports of the project. Again, do not increase the length of the previous progress reports by more than, say, twenty percent, for your customer may suspect that you are trying to "cover up" and divert attention. Most important of all, do not include a set of firm conclusions or ultimate recommendations in a progress report, for you will thereby steal the fire from your final report. At the worst, your customer may collapse the project before its normal completion date.

Ratio of Fact to Procedure

Every progress report should be dynamic; that is, it should honestly reflect the movement, vigor, and advance of the investigation it reports and of the investigators who are doing the work. The energy built into the report will communicate itself to the customer and assure the customer that the researchers are not drowsing on the tails of their spines. Yet there is an equally important necessity for factual content. Support the enthusiasm with a body of information consisting of quantities, numbers, names, dates, specifications, and concrete fact.

Therefore you should watch the ratio of fact to procedure while writing and rewriting a progress report. Twenty percent factual content is about minimum. Some of this will be packed rather densely into some sections of the report; the rest of it will be distributed as opportunity suggests.

Sample Progress Report

The six-page progress report (see Figure 16-2) that begins on page 375 was submitted by a student. Notice that this progress report, as indicated by the centered headings, is organized by the topical plan.

EXERCISES

1. In what ways would a diary of activities (log or journal) benefit the researcher when the time came to compose a progress report? To what other uses might the researcher or a superior put such a diary?

2. Referring to Figure 16-1, can you explain in your own words why the first progress report for a project is often relatively poor? Why the second and third progress reports tend to be very good? Why the progress reports thereafter tend to decline in interest and readability?

3. What section or sections of the usual progress report hold the greatest interest for the project sponsor or customer?

4. In outlining a progress report, what considerations should guide you in selecting the time plan? The task or topics plan?

5. What elements of human psychology enter into the composition of progress reports?

Figure 16-2   Sample Progress Report

RAILROAD CONSULTANTS INCORPORATED
827 OHIO AVENUE
CHICAGO, ILLINOIS 60010

119–F–66–18
9 November 1979

Mr. George C. McClellan
Executive Vice President
Treasurer and General Manager
Bellefonte Central Railroad Company
Bellefonte, Pennsylvania 16823

Dear Mr. McClellan:

On 29 October 1979 RCI received the Bellefonte Central's formal acceptance of proposed study 119–F–66–18 concerning the
profitability of the Bellefonte obtaining a steam locomotive
and conducting weekend excursion trips on its right–of–
way.   Following the signing of the contract, research was
started on the initial phases of the study.   As stipulated in
our agreement, RCI now submits a report on the progress that
has been made during the period from 16 October to 7 November
1979.

We have completed what I consider a successful outline for the
necessary research.   The work has been segregated in the following manner:

> Phase 1: Revenue Problem of Bellefonte Central; Steam
>   Locomotive and Passenger Coach Availability; Route
>   Location and Scheduling

> Phase 2: Public Interest

Work has progressed well on all components of Phase 1.   The
bulk of the work remaining to be done is concerned with public
interest, although preliminary studies have been done on all
parts of the project.

Figure 16-2   (continued)

Mr. George C. McClellan   -2-   9 November 1979

WORK STATEMENT

Purpose

The purpose of project 119-F-66-18 is to determine the feasibility of the Bellefonte Railroad obtaining a steam locomotive and running weekend excursion trips as a means of
supplementing its loss of revenues as a result of volume rate
reductions on lime traffic.

Procedure

To complete the study successfully it will be necessary to determine the exact nature of the Bellefonte Central's revenue
problem, examine the availability of the necessary equipment
for such operations, and analyze the public interest in such
excursion railroads.

REVENUE PROBLEM

Work Completed

The library research and interviewing concerning the revenue
problem of the Bellefonte Central are almost complete.   Our
research team has found that the Bellefonte Central currently
serves two prime functions.   One of these is the haulage of
lime and limestone away from the National Gypsum Company
plant in Buffalo Run Valley.   This is the source of about 90%
of the railroad's tonnage.

Although the volume of traffic handled by the Bellefonte Central increased approximately 30% over 1977, freight revenues
were almost the same as in the previous two years.   This
paradox was a result of further expansion of volume rate con-
cessions affecting a large part of the lime traffic, without
compensatory increased traffic.

The minimum tonnage requirement for the volume rate was originally 50,000 tons from the three plants in the vicinity of
Bellefonte.   Now the requirement refers to 50,000 tons from
anywhere in the United States.   This has been the major cause
in the decreasing revenues of the Bellefonte Central.

Figure 16-2   (continued)

Mr. George C. McClellan   -3-   9 November 1979

Work Remaining

The work remaining in this area is mainly organization of
material concerning the revenue problem into a logical, pre-
sentable order for the final report.

STEAM LOCOMOTIVE AND PASSENGER COACH AVAILABILITY

Work Completed

Several sources have been considered which could possibly
have steam locomotives and passenger coaches suitable for use
on excursion lines.   From the library research thus far it
appears that other shortline railroads who have used excursion trips as a source of revenue have either used old coaches
of their own or purchased them from other railroads or scrap
dealers.

The expense involved in leasing such equipment is high; also
the Pennsylvania Railroad, from whom the Bellefonte Central
could most easily lease equipment, cannot make passenger
equipment available on weekends according to an interview
with Mr. Henry Scott.   The expenses include a dead head
charge between Philadelphia and Bellefonte of over $25 per
passenger car plus a certain charge for running the cars in a
revenue-producing service.

There are some steam locomotives available in Pennsylvania.
Mr. Steven Kovalchik is a scrap dealer with his main operation in Indiana, Pennsylvania.   The Luria Brothers Scrap
Company of Modesta, Pennsylvania is another possible source
of steam locomotives.

ROUTE LOCATION AND SCHEDULING

Work Completed

This is the section of Phase 1 which our research team is currently examining.   Sunset: The Magazine of Western Living
has been particularly helpful in providing material on the
schedules of successful excursion railroads.   From these
schedules tentative schedules for the Bellefonte Central are

Figure 16-2   (continued)

Mr. George C. McClellan   –4–   9 November 1979

being formulated.   After an interview with Dr. Hadley Waters, concerning the points of interest along the Bellefonte
Central's right-of-way, a tentative route was set up which
would have the train leaving State College and making a round
trip only to Waddle, Pennsylvania approximately eight miles
away by rail.   A map of this intended route has been enclosed
for your benefit.

Work Remaining

The final schedules and routes cannot be determined until a
public interest study has been run.   Also an attempt is being
made to obtain a ride over the Bellefonte Central right-of-
way to help determine the most interesting aspects of the
route.

PUBLIC INTEREST

Work Completed

Although work has not been formally started on this phase of
the project, some research has been done on past excursion
trips of the Bellefonte Central.   Thus far, the trips seem to
have had considerable participation.

Work Remaining

The final analysis of public interest will be an incorporation of data on other successful excursion railroads and results of past excursion trips by the Bellefonte Central.

The cost of acquisition of a steam locomotive has not been as-
certained nor have even approximate costs of operation of the
equipment.   Letters of inquiry to the Association of American Railroads and certain successful excursion lines should
aid our research team in this area of the study.

Figure 16-2   (continued)

Mr. George C. McClellan   −5−   9 November 1979

OVERALL APPRAISAL

Because of a lack of cost data concerning steam locomotives by the Bellefonte Central, figures from similar operations such as the Arcadia and Attica Railroad may have to be adapted for our purposes.   The quantity of literature pertaining directly to the problem which is available to our research team is somewhat limited, but the sources which have been located have been very informative.   Although we are awaiting replies to correspondence sent to Dr. Thomas J. Sinclair, Association of American Railroads, and W. D. Midcap, Brotherhood of Locomotive Fireman and Engineman, on vital aspects of the study, the results of interviews with Dr. John Coyle, Dr. Hadley Waters and Mr. George McClellan have been very rewarding.   Considering our findings to date we at RCI are confident that we will meet the proposed schedule for submission of the final report.

— — — — — — — — — — — —

Should there be any questions regarding our research, RCI will gladly aid you by answering them based on the results thus far.

Respectfully yours,

A. B. Orme, Jr.

A. B. Orme, Jr.
Coordinator Class II
and Short Line Department

ABO : wmo

Enclosure

Figure 16-2 (continued)

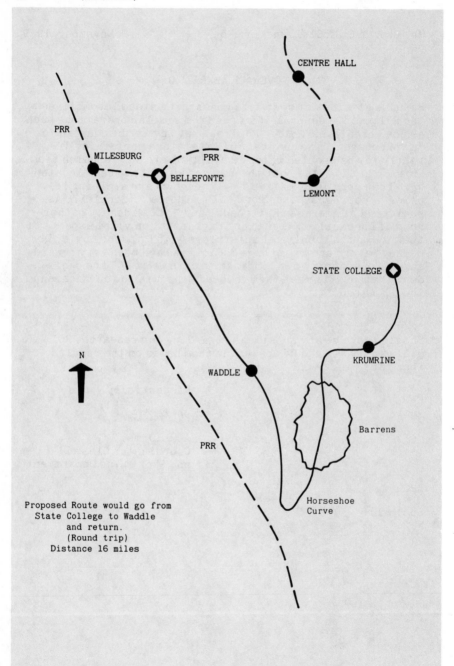

CENTRE HALL

PRR

MILESBURG

PRR

BELLEFONTE

LEMONT

STATE COLLEGE

N

WADDLE

KRUMRINE

Barrens

PRR

Horseshoe
Curve

Proposed Route would go from
State College to Waddle
and return.
(Round trip)
Distance 16 miles

CHAPTER 17

Physical Research Reports

What are the principal traffic and parking problems in your town? What pesticides are most effective in preventing moth damage to clothing stored over winter? What are the best times of year to seed a lawn? Does frequent switching on-and-off of your television set really shorten the life of the picture tube?

To get answers to questions of this kind, you can do two things:

1. You can summarize the answers that previous researchers obtained, by whatever means they used.
2. You can obtain first-hand answers for yourself, by direct physical experimentation and observation.

Of course, you can—and probably should—use these two means in combination. However, in studies and reports of the kind treated in this chapter, the emphasis falls upon physical experimentation.

An illustration may help to clarify our point. Suppose that we have a chunk of glass, crude and irregular, dumped out of the ladle and unmolded. We desire to find the impact strength of the chunk of glass; that is, how many pounds of force will be required to shatter it. We may approach our solution in two ways:

1. We may read up on the chemical makeup of the chunk of glass. We may measure its geometric properties. We may pass white light through it to obtain a reading of its internal structure. By turning to suitable handbooks we may then estimate the minimum impact force required to shatter the chunk.

2. We can whack the chunk with a hammer, hitting harder and harder until
   it shatters. A pressure gauge or similar accessory will tell us how hard we
   had to hit to get the result we wanted. This is the pragmatic test, pure
   and simple.

The physical research study places the emphasis on the second approach.

Physical research—whether or not it is followed by written or
oral reports—is very common in our daily lives. Take the bachelor,
keeping house for himself, as a prime example. Suppose he has just
had a new picture window installed, 5 feet high and 12 feet wide.

Since buying a home six years ago, he has become a faithful
reader of the "Household Hints" column in the *Morning Mirror*. He
tries to recall what the contributors to the column have had to say
about cleaning large windows. What did his parents do? He recalls a
mish-mash of advice: Start at the bottom. Start at the top. Use a
circular cleaning motion. Use broad horizontal strokes. Never clean
with the sun on the glass. Never clean the inside when it's dark outside; you'll be sorry in the morning. Squeaky glass is clean glass.
Scouring powders will dull the surface. For guaranteed results, call
your Handi-Dandi Window Cleaning Service. Helpful or not, this
body of "theory" runs through his head as the time comes to clean the
new picture window.

What cleaning agents should he use?

A commercial window spray?
Detergent and water?
A powder such as *Belle de Jour?*
Ammonia and vinegar diluted with water?

And what should he use to wipe it? His cabinet under the sink holds
an extensive array of likely tools: natural sponges, synthetic sponges,
rolls of cotton gauze, paper towels, old Turkish towels, torn bed
sheets, and chamois skins.

Though he may never state his objectives in so many words, he
senses them keenly. He wants to keep the cleaning costs down. He
wants a window that is clear and sparkling clean. Washing windows
is not his favorite pastime, so he wants the glass to *stay* clean as long
as possible. And he wants to get the job done quickly and efficiently,
avoiding spillage, fuss, and excitation of any allergies he may have.

Being a conscientious homeowner, he now tries out the various
cleaning agents, wiping materials, and application techniques. It's a
long haul. Every cleaning session is followed by hard stares at the
window and by stocktaking and debate. He concludes that anything
is better than nothing but that everything is less than perfect.

During coffee break next day, he may enlighten his colleagues about the pros and cons of cleaning picture windows: what left streaks and what didn't, how much "elbow grease" he had to use, whether lint had to be flicked off the shiny glass, and so on.

In this homely illustration, we have passed through the major stages involved in all physical research, whether it costs a dime or a million dollars:

Introduction	Results
Literature review	Discussion
Materials	Summary and Conclusions
Methods	

These functions constitute the heart of the physical research and the central portion of any written or oral report that may be produced. When put into prose, however, such a report would probably be expanded and then arranged very much as shown here:

Abstract	Methods
List of illustrations	Results
List of symbols	Discussion
Introduction	Summary and Conclusions
Literature review	References
Materials	

The parts that have been added are treated fully elsewhere in this book. Therefore only the parts in the first list warrant detailed treatment here.

The Routine of the Physical Research Report

At hand we have a bona fide physical research report. Its title: "The Electrical Prediction of Local Thunderstorms." For a student report it is fairly long, encompassing seven Roman-numbered preliminary pages and twenty Arabic-numbered pages. Throughout this chapter we shall use this report as a "running" illustration. Although the original report has twelve parts, we shall limit our discussion to the seven parts that constitute the heart of the physical research report.

Introduction

Introductions to physical research reports seldom have subheadings. However, analysis of these reports reveals that the contents of the introduction could be sorted and placed under these headings:

Reasons for the investigation
Some historical background
Purpose and objectives of the investigation
Scope and limitations of the work
Plan of development (if unconventional)

In other words, the introduction provides the reader with a backdrop against which the new research is to be seen and appraised.

Our student researcher devoted three paragraphs to his introduction, all of which we reproduce here:

> The purpose of this project was to determine whether measurement of the changes in air-to-earth potential gradient could be used as a reliable and practical method of predicting local thundershowers.
>
> The gold-leaf electroscope has been familiar to physicists for many years as an instrument for detecting static charges. The development of the vacuum tube has made possible refinement of the electroscope so that it is now capable of detecting the minute currents generated in the atmosphere by the formation of thunderstorms.
>
> This project was concerned only with determining whether a simple version of the electrometer could be used alone to predict thundershowers. No attempt was made to use more complicated electrometers, nor was any attempt made to use the electrometer in conjunction with other meteorological forecasting equipment.

Literature Review

It is a foolish thing to begin any physical research as though the world had been created yesterday. We must build on the past, and then move through the present into the future.

In the Literature Review section you summarize your professional knowledge and gleanings from the literature and bring all of this to bear upon the problem at hand. Mathematical models, algebraic computations, chemical equations, physical laws, prior discoveries, and the like are presented for the reader's illumination and understanding of the present problem.

Our student author (a senior in electrical engineering) therefore went to the literature to find what previous researchers had already learned. His list of references contains 13 items, and his report contains 175 lines of information excerpted and paraphrased from the literature. Here we reproduce one paragraph of his Literature Review section:

The surface of the earth and the highly conducting layer of the ionosphere can be regarded as two charged parallel surfaces. The outer surface has a net positive and the inner one a net negative charge, and the two are separated by the atmosphere. (1:1) The normal potential gradient is one in which the atmosphere is positive with respect to the earth. (2:36) In regions of the atmosphere which do not lie in the direct fields of thunderstorms, only vertical gradients exist normally at ground level. (1:4) Above ground level, the gradient is three-dimensional. (3)

All too commonly, the Literature Review section of physical research reports gives a welter of conflicting opinions and hypotheses. Experts often do disagree, of course, and when this occurs you should never hesitate to say so, as a matter of "truth in reporting." But having cited the conflicts, you have the further duty to try to resolve them, or at least to assess them for relative validity. The guide who takes readers into the jungle must bring them out again.

In ten lines the author of "Electrical Prediction" has presented us with basic information on the electrostatic makeup of the earth and its immediately surrounding atmosphere.

Why is this information pertinent to his research problem? To discover this we have to go back to his introduction, in which he states that his purpose is to determine whether a particular device (the simple electrometer) is capable of detecting the approach of local thunderstorms. It should be clear, now, that the information cited in the paragraph quoted above is critical to his problem, for his electrometer will have to detect and measure changes in the electrostatic forces.

Materials

In the Materials section you describe and specify the equipment you used, and also your reasons for using it. Sometimes you may also describe the equipment you have previously used, might have used alternatively in the present instance, or will use at a later time.

The student who performed the physical research described here drove a van-load of equipment to Burlington, Massachusetts, where he spent ten weeks working with the problem. Although his report uses seven pages to describe his equipment, we shall limit our excerpts to two of his paragraphs:

The radioactive collector consists of a long steel wire suspended about 20 feet above the ground. It is necessary to highly insulate the wire from ground because of the overall high effective impedance (10^{10} ohms) of the collector. (6) A radioactive ionizing unit is attached near the center of the wire. Polonium ionizers with a radioactive intensity of 500 or 1000 microcuries are most commonly used.

The electrometer is an instrument for detecting or measuring a potential by means of the mechanical forces exerted between electrically charged bodies. (9) The vacuum tube electrometer is a modern version of this instrument which detects potentials by means of the electrical forces exerted on electrons flowing in a highly specialized vacuum tube. Tubes of this type have been developed which permit the measurements of currents as low as 50 electrons per second. (9)

Understand that the full original version includes numbers, formulas, and graphs. Although we spare you these details, be sure to realize that physical research reports should leave little or nothing to the reader's imagination. All such research should be reproducible, and to reproduce the work of others you must have all the technical and procedural details spelled out in full.

Methods

In the Methods section you describe the conditions under which you performed the experiment, the safeguards you practiced, the manner in which you set up and operated the equipment, and the way you recorded and analyzed the results.

Having described the hardware he would employ, our student researcher was ready to install his equipment and put it into operation. Here, in two paragraphs, he covered the methods he used:

> The tube and resistors were mounted inside a 2 in. by 2 in. by 4 in. aluminum case, which provided a complete light and electrostatic shield. Light shielding is necessary to prevent photo-electrons from being emitted in the tube, while electrostatic shielding prevents the action of external electric or magnetic fields on the tube current. (5:268)
> The batteries were taped together and connected by shielded cable to the aluminum box. The recording microammeter was also connected to the box by a shielded cable.
>
> After the wiring was completed, the box was immersed in absolute methyl alcohol and scrubbed. This was done to remove any dirt or fingerprints that might have been deposited during construction. Impurities on the surfaces of the tube, insulators, and resistors can cause erroneous meter readings. (11)

To exemplify the precautions he took while making measurements, we reproduce all of his section entitled "Data Compilation."

> After the installation of the electrometer had been completed, the unit was left on for 24 hours with the input short-circuited to ground. It was noted that the output level varied slightly with ambient temperature, but the overall stability of the instrument was satisfactory.

The collector was then connected to the input, and the chart from the recording <u>microammeter</u> was calibrated in time and signal intensity.   Accurate time reference was obtained from a WWV receiver, while a 6.3 volt dry cell attached across the input was used as a polarity and intensity reference. This calibration was repeated daily throughout the entire data-gathering run.

Weather observations were made once a day and logged. Continuous observations were logged during violent weather activity.   The electrometer chart was checked frequently during the afternoons for any unusual fluctuations which might indicate the presence of a nearby thunderstorm.

Results

In the Results section you set down the data you recorded, usually in tabular or graphic form, together with all necessary prose explanation.

From day to day during his ten-week stay in Burlington, our student researcher recorded his findings in log books. Many minor decisions had to be made. His equipment had to be recalibrated daily. The work was delicate and demanding, and yet the results seem sparse. Essentially, this is all he learned:

TABLE I – COMPARISON OF RESULTS

	quantity
Thunderstorms detected by electrometer	5
Thunderstorms detected visually	5
Thunderstorms occurring within 100-mile radius of site during 10-week period	5
Thunderstorms detected by electrometer before being detected visually	2
Chart traces resembling thunderstorm traces	27

There is a saying among physical researchers that nature is reluctant to give up her secrets. The handful of data above, however, slight as it may appear, is noteworthy in that it had never before been obtained as far as the literature reveals.

Discussion

In the Discussion you review and analyze the data presented in the preceding section. The discussion is one of the most important sections of physical research reports—and, unfortunately, often one of the most poorly handled. You should therefore take special pains with this section. Try to step back from your work and findings and reflect upon them before putting the whole story together. Our student researcher takes this objective approach in the paragraphs quoted here:

> Continuous potential gradient measurements were taken at Burlington, Massachusetts, during a ten-week period and were correlated with visual observations of thunderstorms which occurred over Burlington during the same period.
>
> The simple electrometer used to make the measurements was found to be capable of detecting and measuring the currents produced by increases in the negative potential at the site. The electrometer detected all five of the surely bona fide thunderstorms which entered the test area during the period of investigation. Two of these five were detected by the electrometer before they were detected by visual means. In twenty-seven additional instances, the electrometer gave readings indicative of thunderstorms but storms did not develop so that they could be detected by visual means.
>
> It is evident that meteorological disturbances such as rain can cause electrometer indications quite similar to those obtained for thunderstorms.

In physical research reports the discussion sometimes does little more than translate numbers into words. Justification for this practice may lie in the fact that all report users are not equally capable of digesting data presented in tabular or graphic form. At other times, and more commonly, the discussion must do far more than provide a prose recapitulation. Are there blanks in the data? Were some of the data "projected" from actual test data? Were the data subjected to mathematical processing? Do the data contain puzzling irregularities? Were all the data taken with the same equipment? Do the data, when considered in their entirety, lead to a significant conclusion? Were there recognized inaccuracies in measuring and recording? Points such as these must be dealt with in the discussion. Analysis and interpretation are frequently needed to qualify and illuminate "raw" data.

Summary and Conclusions

In the Summary section you extract and extrapolate from the preceding sections of the report what you consider the outcome and final worth of the investigation. You may comment on the reliability and validity of the results, and you may suggest new approaches to the problem for later use. These points are illustrated in this excerpt:

> It should be borne in mind that the test results were obtained under less than optimum conditions: the series of tests was conducted within a single period of ten weeks, no alternate equipment was available, and the peak thunderstorm season had passed.
>
> On the basis of the data obtained, however, it appears that the electrometer used in the present study is not a reliable indicator of approaching thunderstorms, largely because it gives many false indications. Nevertheless, further experimentation with electrometers appears to be warranted. If this is done, electrometers of other types and parameters should be tried, and they should be employed in conjunction with other meteorological devices, including barometers and wind gauges.

Some Additional Considerations
in Reporting Physical Research

Every program of physical research necessitates imaginative and painstaking design from beginning to end. Yet despite the most rigorous planning, mistakes and oversights are almost sure to occur. Every researcher, after finishing work on a problem, would like to start over and do it differently. In the full report on electrical prediction of local thunderstorms there are many intimations that the researcher wished he had been in Burlington at a more favorable time of year, that he had used a more sophisticated electrometer, and that he had taken along other storm-sensing devices. For reasons like these, most research is done not once but several times, to remove the "bugs" from earlier equipment and procedures, to rectify mistakes previously made, and to confirm earlier findings. Like the bachelor in our window-washing illustration, researchers must approach the same task from different angles to achieve the most satisfactory result.

In this chapter on the conduct and reporting of physical research, we have tried to give you general, overall advice, with no particular field in mind. You should understand, nonetheless, that both major and minor adjustments are necessary if you move from testing tire traction on wet pavements for the Department of Transportation, to determining the cause of whirling disease in trout for the Bureau of Fisheries, to analyzing and reporting the reasons for student dissatisfaction in the school cafeteria. In other words, every department or problem has peculiarities that require adaptation on the part of the research-writer. We therefore urge you to carefully examine representative journals, reports, and other literature before entering into physical research in any field with which you are not well acquainted.

You may wish to consult the index of this book, under the following headings, for pages with more information pertaining to physical research reports: Abstract, Discussion, Introduction, List of Illustrations, List of Symbols, Literature Review, Materials, Methods, References, Results, Rhetorical Methods, and Summary.

EXERCISES

1. Physical research as treated in this chapter is primarily concerned with *physical determination* (i.e., fact-finding). Do you see any similarity between physical research and feasibility studies (Chapter 18)? In what major respect are they different?

2. Select a physical research problem with which you are thoroughly familiar. It might be the testing of garden soil for pH or removing stains from upholstery or balancing a checking account. In outline form, and using the seven steps listed on page 383, prepare this material for presentation to someone who is unacquainted with the problem.

3. Referring to Chapter 3 of this book, "Gathering Information," determine as well as you can how the methods one uses to gather information are affected by the nature and purpose of the investigation.

4. Why do reports of physical research usually contain references to the literature on the problem being studied?

CHAPTER **18**

Feasibility Reports

At all levels of human activity, from the individual engrossed in personal and domestic problems to the highest level of policy-making in government, we live in a decision-making society.

After retiring for the night, homeowners may lie awake for an hour or so debating whether to have the roof reshingled this spring or to risk waiting till next year. Should they continue using fuel oil for heating or convert to gas or electricity? Moving up a notch, the town council may debate whether to simply resurface Main Street or to rebuild it from the ground up. On a still higher plane, the state legislature may argue the questions of instituting a state lottery, reapportioning the electoral districts, changing from a flat-rate income tax to one that is graduated. Washington debates the question of whether to recognize a new foreign government when the old one has been overthrown.

Without feasibility studies our world might not come to a screeching halt, but most surely we would waste a lot of time and money. As this chapter was being written, a news article appeared in the morning newspaper (see Figure 18-1) demonstrating that our highly organized technological society demands feasibility studies before time, personnel, and dollars are committed to any sizable project.

Answers to questions like those above cannot be looked up in any book. They have to be threshed out on the basis of available information. The analysis requires logic, hard work, intelligence, and freedom from personal bias. In most instances, also, those who do the research must take human motives and benefits into account.

Figure 18-1   Article Illustrating Need for Feasibility Studies

Await airport study, planners tell airline

STATE COLLEGE – Allegheny Airlines should make no decision about relocating at Mid-State until PennDOT's airport study is complete.

That was the recommendation from Centre Regional Planning Commission Thursday night. PennDOT officials have promised this area first priority in the study with at least preliminary answers in about six months.

Objecting to the proposed transfer of commuter services from University Park Airport to Mid-State Airport in Philipsburg, Charles Hosler, dean of the college of Earth and Mineral Sciences at Penn State, said at least 100 university staff members use the service regularly.

He also said there is a possibility of a major research laboratory locating in the Nittany Valley and availability of commuter service at the University Park Airport would be factor in the location decision.

Waiting for completion of another study was part of a second recommendation from the planners.

The commission went on record saying the area municipalities should take whatever action needed to keep the financially-troubled Fullington bus service in operation until the regional transportation and transit study is completed in about two years.

≈

STATE COLLEGE – Should the Appalachian Thruway go through Bald Eagle Valley, or Nittany Valley, or on the side of Bald Eagle Valley, or nowhere at all in Centre County?

Charles Hosler of Penn State's Earth and Mineral Science Department told the Centre Regional Planning Commission that there is a pollution problem in the Bald Eagle Valley "at least 100 days" of the year and that the problem is acute 50 to 60 days of the year.

He said because the valley is so narrow and so deep, cold air from the Allegheny plateau descends into the valley, fills it up at night and leaves the warm air above the valley. Pollution mixes with the cold air, Hosler said, and since the warm air is above it cannot escape.

The Nittany Valley is wider and not so deep and pollution is dispersed more readily, Hosler said. The lower the point in the Bald Eagle Valley, the worse the problem, he said.

If the thruway must be located in the Bald Eagle corridor it might be better to locate it along the mountain range at a higher elevation where winds and other elements could more easily disperse the pollution, Hosler said.

Commission member Pierce Lewis asked district PennDOT engineer Bruce Speegle if it would be a "viable alternative" to consider no highway at all. Speegle said that would be a possibility.

Meanwhile, PennDOT officials in Harrisburg are preparing to spend approximately $50,000 and one year to study the problem of where to put the thruway.

The federal Environmental Protection Agency must approve PennDOT's final decision before federal funds are granted to the project.

From *Pennsylvania Mirror.*

What, actually, do we mean by *feasibility* studies? We have explored this idea in detail on pages 349–350. In essence, a "feasible" plan is one that can be carried out with a reasonable degree of success. Thus, a "feasibility report" discusses the practicality of a given project in physical and economic terms; in addition, the report considers the *suitability* of the project from the viewpoint of those who will be affected by it.

Of all types of investigation, the feasibility report is the most demanding, because you have to reach a decision. Having reached your decision, you state it as a recommendation that some action be taken or not taken; that some choice be adopted or rejected. You have to supply all the facts required to support your recommendation. The logic of your reasoning must be inescapable from start to finish. You have to be free of any bias that might taint your recommendations. Because the feasibility study is critical from start to finish, you have to be clearheaded at every turn, and the first test comes when you formulate and state the purpose of your study.

Before you do any research, define the precise purpose of all the work you will do. Often a single sentence is ideal for the purpose statement:

> The purpose of this investigation is to determine whether X Company of Old Town should establish a branch plant in New Town.

An announcement such as this may seem easy and self-evident, perhaps superfluous. But many an investigator has floundered around and eventually bogged down simply because he did not clearly and consciously formulate the objective toward which he was striving.

If you do not know what you are trying to accomplish with a feasibility study, you have no sensor to tell you when you are on or off the right track. Get on the right track at the start and keep on it by constantly reminding yourself of the purpose.

Here are some additional examples of purpose statements:

> The purpose of this study is to determine the feasibility of using particle board to sheathe the interior of house boats.

> The purpose of this investigation is to select the best of several promising methods of obtaining the major supply of fresh water for the Pink Sea Horse Motel.

> Our primary objective is to decide whether it would be more profitable to market by-product X-acid in bulk or to convert it on site into commercial fertilizer.

Inexperienced investigators are inclined to resent and reject the notion that their work should have a stateable and stated purpose. They are satisfied, rather, "to learn all they can about such-and-such a subject." This kind of open-ended purpose is sometimes proper for private study, but it will not serve in a practical situation. Here an extended illustration will help to clarify our point.

One student investigator proposed to her instructor in technical writing that she be permitted to investigate the fossil spore content of a cranberry bog. "What bog do you have in mind?" her instructor asked. The student looked pained, but under pressure disclosed that she had no particular bog in mind. Her instructor immediately sent her to the Department of Geology to learn the name and location of some particular cranberry bog certified to contain genuine fossil spores. The Department of Geology was entirely cooperative and within the hour came up with the name and location of a bona fide cranberry bog in the vicinity of New Bedford, Massachusetts. Both student and instructor were pleased, for visible headway had been made.

"Now," the instructor asked, "why do you want to investigate the fossil spore population of this bog?" His student looked more pained than before. After some moments of cogitation, she volunteered this information: "Well, I want to find out what kinds of fossil spores are in this bog and how many there are."

"But what are you going to *determine?*" her instructor pressed.

"Why, just what I said," the student replied, "what kinds are there and how many."

"I'm afraid that won't do," said her instructor. "The effort would be interesting and informative, I'm sure, but you are really describing an activity rather than stating the purpose of that activity. The real question hinges on what you will do, or recommend that someone else do, as a result of what you learn."

It was a sultry August afternoon; minds were working none too well and tempers were brittle. But by 5:15 the student was at last equipped with an acceptable purpose, as follows:

> The purpose of my project is to decide whether Mrs. Rose's cranberry bog on the southern outskirts of New Bedford contains an adequate fossil spore population so that the Department of Geology should acquire exploratory rights over the bog for field research by its graduate students.

Equipped with this "hard" purpose, the student would not waste her time beating around cranberry bushes and plowing knee-deep in watery bogs. Rather, she was committed to making a decision, settling a question. In other words, she was now furnished with an outcome to be accomplished instead of a nebulous description of a messy activity.

Once the purpose of an investigation has been clearly and exactly decided, a fairly tight logical process has been set in motion. Certain other decisions follow with near-inevitability. Ordinarily, the next major decision hangs on this question: Given my stated purpose, what must I do to accomplish it? The term "scope" is often applied to this concept. Whatever it is called, the decision consists of determining the actions to be taken, the range of data to be delved into, the bounds to be set to the problem, the emphasis to be made. To illustrate, let us use a previous example of purpose statement to find what it implies in terms of scope:

Purpose
Should X Company of Old Town establish a branch plant in New Town?

Scope
Does X Company now have, or can it develop, enough business in New Town to justify a branch there?

What would be the advantages and disadvantages of operating a branch plant in New Town?

Does New Town offer adequate physical facilities, utilities, and other services for plant operation there—office space, transportation, communications, etc.?

Can the required staff be obtained, whether by local hiring or moving personnel into the area, or both?

Are local business practices and codes, tax structure, etc., favorable for conducting business there?

What impact, for better and worse, would opening a branch plant in New Town have upon overall company organization, operations, policy, financial condition?

Necessary as it is to generate and compile the chief items of the scope statement early in an investigation, we would be dishonest to pretend either that the process is easy and forthright or that first efforts will produce a scope statement which will hold unchanged throughout the investigation. Devising a scope statement requires insight, foresight, and hindsight—in other words, speculative thinking. But the ability to create serviceable scope statements improves with practice and with experience gained in the subject matter.

In any case, the investigator-reporter should not remain blindly committed to the initial statement, but should reexamine it from time to time, looking for "holes," overlaps, superfluous items, and the like. Frequently, another person, one who is unacquainted with the study to date, is in a far better position than the originator to spot shortcomings and illogicalities in the statement. Thus, someone outside the study should be asked to review and react to the list of scope items compiled by the investigator.

Logic of the Feasibility Report

When well-conceived and well-executed, the feasibility report is inexorable in its logic. Figure 18-2 shows key portions of the logic. In the purpose statement, you state your intentions, and this statement becomes a binding contract between you and your readers. In your scope statement (sometimes combined with the purpose statement), you indicate what you will do and sometimes how you will do it. In other words, you set bounds to the project and spell out the means to be used within these bounds. In the body portion you present the data you have collected, together with tables and figures and your own analysis and explanation. Toward the back of the report you provide the reader with a digest of the contents, either in the form of a list of conclusions (highlights) or a prose summary. In the recommendations you tell the reader (client or customer) what action you believe should be taken based on the data you have presented.

Routine of the Feasibility Report

In its oldest and most traditional arrangement, the feasibility report contains twelve or more parts:

Title page	Introduction
Abstract	Body chapters
Summary	Conclusions
Contents page	Recommendations
List of illustrations	References
Glossary of terms	Appendix(es)

In practice, however, you will encounter several departures from the list above. Sometimes the abstract and summary are combined as an informative summary (see page 187). At times the list of illustrations is omitted because the report contains few illustrations or none at all. At other times the list of illustrations is followed or replaced by a list of symbols, as needed. On occasion, the summary (introductory) takes the place of the conclusions section or follows it, toward the end of the report, as a terminal summary. The appendix is optional.

We wish that we could be more definite in describing these parts, but the fact is that the feasibility report is—and must be—a flexible instrument. Within the range of variation allowed by current practice, we make whatever adaptations are necessary in view of the problem at hand.

Figure 18-2

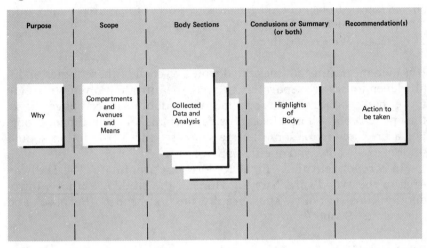

Purpose	Scope	Body Sections	Conclusions or Summary (or both)	Recommendation(s)
Why	Compartments and Avenues and Means	Collected Data and Analysis	Highlights of Body	Action to be taken

We shall confine our discussion to four of the sections listed above: introduction, body chapters, conclusions, and recommendations. The other eight parts are parallel to those described in Chapters 9 and 10.

Introduction to the Feasibility Report

Depending upon the length and complexity of the report, the introduction may range from a single paragraph to several pages. However, the introduction should be limited to essential preliminaries. Avoid dallying with the introduction simply because you are reluctant to get down to work in the body of the report. Readers, understandably, become annoyed if they are "teased" by an unduly extended introduction.

The topics listed here should always be *considered* for inclusion in the introduction:

1. Subject matter of the research and the report
2. Purpose of the investigation and the report
3. Reasons for conducting the research
4. Identification and characteristics of the person or company performing the work
5. Historical background of the problem
6. Extended definition or description of the problem
7. Scope or extent of the work performed
8. Any limitations imposed upon the study

9. Procedures and methods employed in the research
10. Acknowledgement of indebtedness to others (if not given in an earlier preface or foreword)
11. Preview of the report that follows

It is evident that if ten lines are devoted to each of these topics and if all eleven topics are separately treated, the introduction might run to more than a hundred lines or about three double-spaced pages. If so, the entire report should be extensive, perhaps a hundred pages long. For a brief report of some ten pages or so, the introduction should seldom exceed one and one-half pages.

In a report entitled "Final Report on Carpeting of the Hallway Area of the Woodward Nursing Home," the student author's introduction consisted of two paragraphs, the first headed "Purpose" and the second "Scope."

PURPOSE

The purpose of this report is to present the results of a study undertaken by the Textile Consultant Division of Henley Rug Manufacturing Co. to determine the most suitable carpeting for the hallway areas of the Woodward Nursing Home.

SCOPE

The study was designed to establish the criteria the carpeting would have to meet as a result of its placement in the hallway of the nursing home: flammability, wear life, static generation, soil resistance, stain resistance, wet cleanability, and cost. Once the criteria were delineated, the four components of carpets—fiber, construction, backing, and underlay—were analyzed separately to see what types of each of these four elements best enabled the carpet as a whole to meet the criteria. Because the study was concerned with contract carpeting, the research team considered only fibers, construction, backing materials, and underlays typically used in the manufacture of commercial grade carpets.

Breaking down the purpose paragraph, we find that it contains these four elements:

1. Purpose of the report
2. Identification of the company doing the study
3. Purpose of the study
4. Identification of the subject matter

It is difficult to see what the author would have gained by including additional topics under the purpose heading. The prose is businesslike and economical, and yet reads easily. Nothing that appears later in the report would have benefited by expansion of the purpose section.

Notice that the scope paragraph is about two and one-half times the length of the purpose paragraph. Analysis of the paragraph discloses that it incorporates these three elements:

1. Statement of the need to establish criteria
2. Statement of the need to measure the components of carpeting against the criteria
3. Announcement that the study was restricted to "contract" carpeting—a limitation placed on the coverage

In these three steps, the author set forth the rationale of her study. The three sentences are interlocking and progressive. Though the paragraph is headed "Scope," it plainly does double or triple duty in that it depicts the author's logical procedure and also previews the organization used in the body of the report.

A few moments of reflection should be enough to convince you that without the enlightenment provided by the introduction the body chapters might be unnecessarily puzzling upon first reading. Like all good reporters, the student put herself in the reader's place.

Body

The first chapter of the carpeting report supplies these headings:

I. DEFINING THE SUITABILITY OF A CARPET
    A. General Procedure for Determining Suitability
    B. Procedure for Determining the Suitability Criteria for the Woodward Nursing Home
       1. Flammability
       2. Wear Life

3. Static Generation
4. Soil Resistance
5. Stain Resistance
6. Wet Cleanability
7. Cost

The outlined development of this chapter is "classical" in that it moves from the general to the particular. The prose gives a closely reasoned and persuasive explanation of how the seven criteria were arrived at. These characteristics come through clearly in the first paragraph of the chapter:

> The term "suitability" in reference to carpeting is not an absolute. It is meaningless to discuss the suitability of a carpet except in regard to the conditions to which it will be exposed. Each carpet, because of its fiber, construction, backing, and underlay, has certain properties that make its performance and cost under existing conditions different from that of another carpet of dissimilar fiber type, construction, backing, and underlay. In other words, one could never say a woven wool carpet with jute backing and a sponge rubber underlay is inherently better than a tufted nylon carpet with a foam back and an all-hair underlay; superiority depends on whether conditions demand, for example, a rug with a long wear life or one with an elegant appearance.

In this paragraph, as elsewhere, the student was clearheaded and perceptive. Her readings in the subject, fortified by her common sense, enabled her to detect and appreciate the many influences and considerations involved in the selection of such apparently mundane stuff as carpeting. As witness, we quote the paragraph setting up her third criterion:

> 3. Static Generation. The Woodward Nursing Home is located in an area which has a low relative humidity for several months of the year. Carpets with a tendency to generate

static will do so during these months, producing sparks.
Electric sparks are merely annoying in a residence, but they
can be deadly in a building that has highly volatile gases
on hand as a nursing home does (4:65). A nursing home,
therefore, should choose a carpet with antistatic properties
in order to eliminate the danger of explosions.

As already noted, the chapter began with a general section previewing what was to follow. Therefore we should not be surprised to find a "wrapping up" or summation paragraph closing the chapter:

After establishing these seven criteria, the research
team and the nursing home staff agreed that in order to be chosen, all four components of the carpet—fiber, construction,
backing, and underlay—would have to be at least acceptable
in all seven categories (3).

This paragraph also serves to link this chapter to one that follows. The second chapter of the report has these headings:

II. DETERMINING THE SUITABILITY OF VARIOUS FIBERS
    A. Types of Fibers Considered
    B. Acrylic, Nylon, and Wool in Regard to the Criteria

Only two paragraphs, totaling eight lines, appear under the "A" heading, but in this brief section the writer set the stage for all that follows in the chapter. This section makes two major points: (1) only three types of fiber merit consideration in the present study and (2) the type of fiber is crucial in selecting the particular carpet judged by the seven criteria given in the first chapter. Because they act as a critical link in the argument of the report, these two paragraphs are reproduced here:

A. Types of Fibers Considered
Technology has made available a wide range of fibers for
use in carpeting. This study compares only acrylic, nylon,

and wool, because these three fibers are virtually the sole
competitors in the field of contract carpeting with which we
are concerned (6).

If everything else is equal, fiber content is an extremely important factor in comparing two carpets. It influences how well the carpet does in all seven categories.

B. <u>Acrylic, Nylon, and Wool in Regard to the
Criteria</u>

Fiber content is an important determinant of the relative
flammability of carpets. The flame of a match is well over
1000° F, and hot tobacco ashes are over 500° F. Acrylic fibers stick at between 400°–500° F and burn readily. The fire
hazard of this fiber could only be reduced by blending it with
modacrylic fibers which soften at 300° F, but which are flame
resistant. Nylon sticks at around 445° F and melts at around
480° F. Although it does not flame, it smolders and could
reach other material such as drapes and furniture, igniting
them. Wool is the only fiber that burns slowly without smoldering and is self–extinguishing (7:12–3). On the basis of
flammability, nylon is completely unacceptable, acrylic is
acceptable only when modified, and wool is inherently
nonflammable.

The following two and one-half pages of the chapter weigh the
pros and cons of the three types of fiber. These pages require close
reading, for the merits and drawbacks of the fibers must be weighed
in relation to the client's needs. Fortunately for us, the writer did
her best to put the whole story together in this prose passage followed
by a table:

Overall cost analysis indicates that nylon is a poor investment at any price, because it is unacceptable in the area
of flammability. Acrylic and wool fibers are comparably

priced; the added expense of blending acrylic with modacrylic
for the sake of flammability control is counteracted by its
lower maintenance cost.   The following table summarizes the
results of the preceding discussion.

Comparative Behavior of Acrylic, Nylon, and Wool Fibers

	Acrylic	Nylon	Wool
Flammability	*acceptable	unacceptable	acceptable
Wear Life	high	extra high	high
Static Generation	medium	high	high
Soil Resistance	high	medium	high
Stain Resistance	high	high	medium
Wet Cleanability	high	high	high
Overall Cost	reasonable	unreasonable	reasonable

*when blended with modacrylic

Again, as we see here, the writer summarized the main sense of
the chapter rather than leaving this vital task to the reader. Thus
Chapters I and II reinforce the old three-part moral: (1) tell the reader
what you will say, (2) say what you have to say, and (3) reiterate what
you have said. This principle of rhetoric is hackneyed, perhaps, but it
is one of the most useful workhorses of exposition.

The third chapter of our illustrative report is set up in this manner:

III. DETERMINING THE SUITABILITY OF THE AVAILABLE CON-
STRUCTIONS
    A. Defining the Elements of Carpet Construction
    B. Rating the Suitability of Various Constructions

Without the preliminary definitions and explanations provided
by the one-paragraph "A" section given below, most readers could
not make sense of the material under "B." Using the stage-setting
technique relied upon in Chapters I and II, the writer prepared the
reader as follows:

A. Defining the Elements of Carpet Construction
Carpet construction usually refers to the way the carpet

is put together.   It includes primarily the body construction, surface construction, pile density, and pile
height.   Modern machine-made carpets are available in five
body constructions: three woven types (Velvet, Wilton, and
Axminster), a knitted type, and a tufted form.   The surface
construction of a carpet refers to whether the surface pile
has been left as loops or cut, and whether the yarn is twisted
or straight.   Pile density can be measured by the weight of
pile yarn per unit surface area, while the pile height is the
thickness of the wear layer measured from the top of the pile
to the top of the back (7:1-3).

In the next two pages the writer delved into the complicated pros and cons of body construction, pile density, and pile height. She closed her third chapter with the following paragraph of summation, which would be very useful to most readers:

Overall cost analysis suggests that a densely constructed tufted carpet with high, straight yarns would be the
least expensive carpet.   The tufting process helps to lower
the initial cost resulting from the dense, high pile, and
those two factors sharply reduce the long-range cost.   It is
a matter of debate as to whether the reduced long-range cost
of the level loop construction would counteract the reduced
maintenance cost of the cut pile form.   If wool or acrylic
fibers composed the carpet, however, the cut pile construction would lose its advantage, because these two are already
inherently soil resistant and easily cleaned.

Having found an effective pattern of organization, the author used it again in the next-to-last chapter: (1) preview, (2) detailed discussion, and (3) summation.

In the fifth and final chapter our author completed the journey she announced in her introduction—here with consideration of carpet underlays.

As you finish her final chapter you should, theoretically, be able to predict the outcome of her study, for you have been presented with practically all of the evidence available to her. However, it is asking too much of most readers to compel them to integrate all that has previously been said. Only the researcher-author knows the material first-hand; weeks of living with the subject bestows familiarity and confidence that no outsider can possibly have. Without the special help that only the author can give, most readers could not with certainty come up with the logical answer to the selection of carpeting for the Woodward Nursing Home.

Reporting the Outcome

In the body chapters all the pertinent and concrete evidence has been presented. But what does it all add up to? The informed and dedicated readers might be able to draw their own conclusions, but the researcher-author occupies a position of special advantage: complete familiarity with the material and its ramifications. Understand, too, that the author contracted to produce a feasibility study report. By so doing, she assumed the obligation to report the outcome, in the form of conclusions and recommendations. Readers may agree with the outcome as stated, but they expect the author to "go on record."

In its full logical form, the outcome of a feasibility study is reported in three stages:

Factual Summary	Conclusions	Recommendation
Two or more facts given in meaningful context.	Interpretive generalizations.	Action urged as a consequence of conclusions.

Let us turn aside briefly from the carpeting report to investigate the rationale for the framework above. Suppose your parents have bought an isolated cottage, far from public power supply, and need access to common household current to operate a stove, refrigerator, radio, clock, and lights. An electric generator seems like a good solution, and you volunteer to investigate several brands and come up with a recommendation as to the make and model they should purchase. Of course, you have to consider many variables, including out-put wattage, reliability, frequency stability, availability of parts and service, and cost.

Let's take the last item, cost. After consulting catalogs and suppliers' literature you learn these facts:

Model A retails for $425
Model B retails for $385
Model C retails for $465

In the body of the report you might conceivably write for your parents' benefit, you would set down these three facts dealing with cost. In a long report, the three facts might appear in different chapters; in a brief report or letter, the three facts might appear in different paragraphs. Later, however, in reporting the outcome of your study, you should bring the three facts together in a meaningful context:

> *Factual Summary*
> Of the three models studied, Model B, retailing at $385, costs the least.

Factual Summary

In summarizing the facts you would be wise to adhere to these six rules:

1. You can summarize only that information appearing elsewhere in the report. A factual summary must not introduce new or additional information.
2. Every statement should present something you learned, or importantly confirmed, during the course of your own investigation. Do not present generally known and accepted facts, such as that the formula for water is H_2O. However, under the provision "importantly confirmed" you may need to verify facts in dispute or to refine and update existing data. (On this basis, the old nautical mile of 6080 feet was recalculated at 6076 feet.)
3. Every statement should be a genuine assertion of *fact*. That is, it should be free of personal opinion. Unless this rule is observed, you "muddy the waters" and offer an unsure footing for conclusions to follow.
    No: Every American home should have at least one and one-half baths. (Who says so? This statement sets up an arbitrary standard of the author's own making. Avoid preferential and opinionative wording such as *ought, should, better, best.*)
    Yes: In 1970 some 6.9% of the year-round housing in the United States lacked basic plumbing facilities (*World Almanac*, 1972, p. 147).

4. Preferably, every statement should be a *composite* statement of facts. A series of factual assertions, clearly related as you see them, may very well puzzle readers. They may not be able to integrate the assertions to arrive at the net sense. On page 408, for example, we offered three separate but related assertions:

> Model A retails for $425
> Model B retails for $385
> Model C retails for $465

Then, to obtain a composite statement integrating these three assertions, we used this wording:

> Of the three models tested,
> Model B, retailing at $385, costs the least.

Do your readers a kindness. Put it all together for them.

5. Every statement should pave the way for one or more conclusions to follow. The only valid reason for summarizing facts is to use them as a foundation for conclusions.

6. Order your summarizing factual statements so that the first is most basic and remote. Lead your readers ahead in sequential and cumulative fashion. The order in which you pick up facts when you reread the body of your report is often a poor order for presenting them in your factual summary. Probably you will have to reword, rearrange, and otherwise process the raw facts. But don't change their meaning!

In professionally prepared reports, factual summaries (by whatever name they are called) may appear in any one of several different places. Sometimes they are simply written as a few paragraphs of text. Or they may be set up as a series of numbered statements, with one or more sentences per item. Summaries may be placed early in a report, perhaps after the abstract or in place of it. If a report arranged in the "traditional" plan includes anything that might be called a summary, it usually follows immediately after the body chapters. A summary may also be identified as "highlights," "findings of the study," and so on.

Conclusions

As shown in the schematic on page 408, conclusions act as the intermediate step or bridge between the factual summary statements and the recommendation(s) yet to come. Formulating conclusions, given adequate facts, is usually a straightforward process. For the electric generator problem, no mental acrobatics are needed to move from one step to the next:

<table>
<tr><td>

Factual Summary
Of the three
models studied,
Model B, retailing
at $385, costs the
least.

</td><td>

Therefore
⟶

</td><td>

Conclusion
On the basis of
least cost, Model
B is to be pre-
ferred.

</td></tr>
</table>

All conclusions, whether simple or complex, are the products of diagnoses, often guided by previous experience; they are most reliably made by experts with knowledge and experience. Yet, for better or worse, the average citizen is required to draw dozens of conclusions—some both complex and crucial—every day.

The carpeting report, as another example, offers sixteen conclusions. Two are reproduced here:

1. The carpeting for the hallway area of the Woodward Nursing Home must be acceptable in all areas: flammability, wear life, static generation, soil resistance, stain resistance, wet cleanability, and cost. (pp. 2–4)

* * * * * * * *

11. Doubling the pile height alone can double the service life of a carpet, while doubling the density can increase the service life four times. (p. 11)

For the sake of logic and reader comprehension, you should try to arrange your conclusions in a clear and premeditated order. The last conclusion, ideally, should serve as a good stepping-off platform for the first (sometimes only) recommendation to follow.

Do not insist that the 1-2-3 order of events in the factual summary be exactly paralleled in the conclusions. Do not insist that each statement in the factual summary be represented by one and only one conclusion. The truth is that a single conclusion usually embraces two or more factual summary statements—and often not consecutive ones, at that.

In complicated reporting situations, it may prove helpful to group conclusions under headings such as "Major" and "Minor" or "General" and "Particular." This kind of decision may seem harrowing at the time, but it illustrates the process that separates the excellent reporter from those who are mediocre or poor.

Recommendations

In contrast to conclusions and implications, a recommendation is an *action* statement. That is, it urges the customer to take some proposed action or to refrain from taking it. *Most important, the first (or only) recommendation discharges the purpose set forth early in the report.*

If the purpose is to determine whether X Company of Erie, Pennsylvania, should establish a branch in Centre County, Pennsylvania, the recommendation (first or only) should read:

> On the basis of the research performed and reported here we recommend that X Company establish (*or* not establish) a branch plant in Centre County.

It would be futile to note simply that "having one or more branch plants is beneficial to some companies." This generalization would be of no use to X Company.

Any further recommendations usually implement the first. If the first recommendation favors establishment of a branch plant, succeeding recommendations may pinpoint the exact town or give a date for completion of the plant. If the first recommendation opposes the establishment of a branch in Centre County, succeeding recommendations may propose alternative locations or urge that the whole question be reinvestigated several years hence.

Recommendations are opinions: professional judgments. No logic, nor array of evidence—nothing under the sun—can make them anything else. Let them therefore be sound and well considered.

The recommendations offered by the researcher-reporter may or may not be accepted by the customer or report user. However, for the feasibility study report the author is obliged to arrive at one or more recommendations. If the research has been properly conducted, then these recommendations should carry great weight.

Again, our simple generator problem shows how the reporter moves from conclusions to recommendations:

Conclusion		*Recommendation**
On the basis of least cost, Model B is to be preferred.	Therefore ⟶	On the basis of least cost, we recommend Model B.

* The ultimate recommendation, all facts considered, might swing to Model A or Model C.

In a much more complicated situation, the author of the carpeting report made five recommendations, including these two:

The Textile Consultant Division of Henley Rug Manufacturing Co. recommends

1. That the fiber content of the carpeting for the hallway of the Woodward Nursing Home be wool treated for static control.
2. That the Woodward Nursing Home select a carpet by Lees, called "Accomplishment," that meets all of the above requirements and costs $15 per yard.

Some Additional Considerations in Feasibility Reporting

In a feasibility report, and of course in the underlying research, the purpose statement is critical in that it acts as the sensor and censor in keeping the researcher-reporter on the right path and headed toward the predetermined objective. Always, the purpose implies a selection or choice, including these varieties:

To act or not to act
To act here or somewhere else
To act now or sometime later
To act by this means or some other means
To decide who should act
To decide whether existing information warrants further study of the question

In every instance, the purpose statement (an implied question) must be worded with particular care, for it serves as the nucleus of the contract that the researcher-reporter must carry out.

On page 398, we displayed the time-tested and still most popular arrangement of parts for feasibility reports. However, another arrangement is now being increasingly used because it is better adapted to the report user's point of view:

1. Title page
2. Abstract
3. Contents page
4. List of illustrations
5. Glossary of terms
6. Introduction
7. Factual Summary
8. Conclusions
9. Recommendations
   —colored separator page
10. Annexes
11. References

Under this modified arrangement, the summary of the old-style report is renamed "Factual Summary" and appears immediately after the introduction, as shown above. More strikingly, the former body chapters are hauled out of central position and placed as "Annexes" after the colored separator page.

This revised plan, sometimes referred to as the "psychological" or "administrative" plan, is designed to simplify the readers' task. Unless they are particularly conscientious persons, they have to read only a few pages of factual summary rather than fifteen or more pages of detail in the body chapters. Further, the presence of the colored separator page invites them to ignore the annexes.

Figures 18-3 and 18-4 will enable you to directly compare the two plans for feasibility reports.

The administrative plan has both advantages and disadvantages. Certainly, it makes the author's job more difficult, for it focuses the reader's attention upon the few pages that precede the annexes. Therefore, these few pages must be very carefully prepared. Some conscientious readers, too, may be annoyed by having the detailed discussion (annexes) relegated to appendix position, an apparent downgrading of this essential part. Determining which plan you should use in a given situation is basically a matter of audience analysis (see Chapters 4 and 12).

For more information on the various parts of a feasibility report, you may wish to refer to the pages listed under the following headings in the index: Abstract, Conclusions, Introduction, List of Illustrations, List of Symbols, Process, Purpose, Recommendations, References, Scope, Summary. Appendix A is an example of a complete feasibility report.

EXERCISES

1. The feasibility study is a necessary prelude to any important action. In your local newspaper identify any news items that call for feasibility studies prior to action. Is your community considering any actions that should be preceded by feasibility studies? (See Figure 18-1.)

Figure 18-3   Conventional Plan

CONTENTS

2. Prepare a slip like the one shown here, but substitute your own
    subject matter following the four colons:

                                                  Your name
    1. General Field: Meteorology
    2. Specific Topic: Short—term weather forecasting
    3. Purpose: To determine the feasibility of making 24—hour
       forecasts for specific areas to expedite road crews for
       highway storm treatment
    4. Customer: A state Department of Transportation

Figure 18-4   Administrative Plan

CONTENTS

ANNEXES

3. Using the slip you prepared for Exercise 2, determine the "scope" of your treatment:
    a. Topics or compartments of information to be researched.
    b. What methods you will use to acquire the necessary information (see Chapter 3).
    c. Any limits or constraints you feel you should impose upon your investigation.

4. Rough out an introduction to your proposed report (see pages 399–401).

5. How many and what kinds of illustrations do you anticipate for inclusion in your final report (see Chapter 11)?

6. Which of the two plans given on pages 414–415 do you believe would be most appropriate for your use? Why?

CHAPTER **19**

Oral Reports and Group Conferences

You will make oral reports for many reasons. You will have to report committee work, laboratory experiments, and research projects. You will give reports to learned societies. You will instruct, if not in a teacher-student relationship, perhaps in a supervisor-subordinate relationship. You may have to persuade a group that a new process your section has devised is better than the present process. You may have to brief your boss about what your department does to justify its existence. You will participate in group conferences.

In this chapter we divide the speech process into two parts: preparation and delivery. Under *preparation* we tell you how to organize and practice oral reports. Under *delivery* we concentrate on the physical aspects of oral reports—voice and body movement—and upon interacting with your audience. Following our discussion of the speech process, we include material about the use of visual aids in speaking, and about group conferences.

Oral Report Preparation

In many ways preparing an oral report is much like writing a paper. Researching an oral report and a paper are identical processes. You will use the rhetorical forms we explained in Chapters 5 and 6— exposition, narration, description, and argument—as much in speaking as you do in writing.

You will analyze your audience as we described in Chapter 4. In addition, find out as much as you can about the physical layout. Inquire about the size of the room you will speak in, and the size of the audience. If you have to speak in a large area to a large group, will a public address system be available to you? Find out if you will have a lectern for your notes. If you plan to use visual aids, inquire about the equipment. Does the sponsoring group have projectors to show 35 mm slides or vugraphs? More than one speaker has arrived at a hall and found all of his vital visual aids worthless because projection equipment was not available. Find out if there will be a chairman to introduce you. If not, you may have to work your credentials as a speaker into your talk. Consider the time of the day and day of the week. An audience listening to you at 3:30 on Friday afternoon will not be nearly as attentive as an audience earlier in the day or earlier in the week. Feel free to ask the sponsoring group any of the above questions. The more you know beforehand, the better prepared and therefore the more comfortable you will be.

Choosing Delivery Techniques

A vital part of your preparation is choosing your delivery technique. There are four basic techniques, but you really need to think about only two of them. The four are (1) impromptu, (2) speaking from memory, (3) extemporaneous, and (4) reading from a manuscript. Impromptu speaking involves speaking "off the cuff." Such a method is too risky for a technical report where accuracy is so vital.

In speaking from memory, you write out a speech, commit it to memory, and then deliver it. This gives you a carefully planned speech, but we cannot recommend it as a good technique. The drawbacks are (1) your plan becomes inflexible; (2) you may have a memory lapse in one place that will unsettle you for the whole speech; (3) you think of words rather than thoughts, which makes you more artificial and less vital; and (4) your voice and body actions become stylized and lack the vital spark of spontaneity.

We consider the best methods to be extemporaneous speech and the speech read from a manuscript, and we will discuss these in more detail.

The extemporaneous speech. Unlike the impromptu speech, with which it is sometimes confused, the extemporaneous speech is carefully planned and practiced. In preparing for an extemporaneous speech, you go through the planning and organizing steps described in Chapter 7. But you stop when you complete the outline stage. You do not write out the speech. Therefore, you do not commit yourself to any definite phraseology. In your outline, however, include any vital facts and figures that you must present accurately. You will want no lapses of memory to make you inaccurate in presenting a technical report.

Before you give the speech, practice it, working from your outline. Give it several times, before a live audience preferably, perhaps a roommate or a friend. As you practice, fit words to your outlined thoughts. Make no attempt to memorize the words you choose at any practice session, but keep practicing until your delivery is smooth. When you can go through the speech without faltering, you are ready to present it. When you practice a speech, pay particular attention to timing. Depending upon your style and the occasion, plan upon a delivery rate of 120–180 words per minute. Nothing, *but nothing*, will annoy program planners or an audience more than to have a speaker scheduled for thirty minutes go for forty minutes or an hour. The long-winded speaker probably cheats some other speaker out of his or her allotted time. Speakers who go beyond their scheduled time can depend upon not being invited back.

We recommend that you type your outline. Use capitals, spacing, and underlining generously to break out the important divisions. But don't do the entire outline in capitals. That would make it hard to read. As a final refinement, place your outline in a looseleaf ring binder. By so doing you can be sure that it will not become scattered or disorganized.

There are several real advantages of the extemporaneous speech over the speech read from manuscript. With the extemporaneous speech you will find it easier to maintain eye contact with your audience. You need only glance occasionally at your outline to keep yourself on course. For the rest of the time you can concentrate on looking at your audience.

You have greater flexibility with an extemporaneous speech. You are committed to blocks of thought, but not words. If by looking at your audience you see that they have not understood some portion of your talk, you are free to rephrase the thought in a new way for better understanding. If you are really well prepared in your subject you can bring in further examples to clarify your point. Also, if you see you are running overtime, you can condense a block by leaving out some of your less vital examples or facts.

Finally, because you are not committed to words, you retain conversational spontaneity. You are not faltering or groping for words, but neither are you running by your audience like a well-oiled machine.

The manuscript speech. Most speech experts recommend the extemporaneous speech above reading from a manuscript. We agree in general. However, speaking in a technical situation often requires the manuscript speech. Papers delivered to scientific societies are almost always written and then read to the group. Often the society will later publish your paper. Often technical reports contain complex technical information or extensive statistical material. Such reports do not conform well to the extemporaneous speech form and you should plan to read them from a manuscript.

Planning and writing a speech is little different from writing a paper. Follow the advice in Chapters 2 through 8. However, in writing your speech try to achieve a conversational tone. Certainly in speaking you will want to use the first person and active voice. Remember that speaking is more personal than writing. Include phrases like "it seems to me," "I'm reminded of," "Just the other evening, I," and so forth. Such phrases are common in conversation and give your talk extemporaneous overtones. Certainly, prefer short sentences to long ones.

Type the final draft of your speech. Just as you did for the extemporaneous speech outline, be generous with capitals, spacing, and underlining. Plan on about three typed pages per five minutes of speech. Put your pages in order and place them in a looseleaf binder.

When you carry your written speech to the lectern with you, you are in no danger of forgetting anything. Nevertheless, you must practice it, again preferably aloud to a live audience. As you practice, remember that because you are tied to the lectern, your movements are restricted. You will need to depend even more than usual on facial expression, gestures, and voice variation to maintain audience interest. Do not let yourself fall into a sing-song monotone as you read the set phrases of your written speech.

Practice until you know your speech well enough to look up from it for long periods of good eye contact. Plan an occasional departure from your manuscript to speak extemporaneously. This will aid you to regain the direct contact with the audience that you so often lose while reading.

Organizing Report Content

For the most part you will organize your speech as you do your written work. The guidelines we laid down in Chapters 7 and 12 for organizing written work apply here as well. However, the speech situation does call for some differences in organization and even content, and we will concentrate on these differences. We will discuss the organization in terms of introduction, body, and conclusion.

Introduction. A speech introduction should accomplish three jobs: (1) create a friendly atmosphere for you to speak in, (2) interest the audience in your subject, and (3) announce the subject, purpose, scope, and plan of development of your talk.

Be alert before you speak. If you can, mingle and talk with members of the group to whom you are going to speak. Listen politely to their conversation. You may pick up some tidbit that will help you to a favorable start. Look for bits of local color or another means to establish a common ground between you and the audience. When you begin to speak mention some member of the audience or perhaps a previous speaker. If you can do it sincerely, compliment the audience. If you have been introduced, remember to acknowledge the speaker and thank him. Unless it is a very formal occasion, begin rather informally. If there is a chairman and a somewhat formal atmosphere, we recommend no heavier beginning than, "Mr. Chairman, ladies and gentlemen."

Gain attention for your subject by mentioning some particularly interesting fact or bit of illustrative material. Anecdotes are good, if they truly tie in with the subject. But take care with humor. Avoid jokes that really don't tie in with the subject or the occasion. Forget about risqué stories.

Some speakers try to startle their audience, but there are pitfalls to this method. Shortly after World War II, one of the authors of this text began a speech by pretending to be Adolf Hitler for a few minutes. (The purpose of the speech was to expose the wickedness of the big lie in propaganda.) He raged and ranted and denounced Wall Street, Franklin Roosevelt, and the Jews. The audience froze into resentment, and he never got it back. When he stopped his play acting and explained why he had imitated Hitler, most of the audience never even heard him—they were already too angry.

Be careful also about what you draw attention to. Do not draw attention to shortcomings in yourself, your speech, or the physical surroundings. Do not begin speeches with apologies.

Announce your subject, purpose, scope, and plan of development in a speech just as you do in writing. (See pages 192–195.) If anything, giving your plan of development is more important in a speech than in an essay. Listeners cannot go back in a speech to check on your organization the way that a reader can in an essay. So, the more organizational guideposts you give an audience, the better. No one has ever disputed this old truism about speech organization: (1) Tell the audience what you are going to tell them. (2) Tell them. (3) Tell them what you just told them. In instructional situations, some speakers provide their audiences with a printed outline of their talk.

Body. When you organize the body of a speech, you must remember one thing: a listener's attention span is very limited. Analyze honestly your own attention span—be aware of your own tendency to let your mind wander. You listen to the speaker for a moment, and then perhaps you think of lunch, of some problem, or an approaching date. Then you return to the speaker. When you become the speaker, remember that people do not hang on your every word.

What can you do about the problem of the listener's limited attention span? In part, you solve it by your delivery techniques. We will discuss these in the next section of this chapter. It also helps to plan your speech around intelligent and interesting repetition.

Begin by cutting the ground you intend to cover in your speech to the minimum. Build a five-minute speech around one point, a fifteen-minute speech around two. Even an hour-long talk probably should not cover more than three or four points.

The beginning speaker is always dubious about this advice. He thinks, "I've got to be up there for fifteen minutes. How can I keep talking if I only have two points to cover? I'll never make it." Because of this fear he loads his speech with five or six major points. As a result, he lulls his audience into a state of somnolence with a string of generalizations.

In speaking, even more than in writing, your main content should be masses of concrete information—examples, illustrations, little narratives, analogies, and so forth—supporting just a few generalizations. As you give your supporting information, repeat your generalization from time to time. Vary its statement, but cover the same ground. The listener who was out to lunch the first time you said it may hear it the second time or the third. You use much the same technique in writing, but you intensify it even more in speaking.

We have been using the same technique here in this chapter. We began this section on the speech body by warning you that a listener's attention span was short. We reminded you that your listening span was short: same topic but a new variation. We asked you what you can do about a listener's limited span: same topic with only a slight shift. In the next paragraph we told you not to make more than two points in a fifteen-minute speech. We nailed this point down in the next paragraph by having a dubious speaker say, "I've got to be up there for fifteen minutes. How can I keep talking if I have only two points to cover?" In the paragraph just preceding this one we told you to repeat intelligently so that the reader "who was out to lunch the first time you said it may hear it the second time or the third." Here we were slightly changing an earlier statement that "You listen to the speaker for a moment, and then perhaps you think of lunch. . . ." In other words we are aware that the reader's attention sometimes wanders. When you are paying attention we want to catch you. Try the same technique in speaking, where the listener's attention span is even more limited than the reader's.

Creating suspense as you talk is another way to generate interest in your audience. Try organizing a speech around the inductive method. That is, give your facts first and gradually build up to the generalization that they support. (The deductive method states the generalization first and then supports it.) If you do this skillfully, using good material, your audience hangs on wondering what your point will be. If you do not do it skillfully or use dull material, your audience will tune you out and tune into their own private worlds.

Another interest-getting technique is to relate the subject matter to some vital interest of the audience. If you are talking about water pollution, for example, remind the audience that the dirtier their rivers get, the more tax dollars it will eventually take to clean them up.

Visual aids often increase audience interest. Remember to keep your graphics big and simple. No one is going to see typewritten captions from more than three or four feet away. Stick to big pie and bar graphs. If you have tables, print them in letters from two to three inches high. If you are speaking to a large group, put your graphic materials on transparencies and project them on a screen. Prepare your transparencies with care. Don't just photocopy typed or printed pages or graphics from books. No one behind the first row will see them. To work, letters and numbers on transparencies should be at least twice normal size. Large-type typewriters are available. If you need an assistant to help you project visual aids, get one or bring one with you.

Do not display a visual aid until you want the audience to see it. While the aid is up call your listeners' attention to everything you want them to see. Take the aid away as soon as you are through with it. If using a projector, turn it off whenever it is not in use. Be sure to key every visual aid into your speaker's script. Otherwise you may slide right by it. (See also the section, "Visual Aids," pages 430–442.)

Conclusion. In your speech conclusion, as in your essay conclusion, you have your choice of several closes. You can close with a summary, or a list of recommendations including a call for some sort of action, or what amounts to "Good-bye, it's been good to talk to you." We stress again here what we said in Chapter 9 and add two points. In speaking, never suggest that you are drawing to a close unless you really mean it. When you suggest that you are closing, your listeners perk up and perhaps give a happy sigh. If you then proceed to drag on, they will hate you.

Second, remember that audience interest is usually highest at the beginning and close of a speech. Therefore, you will be wise to provide a summary of your key points at the end of any speech. Give your listeners something to carry home with them.

Report Delivery

After you have prepared your speech you must give it. For many people giving a speech is a pretty terrifying business. Before speaking they grow tense, have hot flashes and cold chills, and experience the familiar butterflies in the stomach. Some people tremble before and even during a speech. Try to remember that these are normal reactions, for both beginning and experienced speakers. Most people can overcome them, however, and it is even possible to turn this nervous energy to your advantage.

If your stage fright is extreme, or if you are the one person in a hundred who stutters, or if you have some other speech impediment, seek clinical help. The ability to communicate ideas through speech is one of humanity's greatest gifts. Do not let yourself be cheated. Winston Churchill had a speech impediment as a child. Some of the finest speakers we have ever had in class were stutterers who admitted their problem and worked at it with professional guidance. Remember, whether your problems are large or small, the audience is on your side. They want you to succeed.

The Physical Aspects of Speaking

What are the physical characteristics of good speakers? They stand firmly but comfortably. They move and gesture naturally and emphatically, but avoid fidgety, jerky movements and foot shuffling. They look directly into the eyes of people in the audience, not merely in their general direction. They project enthusiasm into their voices. They do not mumble or speak flatly. We will examine these characteristics in detail—first movement and then voice.

Movement. A century ago a speaker's movements were far more florid and exaggerated than they are today. Today we prefer a more natural mode of speaking, closer to conversation than oratory. To some extent, electronic devices such as amplifying systems, radio, and television have brought about this change. However, you do not want to appear like a stick of wood. Even when speaking to a small group or on television (or, oddly enough, on the radio) you will want to move and gesture. If you are speaking in a large auditorium, you will want to broaden your movements and gestures. From the back row of a 2500-seat auditorium you look about three inches tall.

Movement during a speech is important for several reasons. First, it puts that nervous energy we spoke of to work. The inhibited speaker stands rigid and trembles. The relaxed speaker takes that same energy and puts it into purposeful movement.

Second, movement attracts attention. It is a good idea to emphasize an idea with a pointing finger or a clenched fist; and a speaker who comes out from behind the lectern occasionally and walks across the stage or toward his audience awakens audience interest. The speaker who passively utters ideas deadens the audience.

Third, movement makes you feel more forceful and confident. It keeps you, as well as your audience, awake. This is why good speakers while speaking over the radio will gesture just as emphatically as though the audience could see them.

What sorts of movements are appropriate? To begin with, movement should closely relate to your content. Jerky or shuffling motions that occur haphazardly distract an audience. But a pointing finger combined with an emphatic statement reinforces a point for an audience. A sideward step at a moment of transition draws attention to the shift in thought. Take a step backward and you indicate a conclusion. Step forward and you indicate the beginning of a new point. Use also the normal descriptive gestures that all of us use in conversation: gestures to indicate length, height, speed, roundness, and so forth.

For most people, gesturing is fairly normal. They make appropriate movements without too much thought. Some beginning speakers, however, are body inhibited. If you are in this category, you may have to cultivate movement. In your practice sessions and in your classroom speeches, risk artificiality by making gestures that seem too broad to you. Oddly enough, often at the very point where your gestures seem artificial and forced to you, they will seem the most natural to your audience.

Allow natural gestures to replace nervous mannerisms. Some speakers develop startling mannerisms and remain completely oblivious of them until some brave but kind soul points them out. Some that we have observed include putting glasses off and on; knocking a heavy ring over and over on the lectern; fiddling with a pen, pointer, chalk, cigar, microphone cord, ear, mustache, nose, you name it; shifting from foot to foot in time to some strange inner rhythm; and pointing with the elbows while the hands remain in the pants pockets. Mannerisms may also be vocal. Such things as little coughs or repeating words such as "OK" or "You know" to indicate transitions may become mannerisms.

Listeners are distracted by such habits. Often they will concentrate on the mannerisms to the exclusion of everything else. They may know that a speaker put her glasses on and off 22 times but not have the faintest notion of what she said. If someone points out such mannerisms in your speaking habits, do not feel hurt. Instead, work to remove the mannerisms.

Movement includes facial movement. Do not be a deadpan. Your basic expression should be a relaxed, friendly look. But do not hesitate to smile, laugh, frown, or scrowl when such expressions are called for. A scowl at a moment of disapproval makes the disapproval that much more emphatic. Whatever you do, do not freeze into one expression, whether it be the stern look of the man of iron or the vapid smile of a smoker in a magazine ad.

Voice. Your voice should sound relaxed, free of tension and fear. In a man, people consider a deep voice to be a sign of strength and authority. Most people prefer a woman's voice to be low rather than shrill. If you do not have these attributes, you can develop them to some extent. Here we must refer you to some of the good speech books in our bibliography (Appendix D) where you will find various speech exercises described. If, despite hard work, your voice remains unsatisfactory in comparison with the conventional stereotypes, do not despair. Many speakers have had somewhat unpleasant voices and through force of character or intellect directed their audiences to their ideas and not their voices. Mrs. Eleanor Roosevelt, for example, had a somewhat shrill, distracting voice all of her life, but her warmth and brilliance shone through, and she captivated most members of her audiences.

Many beginning speakers speak too fast, probably because they are anxious to be done and sit down. A normal rate of speech falls between 120 and 180 words a minute. This is actually fairly slow. Generally, you will want a fairly slow delivery rate. When you are speaking slowly, your voice will be deeper and more impressive. Also, listeners have trouble following complex ideas delivered at breakneck speed. Slow up and give your audience time to absorb your ideas.

Of course, you should not speak at a constant rate, slow or fast. Vary your rate. If you normally speak somewhat rapidly, slowing up will emphasize ideas. If you are speaking slowly, suddenly speeding up will suggest excitement and enthusiasm. As you speak, change the volume and pitch of your voice. Any change in volume, whether from low to loud or the reverse, will draw your listeners' attention and thus emphasize a point. The same is true of a change in pitch. If your voice remains a flat monotone and your words come at a constant rate, you deprive yourself of a major tool of emphasis.

Many people worry about their accent. Normally, our advice is *don't*. If you speak the dialect of the educated people of the region where you were raised, you have little to worry about. Some New Englanders, for example, put r's where they are not found in other regional dialects and omit them where they are commonly found. Part of America's richness lies in its diversity, and President Kennedy reminded all of America that an educated man could say "Cuber" and "idear." Accents vary in most countries from one region to another, but certainly not enough to hinder communication.

If, however, your accent is slovenly—"Ya wanna cuppa coffee?"—or uneducated, do something about it. Work with your teacher or seek other professional help. Listen to educated speakers and imitate them. The musical "My Fair Lady" has probably made the need for correct speech, and some of the methods used to attain it, known to most Americans.

Whatever your accent, there is no excuse for mispronouncing words. Before you speak, look up any words you know you must use and about which you are uncertain of the pronunciation. Speakers on technical subjects have this problem perhaps more than other speakers. Many technical terms are jawbreakers. Find their correct pronunciation and practice them until you can say them easily.

Audience Interaction

One thing speakers must learn early in their careers is that they cannot count on the audience hanging on every word. Some years ago an intelligent, educated audience was asked to record its introspections while listening to a speaker. The speaker was an excellent one. Despite his excellence and the high level of the audience, the introspections revealed that the audience was paying something less than full attention. Here are some of the recorded introspections:

> God, I'd hate to be speaking to this group. . . . I like Ben—he has the courage to pick up after the comments. . . . Did the experiment backfire a bit? Ben seems unsettled by the introspective report. . . . I see Ben as one of us because he is under the same judgment. . . . He folds his hands as if he was about to pray. . . . What's he got in his pocket he keeps wriggling around. . . . I get the feeling Ben is playing a role. . . . It is interesting to hear the words that are emphasized. . . . This is a hard spot for a speaker. He really must believe in this research. . . .

> Ben used the word "para-social." I don't know what that means. Maybe I should have copied the diagram on the board. . . . Do not get the points clearly . . . cannot interrupt . . . feel mad . . . More words. . . . I'm sick of pedagogical and sociological terms. . . . Slightly annoyed by pipe smoke in

my face. . . . An umbrella dropped. . . . I hear a funny rumbling noise. . . . I wish I had a drink. . . . Wish I could quit yawning. . . . Don't know whether I can put up with these hard seats for another week and a half or not. . . . My head itches. . . . My feet are cold. I could put on my shoes, but they are so heavy. . . . My feet itch. . . . I have a piece of coconut in my teeth. . . . My eyes are tired. If I close them, the speaker will think I'm asleep. . . . I feel no urge to smoke. I must be interested. . . .

Backside hurts . . . I'm lost because I'm introspecting. . . . The conflict between introspection and listening is killing me. Wish I didn't take a set so easily. . . . If he really wants me to introspect, he must realize himself he is wasting his time lecturing. . . . This is better than the two hour wrestling match this afternoon. . . . This is the worst planned, worst directed, worst informed meeting I have ever attended. . . . I feel confirmation, so far, in my feelings that lectures are only 5% or less effective. . . . I hadn't thought much about coming to this meeting, but now that I am here it is going to be O.K. . . . Don't know why I am here. . . . I wish I had gone to the circus. . . . Wish I could have this time for work I should be doing. . . . Why doesn't he shut up and let us react. . . . The end of the speech. Now he is making sense. . . . It's more than 30 seconds now. He should stop. Wish he'd stop. Way over time. Shut Up. . . . He's over. What will happen now? . . .[1]

As some of the comments reveal, perhaps being asked to record vagrant thoughts as they appeared made some members of the audience less attentive than they normally would have been. But most of us know that we have very similar thoughts and lapses of attention while we attend classes and speeches.

Reasons for audience inattention are many. Some are under the speaker's control; some are not. The speaker cannot do much about such physical problems as hard seats, crowded conditions, bad air, and physical inactivity. The speaker can do something about psychological problems such as the listeners' passivity and their sense of anonymity, their feeling of not participating in the speech.

Even before they begin to speak, good speakers have taken audience problems into account. They have analyzed the audience's education and experience level. They have planned to keep their points few and to repeat major points through carefully planned variations. They plan interesting examples. While speaking they attempt to interest the audience through movement and by varying the speech rate, pitch, and volume.

But good speakers go beyond these steps and analyze their audience and its reactions as they go along. In an extemporaneous speech

---

[1] These introspections were compiled at a session of the National Training Laboratory in Group Development that one of the authors attended in Bethel, Maine. Many of the ideas expressed in this chapter were first developed in the National Training Laboratory.

and even to some extent in a written speech, you can make adjustments based on this audience analysis.

To analyze your audience, you must have good eye contact. You must be looking at Ben, Bob, and Irma. You must not merely be looking in the general direction of the massed audience. Look for such things as smiles, scowls, fidgets, puzzled looks, bored expressions, interested expressions, sleepy eyes, heads nodded in agreement, heads nodded in sleep, heads shaken in disagreement. You will not be 100 percent correct in interpreting these signs. Many students have learned to smile and nod in all the proper places without ever hearing the instructor. But, generally, such physical actions are excellent clues as to how well you are getting through to your audience.

If your audience seems happy and interested, you can proceed with your speech as prepared. If, however, you see signs of boredom, discontent, or a lack of understanding, you must make some adjustments. Exactly what you do depends to some extent on whether you are in a formal or informal speaking situation. We will look at the formal situation first.

In the formal situation you are somewhat limited. If your audience seems bored, you can quickly change your manner of speaking. Any change will, at least momentarily, attract attention. You can move or gesture more. With the audience's attention gained, you can supply some interesting anecdotes or other illustrative material to support better your abstractions and generalizations. If your audience seems puzzled, you know you must supply further definitions and explanations and probably more concrete examples. If your audience seems hostile, you must find some way to soften your argument while at the same time preserving its integrity. Perhaps you can find some mutual ground upon which you and the audience can agree and move on from there.

Obviously, such flexibility during the speech requires some experience. Also it requires that the speaker have a full knowledge of the subject. If every bit of material the speaker knows about the subject is in the speech already, the speaker has little flexibility. But do not be afraid to adjust a speech in midstream. Even the inexperienced speaker can do it to some extent.

Many of the speaker's problems are caused by the speech situation being a one-way street. The listeners sit passively. Their normal desires to react, to talk back to the speaker, are frustrated. The problem suggests the solution, particularly when you are in a more informal speech situation, such as a classroom or a small meeting.

In the more informal situation you can stop when a listener seems puzzled. Politely ask him where you have confused him and attempt to clarify the situation. If a listener seems uninterested, give him an opportunity to react. Perhaps you can treat him as a puzzled listener. Or, you can ask him what you can do to interest him more. Do not be unpleasant. Put the blame for the lack of interest on yourself, even if you feel it does not belong there. Sometimes you may be displeased or shocked at the immediate feedback you receive, but do not avoid it on these grounds. And do not react unpleasantly to it. You will move more slowly when you make speaking a two-way street, but the final result will probably be better. Immediate feedback reveals areas of misunderstanding or even mistrust of what is being said.

In large meetings where such informality is difficult, you can build in some audience reaction through the use of informal sub-groups. Before your talk, divide your audience into small subgroups, commonly called buzz groups. Use seating proximity as the basis for your division if you have no better one. Explain that after your talk the groups will have a period of time in which to discuss your speech. They will be expected to come up with questions or comments. People do not like to seem unprepared, even in informal groups. As a result, they will be more likely to pay attention to your speech in order to participate well in their buzz groups.

Whether you have buzz groups or not, often you will be expected to handle questions following a speech. If you have a chairperson, he or she will field the questions and repeat them, and then you will answer them. If you have no chairperson you will perform this chore for yourself. Be sure everyone understands the question. Be sure you understand the question. If you do not, ask the questioner to repeat it and perhaps to rephrase it.

Keep your answers brief, but answer the questions fully and honestly. When you do not have the answer, say so. Do not be afraid of conflict with your audience. But keep it on an objective basis: talk about the conflict situation, not personalities. If someone reveals through his question that he is becoming personally hostile, handle him courteously. Answer his question as quickly and objectively as you can and move on to another questioner. Sometimes the bulk of your audience will grow restless while a few questioners hang on. When this occurs release your audience and, if you have time, invite the questioners up to the platform to continue the discussion. Above all, during a question period be courteous. Resist any temptation to have fun at a questioner's expense.

Visual Aids

Today, good speakers increasingly use visual aids during their talks. You will use a visual aid for three reasons: (1) *to support* and *expand* the content of your message; (2) to *focus the audience's attention* on a critical aspect of your presentation; and (3) to *clarify* your meaning.[2]

To augment your message. The first purpose of any visual support material, then, is to augment your message—to enlarge on the main ideas and give substance and credibility to what you are saying. Obviously, the material must be relevant to the idea being supported. Too often a speaker gives in to the urge to show a visually attractive or technically interesting piece of information that has little or no bearing on the subject.

Suppose, for the purposes of our analysis, that you were asked to meet with government people to present a case for your company's participation in a major federal contract. Your visual support would probably include information about the company's past performances with projects similar to the one being considered. You would show charts reflecting the ingenious methods used by the company's development people to keep costs down; performance statistics to indicate your high quality standards; and your best conception-to-production times to show the audience how adept you are at meeting target dates.

In such a presentation, before an audience of tough-minded officials, you wouldn't want to spend much of your time showing them aerial views of the company's modern facilities or mug shots of smiling employees, antiseptic production lines, and the company's expensive air fleet. Such material would hardly support and expand your arguments that the company is used to working and producing on a Spartan budget.

To focus attention. Your second reason for using visual aids is *focus of attention*. A good visual can arrest the wandering thoughts of your audience and bring their attention right down to a specific detail of the message. It forces their mental participation in the subject.

When you are dealing with very complex material, as you often will be, you can use a simple illustration to show your audience a single, critical concept within your subject.

For clarity. Finally, the overall purpose of any support—verbal or visual—is to help *clarify* the message. Visual material clarifies by adding information, by expanding on the ideas provided by the speaker verbally. Obviously, to accomplish this the visual material must be designed to provide *further* meaning to the message.

---

[2] The material in this section has been especially prepared for this chapter by Professor James Connolly of the University of Minnesota, based upon his book *Effective Technical Presentations* (St. Paul, Minn.: 3M Business Press, 1968).

Criteria for Visual Aids

But what about the visual aids themselves? What makes one better than another for a specific kind of presentation? Before we consider individual visual aids, let's look at the qualities that make a visual aid effective for the technical speaker.

Visibility. First, a visual aid must be *visible*. If that seems so obvious that it hardly need be mentioned at all, it may be because you haven't experienced the frustration of being shown something the speaker feels is important—and not being able to read it, or even make out detail. To be effective, your visual support material should be clearly visible from the most distant seat in the house. If you have any doubts, sit in that seat and look. Remember this when designing visual material: *Anything worth showing the audience is worth making large enough for the audiience to see.*

Clarity. The second criterion for a good visual is *clarity*. The audience decides this. If they're able to determine immediately what they are seeing, the visual is clear enough. Otherwise, it probably calls for further simplification and condensation. Such obvious mistakes as pictorial material out of focus, or close-ups of a complex device that will confuse the audience, are easy to understand. But what about the chart that shows a relationship between two factors on x- and y-axes when the axes are not clearly designated or when pertinent information is unclear or missing? *Visual material should be immediately clear to the audience, understandable at a glance, without specific help from the speaker.*

Simplicity. The third criterion for good visual support is *simplicity*. No matter how complex the subject, the visual itself should include no more information than absolutely necessary to support the speaker's message. If it is not carrying the burden of the message, it need not carry every detail. Limit yourself to *one* idea per visual—mixing ideas will totally confuse an audience, causing them to turn you off midsentence.

When using words and phrases on a visual, limit the material to key words that act as visual "cues" for you and the audience. So, if a visual communications expert wished to present the criteria for a good visual he might *think* something like this:

"A good visual must manifest visibility.

"A good visual must manifest clarity.

"A good visual must manifest simplicity.

"A good visual must be easy for the speaker to control."

What would he show the audience? If he knows his field as well as he should, he'll offer the visual shown in Figure 19-1.

Figure 19-1   Criteria for Visual Aids

CRITERIA

Visibility

Clarity

Simplicity

Control

The same information is there. The visual is being used appropriately to provide emphasis while the speaker supplies the ideas and the extra words. The very simplicity of the visual has impact and is likely to be remembered by the audience.

Control. The fourth quality a visual aid needs is *control,* speaker control. You should be able to add information or delete it, to move forward or backward to review, and, finally, *to take it away* from the audience to bring their attention back to you.

Some very good visual aids can meet the other criteria, and prove almost worthless to a speaker because they cannot be easily controlled. The speaker, who must maintain a flow of information and some kind of rapport with his audience while he is doing it, can't afford to let his visual material interfere with his task. Remember, visual material is meant to *support* you as a speaker, not to replace you.

Visual Design—What and How?

So far we have discussed the *why* of visual support material. The remaining two questions of concern to you are What do I use? How do I use it? Let's consider them in that order, applying the criteria already established as we go. Types of visual material can be roughly classified in six categories: (1) graphs, (2) tables, (3) representational art, (4) words and phrases, (5) cartoons, and (6) hardware. Graphs, tables, and representational art are discussed in detail in Chapter 11. You'll want to apply the suggestions made there to the visual materials you use when speaking. In the next few pages, we'll discuss using words and phrases, cartoons, and hardware.

Words and phrases. There will always be circumstances where you will want to emphasize key words or phrases by visual support.

This type of visual can be effectively used in making the audience aware of major divisions or subdivisions of a topic, for instance.

There is danger, however, in the overuse of words—too many with too much detail. Some speakers tend to use visuals as a "shared" set of notes for their presentation, a self-limiting practice. Audiences who are involved in reading a long, detailed piece of information won't recall what the speaker is saying.

With technical presentations, there is still another problem with the use of words. Too often, because they may be parts of a specialized vocabulary, they do more to confuse the audience than increase their understanding. Such terms should be reserved for audiences whose technical comprehension is equal to the task of translating them into meaningful thoughts.

Cartooning. Cartooning is no more than illustrating people, processes, and concepts with exaggerated, imaginative figures—showing them in whatever roles are necessary to your purpose. (See Figure 19-2.) Not only does it heighten audience interest, but cartooning can be as specific as you want it to be in terms of action or position.

Some situations in which you might choose to use cartooned visual material are these:

1. When dealing with subjects that are sensitive for the audience.
2. When showing people-oriented action in a stationary medium—any visual aid outside the realm of motion pictures.

The resourceful speaker will use cartoons to help give additional

Figure 19-2   Cartoon Used to Illustrate a
Computer Process

meaning to other forms of visual support. The use of cartoons as elements in a block diagram tends to increase viewer interest.

Cartooning, like any other technique, can be overdone. There are circumstances where the gravity of the situation would suggest that you consider only the most formal kinds of visual support material. On other occasions, cartooning may distract the audience, call too much attention to itself. Your purpose is not to entertain, but to communicate.

Hardware. After all this analysis of visual support material, you may wonder if it wouldn't be somewhat easier to show the real thing instead.

Certainly there will be times when the best visual support you can have is the actual object. Notably, the introduction of a new piece of equipment will be more effective if it is physically present to give the audience an idea of its size and bulk. And, if it is capable of some unique and important function, it should definitely be seen by the audience. (The greatest difficulty with the use of actual hardware is control. The device that is small enough for you to carry conveniently may be too small to be seen from the audience.)

Even when the physical presence of a piece of equipment is possible, it is important to back it with supplementary visual materials. Chances are the audience will not be able to determine what is happening inside the machine, even if they understand explicitly the principle involved. With this in mind, you will want to add information with appropriate diagrams, graphs, and scale drawings.

In this discussion of visual support, the points of visibility, clarity, simplicity, and control have been stressed over and over. The reason for this is their importance to the selection and use of visual support by the technical communicator. In the end, it is you who can best decide which visual support form is required by your message and your audience.

You are also faced with the choice of visual tools for presenting your visual material. The next section will deal with popular visual tools, their advantages, disadvantages, and adaptability to the materials we've already discussed.

The Tools of Visual Presentation

The major visual tools used today are these:

1. Chalkboards
2. Charts
3. Slides
4. Movies
5. Overhead projection

In the next few pages we'll discuss them individually, with an eye on the advantages each one offers the technical speaker.

The chalkboard. Anyone who has attended school in the past half century is familiar with the chalkboard. It has been a standard source of visual support for much longer than that, and often the only means of presenting visual information available to the classroom teacher.

As a visual aid, it leaves something to be desired. In the first place, preparing information on a chalkboard, especially technical information where every sliding scrawl can have significance, takes time. And, after the material is in place, it cannot be removed and replaced quickly.

Secondly, the task of writing on a surface that faces the audience requires that you turn your back toward them while you write. And people don't respond well to backs. They want you to face them while you're talking to them.

Add to these problems the difficulties of moving a heavy, semipermanent chalkboard around, and you begin to wonder why anyone bothers.

Low cost and simplicity are the reasons. The initial cost of a chalkboard is higher than you may think; but the cost of erasers and chalk is minimal. And, in spite of its drawbacks, a chalkboard is easy to use. It may take time, but there's nothing very complicated about writing a piece of information on a chalkboard. This simplicity, of course, gives it a certain flexibility, makes it essentially a spontaneous visual aid on which the speaker can create his visual material as he goes.

There are specific techniques for using a chalkboard which make it a more effective visual tool and help overcome its disadvantages. Let's consider them one at a time.

1. Plan ahead. Unless there is a clear reason for creating the material as you go, prepare your visual material before the presentation. Then cover it. Later, you can expose the information for the audience at the appropriate point in your speech.

Interrupting your presentation and the flow of ideas to write on the board can ruin the impact of your material. For this reason, it's a good idea to prepare the board before you begin your speech. Then, because you normally won't want the audience staring at the information you've written until the right moment during your speech, cover it with a cloth or sheet of paper. In certain instances, the mystique of the hidden information will actually work to your advantage, whetting the audience's appetite and sharpening their curiosity.

2. Be neat and keep the information simple and to the point. If your material is complex, find another way of presenting it.

3. Prime the audience. Before showing your information, tell them what they are going to see and why they are going to see it.

This last point is especially important when you are creating your visual support as you go. Priming your listeners will allow you to maintain the flow of information and, at the same time, prepare them mentally to understand and accept your information.

Charts. Charts take a couple of forms. The first is the individual *hardboard* chart, rigid enough to stand by itself and large enough to be seen by the audience from wherever they might be seated in the room. It is always prepared before the presentation, sometimes at considerable cost.

The second chart form is the *flip* chart, a giant-sized note pad which may be prepared before or during the presentation. When you have completed your discussion of one visual, you simply flip the sheet containing it over the top of the pad as you would the pages of a tablet.

The two types of charts have a common advantage. Unlike the chalkboard, they allow for reshowing a piece of information when necessary—an important aid to speaker control.

The techniques listed below will help you use charts more effectively during your presentation. They're really rules of usage, to be followed each time you choose this visual form for support.

1. Keep it simple. Avoid complex, detailed illustrations on charts. A 3 ft × 5 ft chart is seldom large enough for detailed visibility.
2. Ask for help. Whenever possible, have an assistant on one side of your charts to remove each one in its turn. This avoids creating a break in your rapport with the audience while you wrestle with a large cardboard chart or a flimsy flip sheet.
3. Predraw your visuals using a very light-colored crayon or chalk. During the presentation, you can simply draw over the original lines in darker crayon or ink. This allows you to create an accurate illustration a step at a time for clarity.
4. Prime the audience. Tell them what they are going to see and why before you show each visual, for the same reason you would do it with the chalkboard.

Slides. The 35 mm slide, with its realistic color and photographic accuracy, has always been a popular visual tool for certain types of technical presentations. Where true reproduction is essential, there is no better tool available.

Modern projectors have two notable advantages over their predecessors. First, the introduction of slide magazines has made it possible for you to organize your presentation and keep it intact; and, second, remote controls allow you to operate the projector—even to reverse the order of your material—from the front of the room.

In order to use slide projectors effectively, however, you must turn off the lights in the room. And any time you keep your audience in the dark, you risk damaging the direct speaker-listener relationship on which communication hinges. In a sense, it takes the control of the presentation out of your hands. Long sequences of slides tend to develop a will and a pace of their own. They tire an audience and invite mental absenteeism.

There are ways to handle the built-in problems of a slide presentation, simple techniques that can greatly increase audience attention and the effectiveness of your presentation.

1. When using slides in a darkened room, light yourself.

A disembodied voice in the dark is little better than a tape recorder; it destroys rapport and allows the audience to exit into its own thoughts. To minimize this effect, arrange your equipment so you may stay in the front of the room and use a podium light or some other soft, nonglaring light to make yourself *visible* to the audience.

2. Break the presentation into short segments of no more than five or six slides.
3. Always tell the audience what they're going to see, and what they should look for.

Everything considered, slides are an effective means of presenting visual material. But, like any visual tool, they require control and preparation on your part. The important thing to remember is that they are there only to support your message—not to replace you.

Movies. Whenever motion and sound are important to the presentation, movies are the only visual form available to the speaker that can accomplish the effect. Like slides, they also provide an exactness of detail and color that can be critical to certain subjects. There is really no other way an engineer could illustrate the tremendous impact aircraft tires receive during landings, for instance. The audience would understand the subject only if they were able to view, through the eye of the camera, the distortion of the rubber when the plane touches down.

But movies *are* the presentation. They cannot be considered visual support material at all in the sense of the term developed in this book. They simply replace the speaker as the source of information, at least for the duration of the film. If the film is long, say the major part of the presentation, the speaker is reduced to a master of ceremonies with little more to do than introduce and summarize the content of the film.

This makes the movie the most difficult visual form to control. Yet it can be controlled and, if it is to perform the support functions we've outlined, it must be. Some effective techniques are given here.

1. Prepare the audience. Explain the significance of what you are about to show them.
2. If a film is to be used, it should make up only a small part of the total presentation.
3. Whenever possible, have the film cut into short three or four minute sections—preferably separated by a few seconds of blank film. During the breaks introduced by these periodic insertions, you can reestablish rapport with the audience by summarizing what they have seen and refocusing attention on the important points in the next segment.

Overhead projection. Throughout this discussion of visual support, we've intentionally stressed the importance of maintaining a good speaker-audience relationship. It's an essential in the communication process. And it's fragile. Any time you turn your back to the audience, or darken the room, or halt the flow of ideas for some other reason, this relationship is damaged.

The overhead projector alone effectively eliminates all of these rapport-dissolving problems. The image it projects is bright enough and clear enough to be used in a normally lighted room without noticeable loss of visibility. And, just as important from your point of view, it allows you to remain in the front of the room *facing* your audience throughout your presentation. The projector itself is a simple tool, and like all simple tools it may be used without calling attention to itself.

Visual material for the overhead projector is prepared on transparent sheets the size of typing paper. In recent years, the methods for preparing these transparencies have become so simplified and made so inexpensive that the overhead has become a universally accepted visual tool in both the classroom and industry.

Perhaps the most important advantage of the overhead projector is the total speaker control it affords. With it, you may add information or delete it in a variety of ways, move forward or backward to review at will, and *turn it off* without altering the communicative situation in any way. It is the last capability that makes the overhead projector unique among visual tools. By flipping a switch, you can literally "remove" the visual material from the audience's consideration, bringing their attention back to you and what you are saying. And, since the projector is used in a fully lighted room, this on-and-off process seldom distracts the audience or has any effect on the speaker-audience relationship.

There are three ways to add information to a visual while the audience looks on—an important consideration when you want your listeners to receive information in an orderly fashion. In order of their discussion, they are (1) overlays, (2) revelation, and (3) writing on the visual itself.

The *overlay* technique (Figure 19-3) combines the best features of preparing your visuals in advance and creating them at the moment of their need. It is the simple process of beginning with a single positive transparency and adding information with additional transparencies by "overlaying" them, placing each one over the first so they may be viewed by the audience as a single, composite illustration. Ordinarily, no more than two additional transparencies should be used this way, but it's possible to include as many as four or five.

The technical person, who must usually present his more complex concepts a step at a time to ensure communication, can immediately see the applications of such a technique.

The technique of *revelation* is simpler. (See Figure 19-4.) It is the process of masking off the parts of the visual you don't want the audience to see. A plain sheet of paper will work. By laying it over the information you want to conceal for the moment, you can block out selected pieces of the visual. Then, when you're ready to discuss this hidden information, you simply remove the paper. The advantage is clear enough. If you don't want the audience to read the bottom line on the page while you're discussing the top line, this is the way to control their attention.

Figure 19-3   Using Overlays with an Overhead Projector

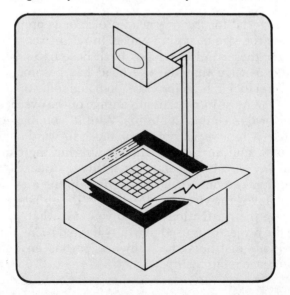

Figure 19-4   The Technique of Revelation

Writing on transparencies. Writing information on an over-head-projection transparency (Figure 19-5) is nearly as easy as writing on a sheet of paper at your desk. Several felt-tipped pens available for this purpose may be used to create visual material in front of the audience. Often you can achieve your purposes by simply underlining or circling important parts of your visual—a means of focusing audience attention on the important aspects of your message.

Figure 19-5   Writing on Transparencies

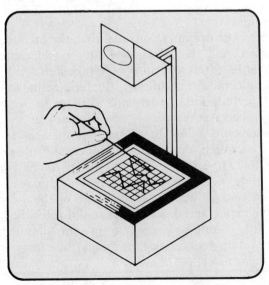

A final way of directing audience attention with overhead projection transparencies is simply to use your pencil as a pointer. The "profile" shadow of the pencil will be seen on the screen, directing the audience's attention to the proper place.

Summary

When considering a visual aid, there are certain things you must keep in mind. First, does it satisfy the fundamental purposes of a visual aid? Will it *augment* the message? Will it provide a *focus of attention* for the audience? And will it *clarify* the subject it is supporting?

If it does these things effectively, and it meets the four criteria of a good visual aid—visibility, clarity, simplicity, and control—you have a legitimate piece of visual support material.

Group Conferences

Most industries and government agencies today use the group conference for training and for problem solving. Therefore, the group conference is another speech situation you may find yourself in.

Group conferences do have their limitations. They are time consuming, and sometimes they spread responsibility too thin. They sometimes fail at truly being a conference when one member of the group has direct control over some of the other members. And if you do not have skilled conferees, you end up pooling ignorance rather than knowledge.

However, most organizations feel that the advantages outweigh the limitations. And the advantages are considerable. Conferences share judgments. Often several people together will have an insight that they would miss individually. Conference judgments are more temperate than individual judgments. Perhaps the biggest advantage of the conference is that people are more eager to carry out and support actions that they have planned themselves. Leaders can make decisions much more quickly alone than in a conference. But they will quickly lose the saved time if they have to spend hours or even days selling their decisions to their subordinates.

In this section we will give you some of the *do's* and *don'ts* of conference behavior and describe the useful roles that conferees can play. We will conclude by describing in some detail the methods of the problem-solving conference, probably the most common type of conference in industry.

Conference Behavior

A good group conference is a pleasure to observe. A bad conference distresses conferees and observers alike. In a bad group conference the climate is defensive. Conferees feel insecure, constantly fearing a personal attack and preparing to defend themselves. The leader of the bad conference cannot talk without pontificating; advice is given as though from on high. The group punishes those members who deviate from the majority will. As a result the ideas offered are tired and trite. Creative ideas are rejected. People compete for status and control, and they consider the rejection of their ideas as a personal insult. They attack those who reject their contributions. Everyone goes on the defensive, and energy that should be focused on the group's task flows needlessly in bitter debate. The group accomplishes little. As a rule, the leader ends up dictating the solutions, perhaps what was wanted all along.

In a good group conference the climate is permissive and supportive. Members truly listen to one another. People assert their own ideas, but they do not censure the opinions of others. The general attitude is, "We have a problem. Let's put our heads together and solve it." Members reward each other with compliments for good ideas, and do not reject ideas because they are new and strange. When members do reject an idea they do it gently with no hint of a personal attack on its originator. People feel free to operate in such a climate. They come forward with more and better ideas. They drop the defensive postures that waste so much energy and put the energy instead into the group's task.

How do the members of a group arrive at such a supportive climate? To simplify things we will present a list of *do's* and *don'ts*. Our rules cannot, of course, guarantee a good conference, but if they are followed they can help contribute to a successful outcome.

Do's

1. Do be considerate of others. Stimulate people to act rather than pressuring them.
2. Do be loyal to the conference leader without saying "yes" to everything. Do assert yourself when you have an idea or when you disagree.
3. Do support the other members of the group with compliments and friendliness.
4. Do be aware that other people have feelings. Remember that conferees with hurt feelings will drag their feet or actively disrupt a conference.
5. Do have empathy for the other conferees. See their point of view. Do not assume you know what they are saying or are going to say. Really listen and *hear* what they are saying.

6. Do conclude contributions you make to the group by inviting criticism of them. Detach yourself from your ideas and see them objectively as you hope others do. Be ready to attack your own ideas.

7. Do understand that communication often breaks down. Do not be shocked when you are misunderstood or when you misunderstand others.

8. Do feel free to disagree with the ideas of other members, but never attack people personally for the ideas they hold.

9. Do remember that most ideas that are not obvious seem strange at first. Yet they may be the best ideas.

10. Do remember that trying to monopolize or dominate a conference is the mark of the insecure "little person." The confident person feels secure and is willing to listen to the ideas of others. Being secure, confident people do not fear to adopt the ideas of others in preference to their own, giving full credit when they do so.

Don'ts

1. Don't continually play the expert. You will annoy other conferees with constant advice and criticism based upon your expertise.

2. Don't pressure people to accept your views.

3. Don't make people pay for past mistakes with continuing punishment. Instead try to change the situation to prevent future mistakes.

4. Don't let personal arguments foul a meeting. Stop arguments before they reach the personal stages by rephrasing them in an objective way.

Perhaps the rule "Do unto others as you would have them do unto you" best summarizes all these *do's* and *don'ts*. When you speak you want to be listened to. Listen to others.

Group Roles

You can play many roles in a group conference. Sometimes you bring new ideas before the group and urge their acceptance. Perhaps at other times you serve as information-giver and at still others as harmonizer, resolving differences and smoothing ruffled egos. In this section we will discuss the useful roles that you as a conference leader or member can play. We purposely do not distinguish between leader and member roles. In a well-run conference, an outsider would have difficulty knowing who the leader is. We divide the roles into two groups: (1) *Task roles:* roles designed to move the group toward the accomplishment of its task; and (2) *Group maintenance roles:* roles designed to maintain the group in a harmonious, working condition.

We pause here to point out that it does not pay to analyze too closely *why* any particular group member plays some particular role.

Few of us have perfectly pure motives for any role we play. Members introducing new ideas may have many motives. They may be seeking power: people often think of the idea-giver as the leader of a group. They may be acting from a sincere desire to help the group solve its problem. They may also be building up their own egos or seeking promotions. At any one time you will be aware of only a few of a person's motives—including your own. So long as a person's role or roles are useful ones, do not turn the spotlight of analysis too brightly. Few of us can stand the glare. Now, on to the roles.

Task roles. When you play a task role you help the group accomplish its set task. Some people play one or two of these roles almost exclusively, but most people slide easily in and out of most of them.

1. *Initiators.* The initiators are the idea-givers, the starters. They get the group moving toward its task, perhaps by proposing or defining the task, or by suggesting a solution to a problem, or a way of arriving at the solution.

2. *Information-seekers.* The information-seekers see where needed facts are thin or missing. They solicit the group for facts relevant to the task at hand.

3. *Information-givers.* The information-givers provide data and evidence relevant to the task. They may do so on their own or in response to the information-seekers.

4. *Opinion-seekers.* The opinion-seekers canvass group members for their beliefs and opinions concerning the problem. They might encourage the group to state the value judgments that form the basis for the criteria of the problem solution.

5. *Opinion-givers.* The opinion-givers volunteer their beliefs, judgments, and opinions to the group or respond readily to the opinion-seekers. They help set the criteria for the problem solution.

6. *Clarifiers.* The clarifiers act when they see the group is confused about a conferee's contribution. They attempt to clear away the confusion by restating the contribution or by supplying additional relevant information, opinion, or interpretation. They may restate and thus clarify a problem or furnish a needed definition.

7. *Elaborators.* The elaborators further develop the contributions of others. They give examples, analogies, and additional information. They might carry a proposed solution to a problem into the future and speculate about how it would work.

8. *Summarizers.* The summarizers draw together the ideas, opinions, and facts of the group into a coherent whole. They may state the criteria that the group has set or the solution to the problem agreed upon. Often, after a summary, they may call for the group to move on to the next phase of the task.

Group maintenance roles. When you play a group maintenance role, you help to build and maintain the supportive group climate. Some people are so task oriented that they ignore the feelings of others as they push for a solution. Without the proper climate in a group, the members will often fall short of completing their task.

1. *Encouragers.* The encouragers respond warmly to the contributions of others. They express appreciation for ideas and reward conferees by complimenting them. They go out of their way to encourage and reward the reticent members of the group when they do contribute.
2. *Feeling-expressers.* The feeling-expressers sound out the group for its feelings. They sense when some members of the group are unhappy and get their feelings out into the open. They may do so by stating the unhappiness as their own and thus encourage the others to come into the discussion.
3. *Harmonizers.* The harmonizers step between warring members of the group. They smooth ruffled egos and attempt to lift conflicts from the personality level and objectify them. With a neutral digression, they may lead the group away from the conflict long enough for tempers to cool, allowing people to see the conflict objectively without further help from them.
4. *Compromisers.* The compromisers voluntarily withdraw their ideas or solutions in order to maintain group harmony. They freely admit error. With such actions they build a climate within which conferees do not think their status is riding on their every contribution.
5. *Gate-keepers.* The gate-keepers are alert for blocked-out members of the group. They subtly swing the discussion away from the forceful members to the quiet ones and give them a chance to contribute.

The Problem-Solving Conference

The type of conference most commonly held in industry is the problem-solving conference. Certainly, it is also the conference type most likely to create tension and argument among the conferees. For both these reasons we will discuss the problem-solving conference in some detail.

The problem-solving process involves five steps:

1. Defining the problem
2. Creating solutions
3. Testing solutions
4. Choosing a solution
5. Planning for action

We will consider each of these steps in detail.

Defining the problem. At the beginning of the conference, the group must define the problem as objectively as it can. Move slowly at this point. Particularly avoid the temptation to assume that everyone understands the problem, and that the group can rush on to the solution stages. Often, some members of the group may be part of the problem or even a cause of the problem. Avoid statements that punish them. Do not drive them from the group. If not driven to the defensive, they may be in a better position to define and solve the problem than anyone else.

As you attack the problem, do not oversimplify. The problem may consist of many parts. For example, in a common industrial situation, some shop in a large plant may not be meeting its production quota. The low output is only the surface problem. The real one lies underneath and may involve such things as outmoded equipment, poor lighting, faulty training, inadequate rest periods, or many other factors. You cannot attempt solutions until you know what the real problem or problems are. Often, finding the true obstacles will make the solution step quite easy. Therefore, you must never hurry the problem-defining step.

Take your time and gather as much useful information as you can about each of the subproblems. In the production-quota problem, find out the educational level of the workers. Who trains them? Are they trained in classes or on the job? What are their incentives to learn their jobs thoroughly, and so forth? Somewhere in the mass of data gathered may lie the clue needed to suggest the solution.

In the problem-definition stage you will also set the criteria and limitations within which you must solve the problem. You may have some control over the criteria and limitations, or they may be forced on you from above. In either case the group must thoroughly understand them before it can suggest solutions. As a criterion for the production problem we have been using, you might have the production quota itself. The shop must turn out x number of units per hour or else the rest of the plant slows down. As a limitation you may have the stipulation that at this time no additional workers may be hired.

When the group has thoroughly defined the problem and has explored the criteria and limitations, it may move on to the next stage.

Creating solutions. There are several good reasons for keeping the step of creating solutions separate from the steps of testing solutions and choosing a solution. First of all the separation slows the process down. The slower pace keeps the group from grasping at the first plausible-sounding solution and rushing into action with it. The best solution may not come until many suggestions have been put forth. The best solution may be a combination of several ideas.

Second, separating the steps reduces the threat of ridicule for the unusual solution, the solution that at first glance may seem too far out even to be considered. Remember that if the problem has existed long enough to reach the conference stage most of the obvious solutions have most likely already been tried. The answer may lie in the odd or the unusual. The nonconformists of the group should be encouraged, not discouraged.

The solutions offered should stem from the data gathered during the problem-defining period. In our production-quota problem, discussion may have revealed that the instruction manuals used in the shop were written in complicated language with little regard for the educational level of the audience. Suggested solution: Rewrite the manuals. This, of course, is a fairly obvious solution given such a well-defined problem. But a group with a success-oriented attitude doesn't let the idea drop with its first statement. The members pick it up and embellish it. Why not instructional movies instead of manuals? Put the manuals in comic book form. Use 35 mm slides. Use closed-circuit TV in the training. How about programmed-learning texts? Once started, the ideas will flow. Some will be practical; some will be impractical. But don't stop the flow at this point by testing them.

As a practical matter someone should keep track of the ideas. The group should appoint a recorder. If a chalkboard is available, record on it. Everyone can see the solutions in this way. The ideas are out in the open; this greatly aids the process of one conferee building on the ideas of another.

By taking solution testing out of this step, you avoid the fear of punishment. But do not overlook the possibility of reward. When someone advances an idea that sounds good to you, say so. A warm, supportive atmosphere generates ideas; a cold, formal one freezes them out.

When the group has obviously exhausted the solutions it has to offer, it is time to move on to the next stage.

Testing solutions. In the problem-definition stage, the group sets up its criteria and limitations. In the testing stage, the group applies the criteria and limitations to the proposed solutions. Very often this step immediately knocks out a number of the proposed solutions. Probably, the group will find it tougher to test the ones that remain. At this point several blocks to the testing process arise.

Although at this stage the group is supposedly only testing solutions, some members are already looking forward to choosing the solution. Personal interests begin to get in the way of the problem-solving process. People seeking status will find favorable arguments for solutions that increase their status and find fault with solutions that decrease their status. Some conferees feel bound to defend their own solutions. People will attack solutions that threaten them or their friends with change. People may even favor a solution that punishes an enemy.

Most of these attitudes result in bitter debate. Oddly enough, a block to solution testing can also arise from a lack of conflict. If the conferees are determined to maintain an atmosphere of sweetness and light, they will not attack ideas sufficiently to test them. Without proper testing of the solution, the group will flounder at the next stage of choosing a solution or the last stage of planning the action.

The group must argue about the solutions. But it must argue on objective and not on personal grounds. Insofar as humanly possible, personalities and personal interests must not influence the testing and the later decision. Such objectivity is difficult, if not impossible, in a group of human beings. Therefore, those members of the group who do not have personal axes to grind must be particularly alert for the arguments that camouflage personal interests. When such arguments arise, the impartial member must steer them as quickly as possible to more objective ground.

Consider whether a solution will create new problems. Suppose that in our production-quota problem, the solution definitely seems to lie in improving the worker's training and instruction. One of the proposed solutions is that instructions be put in comic book form. On the surface this sounds like a good solution. The combination of pictures and simple prose will enable the workers to understand the involved processes. The cost of the books does not seem prohibitive. "But," someone says, "the average worker in this section of the plant has a high school diploma. Will we insult the workers with a comic book approach?" The group kicks this idea around. In objective arguments like this, solutions gain or lose ground. If most of the group agree that comic books will insult the workers, the comic book approach will fade away. Unless someone in the group is attached to the idea for personal reasons, there will be little argument. The idea has been tested and found wanting. In this subtle way, the stage of testing solutions moves easily into the next stage, that of choosing a solution.

Choosing a solution. As solutions are tested, most of them will gradually lose favor and drop from the discussion. Perhaps only one will remain and the group has therefore made the decision. But often several solutions will survive the testing period. How does the group decide which is best?

Essentially, there are two yardsticks to use in the final measurement before a choice.

1. *The objective test:* Is the solution objectively best? Does it best fit the facts of the situation? Which solution will do the most for the least cost? Which solution will gain the most support from the workers? Which solution will disrupt current practices the least? Such questions decide the objective worth of a solution. When the members of the conference do not have to carry out the solution personally, the objective approach is the best.

2. *The subjective approach:* Often some of the members of the conference must carry out the decision. When this is the case, the other members must pay special attention to their preferred solution. The group should choose the solution the implementers favor unless it is obviously much poorer than the best objective solution. When the implementers favor a solution, they will work harder and more enthusiastically at it. This harder work and enthusiasm may very well make their solution the best.

Whether it uses the objective or subjective method, the group should take its time. The group should be patient with those who hold out against the majority. Sometimes, given enough time, the minority will prove to be right. In any event, don't rush the decision by voting. The final decision should be one that, if at all possible, the whole group approves. The term for this general approval is *consensus*. Consensus will be difficult to reach if the group takes an early vote. Once conferees commit themselves to a solution by a vote, they find it more difficult to change their mind later.

Finally, however, the group must reach its decision. (Remember that indecision is itself a form of decision.) After the group reaches its decision it must plan its action.

Planning the action. What action the group plans depends entirely on the responsibilities of the group. If the group does not have the responsibility for carrying out the decision, probably the only action needed is a report to higher authority. In this case, you will find the rhetorical forms for Argument (pages 138–143) and Comparison and Contrast (pages 107–110) useful to you.

If the group has the responsibility for carrying out the action, then it must begin detailed planning. Primarily, the planning will involve assigning responsibilities to different members of the group. We cannot go much beyond this step in advising you how to plan. The content and extent of the planning depend on the situation and the responsibilities of the group.

We will close with one final hint. Do not be surprised if you run into blocks while planning the action. Failure to test the solution thoroughly or failure to obtain a true consensus while choosing a solution may slow you up at this stage. When such blocks occur, back up to the stage where you failed and work through it again.

EXERCISES ██

1. Deliver a speech in one of the following speech situations:
    a. You are an instructor at your college. Prepare a short extemporaneous lecture on a technical subject. Your audience is a class of about 20 upperclassmen.
    b. You are the head of a team of engineers and technicians who have developed a new mechanism or process. Your job is to persuade a group of senior managers from your own firm to accept the process or mechanism for company use. Assume these managers to have a layman's knowledge about your subject. Speak extemporaneously.
    c. You are a known expert on your subject. You have been invited to speak about your subject at the annual meeting of a well-known scientific association. You are expected to write out and read your speech. You are to inform the audience, which is made up of knowledgeable research scientists and college professors from a diversity of scientific disciplines, about your subject or to persuade it to accept a decision you have reached.

2. Change one of your written reports into an oral report. Deliver the report extemporaneously. Prepare several visual aids to support major points.

3. Focus upon some problem at your school and have a problem-solving conference concerning it. You may choose some technical problem such as parking, inadequate cafeteria space, poor lighting in classrooms or dormitories, or inadequate study facilities in the library. Or you may choose a more abstract issue such as required class attendance, student representation in school decisions, lower tuition rates, over- (or under-) emphasis on athletics, and so forth. You might even have a conference on the subject of *What Problem Shall We Consider?*

    Often it is a valuable experience—particularly at early conferences—to assign roles. Before the conference begins, write the names of the various roles explained on pages 444–446 on slips of paper. Fold the slips and have every conferee choose one or more of them. The conferees will not tell anyone else what roles they have picked, but they will play the roles as defined.

PART 4

Handbook

Any language is a growing, flexible instrument with rules that are constantly changing by virtue of the way it is used by its live, independent speakers and writers. Only the rules of a dead language are unalterably fixed.

Nevertheless, at any point in a language's development certain conventions are in force. Certain constructions are considered errors, and the person who uses them is considered illiterate or uneducated. It is with these conventions and constructions that this Handbook deals.

Because we wished to use continuous numbering, we have not divided Part 4 into chapters. But we have divided it into two sections. The first deals with eleven common errors. The second deals with the conventions of punctuation, italicization, capitalization, numbers, and abbreviations.

Each error and convention dealt with has a number and an abbreviated reference tag. The tags are reproduced on the back endpapers along with references to Chapter 8 on style and some of the more important proofreading symbols. If you are currently in a college writing course, your instructor may use some combinations of these tags and symbols to indicate revisions needed in your papers.

Common Errors

1. Fragmentary Sentence (Frag)

Rewrite or repunctuate the sentence to make it a complete sentence, or to join it to a complete sentence.

Most fragmentary sentences are either verbal phrases or subordinate clauses that the writer mistakes for a complete sentence.

A verbal phrase has in the predicate position a participle, gerund, or infinitive, none of which functions as a complete verb.

> Norton, *depicting* the electromagnetic heart. *(participle)*
> The *timing* of this announcement about Triptycene. *(gerund)*
> Braun, in order *to understand* tumor cell growth. *(infinitive)*

When your fragment is a verbal phrase, either change the verb to a complete verb or repunctuate the sentence so that the phrase is joined to the complete sentence of which it is actually a part.

Fragment
Norton, depicting the electromagnetic heart. She made a mockup of it.

Rewritten
Norton depicted the electromagnetic heart. She made a mockup of it.
Norton, depicting the electromagnetic heart, made a mockup of it.

Subordinate clauses are distinguishable from phrases in that they have complete subjects and complete verbs (rather than verbals) and are introduced by relative pronouns *(who, which, that)* or by subordinating conjunctions *(because, although, since, after, while).*

The presence of the relative pronoun or the subordinating conjunction is a signal that the clause is not independent but is part of a more complex sentence unit. Any independent clause can become a subordinate clause with the addition of a relative pronoun or subordinating conjunction.

Independent Clause
Early transistors were only the size of a pea.

Subordinate Clause
Although early transistors were only the size of a pea.

Repunctuate subordinate clauses so that they are joined to the complex sentence of which they are a part.

Fragment
Although early transistors were only the size of a pea. They were enormous compared to modern transistors.

Rewritten
Although early transistors were only the size of a pea, they were enormous compared to modern transistors.

Various kinds of elliptical sentences minus a subject and/or a verb do exist in English, for example, "No!" "Oh?" "Good shot." "Ouch!" "Well, now." These constructions may occasionally be used for stylistic reasons, particularly the representation of conversation, but they are seldom needed in technical writing. If you do use such constructions, use them sparingly. Remember that major deviations from normal sentence patterns will probably jar your readers and break their concentration upon your thought, the last thing that any writer wants.

2. Comma Splice or Fused Sentence (CS or FS)

Rewrite or repunctuate the sentence to make it a more effective sentence and to remove the comma splice or fused sentence error.

Punctuate two independent clauses placed together with one of the following marks of punctuation: period, semicolon, or a comma and a coordinating conjunction *(and, but, for, nor, or yet).* Infrequently the colon or dash is used also. (In modern usage, there are some exceptions to these rules. See Rule 16-1.)

Check the hydraulic pressure. If it reads below normal, do not turn on the aileron boost.

We will describe the new technology in greater detail; however, first we will say a few words about the principal devices found in electronic circuits.

If the underlined places in the sentences above are punctuated only with commas, the error is called a *comma splice*. If even the comma is omitted, the error is called a *fused sentence*.

Writers most frequently commit comma splices when they mistake conjunctive adverbs for coordinating conjunctions. The most common conjunctive adverbs are *also, anyhow, besides, consequently, furthermore, hence, however, moreover, nevertheless, therefore,* and *too.*

When a conjunctive adverb is used to join two independent clauses, the mark of punctuation most often used is a semicolon (a period is used infrequently).

Ice fish are nearly invisible; however, they do have a few dark spots on their bodies.

Often the sentence will be more effective if it is rewritten completely, making one of the independent clauses a subordinate clause or a phrase.

Comma Splice
The students at the university are mostly young Californians, most of them are between the ages of 18 and 24.

Rewritten
The students at the university are mostly young Californians between the ages of 18 and 24.

3. Verb-Subject Agreement (V/ag)

Make the verb and subject agree in number and person.

Most of the time verb-subject agreement presents no difficulty to the writer. For example, in the sentence, "He speaks for us all," only a child or a foreigner learning English might say, "He speak for us all." However, various constructions exist in English that do present agreement problems even for the adult, educated, native speaker of English. These troublesome constructions are examined below.

The following words take singular verbs: *each, everyone, either, neither, anybody, somebody.*

Writers rarely have trouble with a sentence such as "No one is going to the game." Problems arise when, as is often the case, a prepositional phrase with a plural object is interposed between the simple subject and the verb as in this sentence: *"Each of the freshmen is going to the game."* In this sentence the temptation is to let the object of the preposition, *freshmen*, govern the verb and write *are*.

Divided usage: The pronouns *any* and *none* are now commonly treated as plural subjects, particularly when they are followed by a prepositional phrase with a plural object. Such usages as *"None of the freshmen are going to the basketball game"* or *"Are any of the freshmen going to the basketball game?"* are now respectable and even preferable.

When a compound subject is joined by *or* or *nor* the verb agrees with the closer noun or pronoun.

> Either the designer or the builders are in error.
> Either the builders or the designer is in error.

In informal and general usage one might now commonly hear, or see, the second sentence above as "Either the builders or the designer are in error." In writing you should hold to the more formal usage of the example.

Parenthetical expressions introduced by such words as *accompanied by, with together with,* and *as well as* do not govern the verb.

> Mr. Roberts, as well as his two assistants, is working on the experiment.

Two or more subjects joined by *and* take a plural verb. Inverted word order does not affect this rule.

> Close to the Academy are Cathedral Rock and the Rampart Range.

Collective nouns such as *team, group, class, committee,* and many others take either plural or singular verbs, depending upon the meaning of the sentence. The writer must be sure that any subsequent pronouns agree with the subject and verb.

> The team is going to receive its championship trophy tonight.
> The team are going to receive their football letters tonight.

Note well: When the team was considered singular in the first example the subsequent pronoun was "its." In the second example the pronoun was "their."

4. Pronoun-Antecedent Agreement (P/ag)

Make the pronoun agree with its antecedent.

Pronoun-antecedent agreement is closely related to Rule 3. For example, the problem area concerning the use of collective nouns explained above is closely related to the proper use of pronouns. When a collective noun is considered singular it takes a singular pronoun as well as a singular verb. Also, such antecedents as *each, everyone, either, neither, anybody, somebody, everybody,* and *no one* take singular pronouns as well as singular verbs.

One additional problem presents itself, however. Often such nouns as *man* or *woman* or *student* are used in their generic sense; that is, they are used to stand for all mankind or womankind, or all students. When they are used generically the temptation is to think of them as plural and follow them with plural pronouns. Do not do so. Only the form of the following example is correct: "One of the characteristics of twentieth-century man is *his* wallet filled with credit cards."

5. Improper Verb Form (Vb)

Use the proper verb form.

This marking on your paper can cover a wide variety of linguistic errors ranging from such nonstandard usages as "He seen the show" for "He saw the show" to such esoteric errors as "He was hung by the neck until dead" for "He was hanged by the neck until dead." Normally a few minutes spent with any collegiate dictionary will show you the correct verb form. College level dictionaries list the principal parts of the verb after the verb entry.

6. Improper Pronoun Form (Pron)

Use the proper pronoun form.

Almost every adult can remember being constantly corrected by parents and elementary school teachers in regard to pronoun form. The common sequence is for the child to say, "Me and Johnny are going swimming," and for the teacher or parent to say patiently, "No, dear, 'Johnny and *I* are going swimming.'" As a result of this conditioning, all objective forms are automatically under suspicion in many adult minds, and the most common pronoun error is for the speaker or writer to use a subjective case pronoun such as *I, he*, or *she*, when an objective case pronoun such as *me, him*, or *her* is called for.

Whenever a pronoun is the object of a verb or the object of a preposition, it must be in the objective case.

> It occurred to my colleagues and *me* to check the velocity data on the earthquake waves.
>
> Just between you and *me*, the man who says, "He grabbed he and I and shook us," is ludicrous.

Use a subjective case pronoun in the predicate nominative position.

This rule slightly complicates the use of pronouns after the verb. Normally the pronoun position after the verb is thought of as objective pronoun territory, but when the verb is a linking verb (chiefly the verb "to be") the pronoun is called a "predicate noun" rather than an object and is in the subjective case.

> It is *she*.
>
> It was *he* who discovered the mutated fruit fly.

7. Spelling Errors (Sp)

Correct your spelling error.

No easy road to spelling perfection exists. The condition of English spelling is chaotic and likely to remain so. George Bernard Shaw once illustrated this chaos by spelling *fish, ghoti.* To do so he took the *gh* from *rough,* the *o* from *women,* and the *ti* from *condition.* You may obtain some help from the spelling section in a collegiate dictionary where the common rules of spelling are explained. You can also buy, rather inexpensively, books that explain the various spelling rules and provide exercises to fix the rules in your mind. Certainly, you should memorize the spelling of words you commonly use and learn to look up in your dictionary difficult words that you use only rarely.

8. Dangling Modifier (DM)

Give the modifier something to modify.

Many curious sentences result from the failure to provide the modifier something to modify.

Having finished the job, the tarpaulins were removed.

In this example it seems as though the tarpaulins have finished the job. As is so often the case, a passive voice construction has caused the problem. If we recast the sentence in active voice we remove the problem:

Having finished the job, the workmen removed the tarpaulins.

9. Misplaced Modifier (MM)

Place the modifier as close as possible to the element modified.

As in the case of dangling modifiers (Rule 8), curious sentences result from the modifier's not being placed next to the element modified.

An engine may crack when cold water is poured in *unless it is running.*

Probably, with a little effort, no one will misread this example; but, undeniably, it says that the engine will crack unless the water is running. Move the modifier and the sense is clear.

Unless it is running, an engine may crack when cold water is poured in.

It should be apparent from the preceding examples that a modifier may be in the wrong position to convey one meaning but in the perfect position to convey a different meaning. In the next example, the placement of *for three years* is either right or wrong. It is in the right position to modify *to work* but in the wrong position to modify *have been trying.*

I have been trying to place him under contract to work here for three years. *(three-year contract)*

As the examples should suggest, correct placement of modifiers sometimes amounts to more than mere nicety of expression. It can mean the difference between stating falsehood and truth, between saying what you mean and saying something else.

A *which* clause that must look over several intervening words before finding its antecedent is sometimes called a "squinting *which* clause."

We found a pile of picture frames in the corner, which had not been dusted for years.

This construction is objectionable only when the intervening words include a substantive (that is, a noun or noun phrase) that could logically be the antecedent of *which*. Usually it is a simple matter for the writer to revise and find a way out of the difficulty:

> In the corner we found a pile of picture frames, which had not been dusted for years.

Of course, the sentence was all right to begin with if a dusty corner is being referred to.

10. Parallelism (Paral)

Make the elements in a series grammatically parallel.

When you link elements in a series, they must all be in the same grammatical form. Link an adjective with an adjective, a noun with a noun, a clause with a clause, and so forth. Look at the italicized portion of the sentence below:

> A good test would *use small amounts of plant material, require little time, simple to run, and accurate.*

The series begins with the verbs *use* and *require* and then abruptly switches to the adjectives *simple* and *accurate*. All four elements must be based on the same part of speech. In this case, it's simple to change the last two elements:

> A good test would use small amounts of plant material, require little time, *be simple to run, and be accurate.*

Always be careful when you are listing to keep all the elements of the list parallel. In the following example, the third item in the list is not parallel to the first two:

> The process has three stages: (1) the specimen is dried, (2) all potential pollutants are removed, and (3) atomization.

The error is easily corrected:

> The process has three stages: (1) the specimen is dried, (2) all potential pollutants are removed, and (3) the specimen is atomized.

When you start a series, keep track of what you are doing and finish the series the same way you started it. Nonparallel sentences are at best awkward and off-key. At worst, they can lead to serious misunderstandings.

11. Diction (D)

Choose a word or words that are more accurate, effective, or appropriate to the situation.

Many different kinds of linguistic sins are covered by the general term *diction*. All involve a faulty choice of words. Poor diction can involve a choice of words that are too heavy or pretentious: *utilize* for *use, finalize* for *finish, at this point in time* for *now*, and so forth.

Tired old clichés are poor diction: *with respect to, with your permission, with reference to,* and many others. We talk about such language in Chapter 8, particularly in the section on pomposity (pages 174–178). Sometimes the words chosen are simply too vague to be accurate: *inclement weather* for *rain, too hot* for *600C*. See the section *Specific Words* (pages 172–174) in Chapter 8 for more on this subject.

Poor diction can mean an overly casual use of language when some degree of formality is expected. One of the many synonyms for *intoxicated,* such as *bombed, stoned,* or *smashed,* might be appropriate in private conversation but totally wrong in a police or laboratory report.

Poor diction can reflect a lack of sensitivity to language—to the way one group of words relates to another group. Someone who writes that "The airlines are beginning a crash program to solve their financial difficulties" is not paying attention to relationships. The person who writes that the "Steelworkers' Union representatives are getting down to brass tacks in the strike negotiations" has a tin ear, to say the least.

Make your language work for you, and make it appropriate to the situation.

Conventions

12. Punctuation (Pn)

The chief marks of punctuation total thirty-six. If from that list we exclude the marks that are technical, foreign, and diacritical, we have thirteen.

Apostrophe	Hyphen
Brackets	Parentheses
Colon	Period
Comma	Question mark
Dash	Quotation Marks
Ellipsis	Semicolon
Exclamation point	

These few marks satisfy all the constantly recurring demands for punctuating page after page of normal prose.

13. Apostrophe (Apos)

There are three chief uses for the apostrophe: (1) to form the possessive, (2) to stand for missing letters or numbers, and (3) to form the plural of certain expressions.

13.1 Possessives

Add an apostrophe and *s* to form the possessive of all nouns, both singular and plural, unless (a) the possessive singular would end in an objectionable *s* or *z* sound or (b) the plural already ends in an *s* or *z* sound.

Singular		**Plural**	
Nom.-obj.	*Poss.*	*Nom-obj.*	*Poss.*
man	man's	men	men's
woman	woman's	women	women's
spectator	spectator's	spectators	spectators'
agent	agent's	agents	agents'
Matthew	Matthew's	Matthews	Matthews'
Matthews	Matthews' (Matthews's)	Matthewses	Matthewses'
Xerxes	Xerxes'		
Moses	Moses'		

To show joint possession, add the apostrophe and *s* to the last member of a compound or group; to show separate possession, add an apostrophe and *s* to each member:

> Gregg and Klymer's experiment astounded the class.
> Gregg's and Klymer's experiments were very similar.

Of the several classes of pronouns, only the indefinite pronouns use an apostrophe to form the possessive.

Possessive of Indefinite Pronouns	**Possessive of Other Pronouns**
anyone's	my (mine)
everyone's	your (yours)
everybody's	his, her (hers), its
nobody's	our (ours)
no one's	their (theirs)
other's	whose
neither's	

13.2 Missing Letters or Numbers

Use an apostrophe to stand for the missing letters in contractions and to stand for the missing letter or number in any word or set of numbers where for one reason or another a letter or number is omitted.

> can't, don't, o'clock, it's (it is), etc.
> We was movin' downriver, listenin' to the birds singin'.
> The class of '49 was Colgate's best class in years.

13.3 Plural Forms

An apostrophe is often used to form the plural of letters and numbers, but this style is gradually dying, particularly with numbers.

> 6's and 7's (but also 6s and 7s)
> A's and b's

14. Brackets (Brackets)

Brackets are chiefly used when a word or comment is inserted into a quotation.

> "The result of this [disregard by the propulsion engineer] has been the neglect of the theoretical and mathematical mastery of the engine inlet problem."
> "An ideal outlet require [*sic*] a frictionless flow."
> "Last year [1979] saw a partial solution to the problem."

Sic, by the way, is Latin for *thus*. Inserted in a quotation, it means that a mistake is the original writer's, not yours. Use it with discretion.

15. Colon (Colon)

The colon is chiefly used to introduce quotations, lists, or supporting statements and between clauses when the second clause is an example or amplification of the first. Additionally it is used in certain conventional ways with numbers and in correspondence and bibliographical entries.

15.1 Introductory

Place a colon before a quotation, or a list, or supporting statements and examples that are formally introduced.

Mr. Smith says the following of wave generation:

The wind waves that are generated in the ocean and which later become swells as they pass out of the generating area are products of
storms. The low pressure regions that occur during the polar winters of
the Arctic and Antarctic produce many of these wave-generating
storms.

The various forms of engine which might be used would operate within
the following ranges of Mach number:

M-0 to M-1.5	Turbojet with or without precooling
M-1.5 to M-7	Reheated turbojet, possibly with precooling
M-7 to M-10+	Ramjet with supersonic combustion

Engineers are developing three new engines: turbojet, reheated turbojets, and ramjets.

Do not place a colon between a verb and its objects or a linking
verb and the predicate nouns.

Objects
The engineers designed turbojets, reheated turbojets, and ramjets.

Predicate Nouns
The three engines the engineers are developing are turbojets, reheated
turbojets, and ramjets.

Do not place a colon between a preposition and its objects.

The plane landed at Detroit, Chicago, and Rochester.

15.2 Between Clauses

If the second clause consists of an example or an amplification of
the first clause, then the colon may replace the comma, semicolon, or
period.

The docking phase involves the actual "soft" contact: the securing of
lines, latches, and air locks.
The difference between these two guidance systems is illustrated in Figure 2: The paths of the two vehicles are shown to the left and the motion
of the ferry as viewed from the target station is shown to the right.

You may follow a colon with a capital or a small letter. Generally,
a complete sentence beginning after a colon is given a capital.

15.3 Styling Conventions

Place a colon after a formal salutation in a letter, between hour and minute figures, between the elements of a double title, and between chapter and verse of the Bible.

Dear Mrs. Jones:
3:45
at 7:15 p.m.
Mary Russell Mitford: Her Life and Writing
I Samuel 7:14–18

16. Comma (C)

The most used—and misused—mark of punctuation is the comma. Writers use commas to separate words, phrases, and clauses. Generally, commas correspond to the pauses we use in our speech to separate ideas and to avoid ambiguity. As writing style has grown plainer, using shorter sentences and simpler constructions, the use of commas has lessened. In many places where your parents would have used a comma you will not. Nevertheless, you will use the comma often: about two out of every three marks of punctuation you use will be commas. Sometimes your use of the comma will be essential for clarity; at other times you will be honoring grammatical conventions. (See also Rule 2.)

16.1 Main Clauses

Place a comma before a coordinating conjunction *(and, but, or, nor, for, yet)* that joins two main (independent) clauses.

During the first few weeks we felt a great deal of confusion, but as time passed we gradually fell into a routine.
We could not be sure that the plumbing would escape frost damage, nor were we at all confident that the house could withstand the winds of almost hurricane force.

If the clauses are short, have little or no internal punctuation, or are closely related in meaning, then you may omit the comma before the coordinating conjunction.

The wave becomes steeper but it does not tumble yet.

In much published writing there is a growing tendency to place two very short and closely related independent clauses (called contact clauses) side by side with only a comma between.

The wind starts to blow, the waves begin to develop.

Sentences consisting of *three* or more equal main clauses should be punctuated uniformly.

We explained how urgent the problem was, we outlined preliminary plans, and we arranged a time for discussion.

In general, identical marks are used to separate equal main clauses. If the equal clauses are short and uncomplicated, commas usually suffice. If the equal clauses are long, internally punctuated, or if their separateness is to be emphasized, semicolons are either preferable or necessary.

16.2 Clarification

Place a comma after an introductory word, phrase, or clause that might be over-read or that abnormally delays the main clause.

As soon as you have finished polishing, the car should be moved into the garage. *(Comma to prevent over-reading)*

Soon after, the winds began to moderate somewhat, and we were permitted to return to our rooms. *(First comma to prevent over-reading)*

If the Polar ice caps should someday mount in thickness and weight to the point that their combined weight exceeded the Equatorial bulge, the earth might suddenly flop ninety degrees. *(Introductory clause abnormally long)*

After a short introductory element (word, phrase, or clause) where there is no possibility for ambiguity, the use of the comma is optional. Generally let the emphasis you desire guide you. A short introductory element set off by a comma will be more emphatic than one that is not.

16.3 Nonrestrictive Modifiers

Enclose or set off from the rest of the sentence every nonrestrictive modifier, whether a word, a phrase, or a clause. How can you tell a nonrestrictive modifier from a restrictive one? Look at these two examples:

Restrictive
A runway *that is not oriented with the prevailing wind* endangers the aircraft using it.

Nonrestrictive
The safety of any aircraft, *whether heavy or light,* is put in jeopardy when it is forced to take off or land in a crosswind.

The restrictive modifier is necessary to the meaning of the sentence. Not just any runway but "a runway that is not oriented with the prevailing wind" endangers aircraft. The writer then has *restricted* the many kinds of runways he could talk about to one particular kind. In the nonrestrictive example, the modifier merely adds descriptive detail. The writer does not restrict *aircraft* with the modifier, but rather makes the meaning a little clearer.

Restrictive modifiers cannot be left out of the sentence if it is to have the meaning the writer intends; nonrestrictive modifiers can be left out.

16.4 Nonrestrictive Appositives

Set off or enclose every nonrestrictive appositive. As used here the term *appositive* means any element (word, phrase, or clause) that parallels and repeats the thought of a preceding element. According to this view, a verb may be coupled appositively with another verb, an adjective with another adjective, and so on. An appositive is usually more specific or more vivid than the element that it is an appositive to; an appositive makes explicit and precise something that has not been clearly enough implied.

Some appositives are restrictive, and therefore are not set off or enclosed.

Nonrestrictive
A crosswind, *a wind perpendicular to the runway,* causes the pilot to make potentially dangerous corrections just before landing.

Restrictive
In some ways, Eisenhower *the President* had to behave differently from Eisenhower *the general.*

In the nonrestrictive example, the appositive merely adds clarifying detail. The sentence makes sense without it. In the restrictive example, the appositives are essential to the meaning. Without them we would have, "In some ways, Eisenhower had to behave differently from Eisenhower."

16.5 Series

Use commas to separate members of a coordinate series of words, phrases, or clauses if *all* the elements are not joined by coordinating conjunctions.

> Instructions on the label state clearly how surfaces should be prepared, how the contents are to be applied, and how to clean and polish the mended article.
>
> To mold these lead figures you will need a hot flame, a two-pound block of lead, the molds themselves, a file or a rasp, and an awl.
>
> Under the microscope the sensitive, filigree-like mold appeared luminous and faintly green.

16.6 Other Conventional Usages

Date
On November 26, 1950, all the forces of nature seemed to combine to wreak havoc upon the Middle Atlantic States.

Note: When you write the month and the year without the day, it is common practice now to omit the comma between them—as in June 1979.

Geographical Expression
During World War II Middletown, Pennsylvania, was the site of a huge military airport and supply depot.

Title After Proper Name
A card in yesterday's mail informed us that Carleton Williams, M.D., Pediatrician, would soon open new offices on Beaver Street, State College.

Noun of Direct Address
Clifton, do you suppose that we can find our way back to the cabin before nightfall?

Informal Salutation
Dear Jane,

17. Dash (Dash)

In technical writing, you will use the dash almost exclusively to set off parenthetical statements. You may, of course, use commas or parentheses for the same function, but the dash is the most emphatic.

separator of the three. You may also use the dash to indicate a sharp transition, but some editors may object to this use. On the typewriter you make the dash with two hyphens. You do not space between the words and the hyphens or between the hyphens themselves.

Typewriter

```
    The first phase in rendezvous--sighting and rec-
ognizing the target--is so vital that we will treat it
at some length.
```

Typeset
The target must emit or reflect light the pilot can see—but how bright must this light be?

18. Ellipsis (Ell)

Use three spaced periods to indicate words omitted within a quoted sentence, four spaced periods if the omission occurs at the end of the sentence.

"As depth decreases, the circular orbits become elliptical and the orbital velocity . . . increases as the wave height increases."
"As the ground swells move across the ocean they are subject to head-winds or crosswinds. . . ."

19. Exclamation Point (Exc)

Place an exclamation point at the end of a startling or exclamatory sentence.

On 9 January 1952, the *S.S. Pennsylvania* sank with no survivors in the North Pacific in a storm that produced 50-knot winds for 33 hours and waves up to 48 feet high!

Obviously, with the emphasis in technical writing on objectivity and calmness, you will seldom use the exclamation point.

20. Hyphen (Hyphen)

Hyphens are used to form various compound words and in breaking up a word that must be carried over to the next line.

20.1 Compound Numbers

See sections 28.1.3 and 28.3.

20.2 Common Compound Words

Observe dictionary usage in using or omitting the hyphen in compound words.

governor-elect	acid-resistant
ex-treasurer	Croesus-like
Russo-Japanese	drill-like
pro-American	self-interest

But:

glasslike	wrist watch
neophyte	sweet corn
newspaper	weather map
newsstand	sun lamp
housewife	prize fight

20.3 Compound Words as Modifiers

Use the hyphen between words joined together to modify other words.

a half-spent bullet
an eight-cylinder engine
their too-little-and-too-late methods

Be particularly careful to hyphenate when omitting the hyphen may cause ambiguity.

two-hundred-gallon drums
two hundred-gallon drums
a pink-skinned Buddha

Sometimes you have to carry a modifier over to a later word, creating what is called a "suspended hyphen."

GM cars come with a choice of four-, six-, or eight-cylinder engines.

20.4 Word Division

Use a hyphen to break a word that must be carried over to the next line.

Words compounded of two roots or a root and an affix are divided at the point of union.

self- / important	wind / jammer
cross- / pollination	desir / able
bladder / wort	anti / dote
summer / time	manage / able

Note: The first two words in this list are always spelled with a hyphen; the remaining words use the hyphen only when the word is syllabicated at the end of a line.

In general, noncompound words of more than one syllable are divided between any two syllables but only between syllables.

soph / o / more	con / clu / sion
sat / is / fac / tion	sym / pa / thy

A syllable of one letter is never set down alone; a syllable of only two letters is seldom allowed to stand alone unless it is a prefix or a suffix, and then only if it is pronounced as spelled.

hello /	elec / tro / type
method /	de / mand
pilot /	ac / cept
many /	walked /
saga /	start / ed

If a consonant is doubled because a suffix is added, include the second consonant with the suffix.

spin / ner	slip / ping
stir / ring	slot / ted

But:

stopped /	pass / ing
lapped /	stall / ing

Because *-ped* is not pronounced as a syllable it should not be carried over. In words like *passing* and *stalling* both consonants belong to the root and therefore only the suffix *-ing* is carried over.

21. Parentheses (Paren)

Parentheses are used to enclose supplementary details inserted into a sentence. Commas and dashes may also be used in this role, but with some restrictions. You may enclose a complete sentence or several complete sentences within parentheses. But such enclosure would confuse the reader if only commas or dashes were used for the enclosure.

> The violence of these storms can scarcely be exaggerated. (Typhoons and hurricanes generate winds over 75 miles an hour and waves fifty feet high.) The study. . . .

21.1 Lists

Parentheses are also used to enclose numbers or letters used in listing.

> This general analysis consists of sections on (1) wave generation, (2) wave propagation, (3) wave action near a shoreline, and (4) wave energy.

21.2 In Sentences

Within a sentence place no mark of punctuation before the opening parenthesis. Place any marks needed in the sentence after the closing parenthesis.

> If a runway is regularly exposed to crosswinds of over 10 knots (11.6 mph), then the runway is considered unsafe.

Do not use any punctuation around parentheses when they come between sentences. Give the statement *inside* the parentheses any punctuation it needs.

22. Period (Per)

Periods have several conventional uses.

22.1 End Stop

Place a period at the end of any sentence that is not a question or an exclamation.

> Find maximum average daily temperature and maximum pressure altitude.

22.2 After Abbreviations

Place a period after abbreviations.

M.D.	etc.
Ph.D.	Jr.

22.3 Decimal Point

Use the period with decimal fractions and as a decimal point between dollars and cents.

.4	$5.60
.05%	$450.23

23. Question Mark (Ques)

Place a question mark at the end of every sentence that asks a direct question.

What is the purpose of this report?

A request that you politely phrase as a question may be followed by either a period or a question mark.

Will you be sure to return the experimental results as soon as possible. (*or ?*)

When you have a question mark with quotation marks, you need no other mark of punctuation.

"Where am I?" he asked.

24. Quotation Marks (Quot)

In technical writing you will use quotation marks to set off short quotations and certain titles. You will have little other need for them.

24.1 Short Quotations

Use quotation marks to enclose quotations that are short enough to work into your own text (normally, less than three lines).

According to Dr. Smith, "Magnetic tape is now so reliable, it is accepted as courtroom evidence."

When quotations are longer than three lines, set them off by single spacing and indenting them. See 15.1 for an example of this style. Do not use quotation marks when quotations are set off and indented.

24.2 Titles

Place quotation marks around titles of articles from journals and periodicals.

Beller's article "Air Breathing Boosters Gain Favor" appeared in *Missiles and Rockets*.

24.3 Within Quotes

When you must use quotation marks within other quotation marks use single marks (the apostrophe on your typewriter).

"Do you have the same trouble with the distinction between 'venal' and 'venial' that I do?" asked the copy editor.

24.4 With Other Punctuation Marks

The following are the American conventions for using punctuation with quotation marks:

Commas and periods: Always place commas and periods inside the quotation marks. There are no exceptions to this rule.

G. D. Brewer wrote "Manned Hypersonic Vehicles."

Semicolons and colons: Always place semicolons and colons outside the quotation marks. There are no exceptions to this rule.

As Dr. Damron points out, "The inlet designer has to satisfy both the aircraft designer and the propulsion engineer"; the aircraft designer need satisfy only himself.

Question marks, exclamation points, and dashes: Place question marks, exclamation points, and dashes inside the quotation marks when they apply to the quote only *or to the quote and the entire sentence at the same time.* Place them outside the quotation marks when they apply to the entire sentence only.

Inside
When are we going to find the answer to the question, "What causes clear air turbulence?"

Outside
Did you read Kelly's "Scientists Consider Aerospace Transporter"?

25. Semicolon (Semi)

The semicolon lies halfway between the comma and the period in force. Its use is quite restricted. (See also Rule 2.)

25.1 Main Clauses

Place a semicolon between two closely connected main clauses that are not joined by a coordinating conjunction: *and, but, or, nor, for,* or *yet.*

The expanding gases formed during burning drive the turbine; the gases are then exhausted through the nozzle.

If the clauses are long, have internal punctuation, or if separate emphasis is desired, then the comma before the coordinating conjunction may be increased to a semicolon.

The front lawn has been planted with a Chinese Beauty Tree, a Bechtel Flowering Crab, a Mountain Ash, and assorted small shrubbery, including barberry and cameo roses; but so far nothing has been done to the rear beyond clearing and rough grading.

25.2 Series

When a series contains commas as internal punctuation within the parts, use semicolons between the parts.

Included in the experiment were Peter Moody, a freshman; Jesse Gatlin, a sophomore; Burrel Gambel, a junior; and Ralph Leone, a senior.

26. Italicization (Ital)

Italic print is a distinctive type face, like this sample: *Scientific American.* When you type or write, you represent italics by underlining, like this:

Scientific American

26.1 Foreign Words

Italicize foreign words that have not yet become a part of the English language.

We suspected him always of holding some *arrière pensée*.
Karl's everlasting *Weltschmerz* makes him a depressing companion.

Also italicize Latin scientific terms.

cichorium endivia (endive)
Percopsis omiscomaycus (trout-perch)

But do not italicize Latin abbreviations or foreign words that have become a part of the language (including such scholarly abbreviations as "ibid." and "op. cit.").

etc.	bourgeois
vs.	status quo

Your college dictionary will normally indicate which foreign words are still italicized and which are not.

26.2 Words, Letters, and Numbers Used as Such

The words *entrance* and *admission* are not perfectly interchangeable.
Don't forget the *k* in *picnicking*.
His 9's and 7's descended below the line of writing.

26.3 Titles

Italicize the titles of books, plays, magazines, newspapers, ships, and artistic works.

Webster's New World Dictionary	*The Free Press*
Othello	*S.S. Pennsylvania*
Scientific American	*Mona Lisa*

27. Capitalization (Cap)

We provide the more important rules of capitalization. For a complete rundown see your college dictionary.

27.1 Proper Nouns

Capitalize all proper nouns and their derivatives.

Places

| America | American | Americanize | Americanism |

Days of the Week and Months

| Monday | Tuesday | January | February |

But not the seasons.

| winter | spring | summer | fall |

Organizations and Their Abbreviations

American Kennel Club (AKC)

United States Air Force (USAF)

Capitalize *geographic areas* when you refer to them as areas.

The Andersons toured the Southwest.

But do not capitalize words that merely indicate direction.

We flew west over the Pacific.

Capitalize the names of *studies* in a curriculum only if the names are already proper nouns or derivatives of proper nouns or if they are part of the official title of a department or course.

Department of Geology

English Literature 25

the study of literature

the study of English literature

Note: Many nouns (and their derivatives) that were originally proper have been so broadened in application and have become so familiar that they are no longer capitalized: *boycott, macadam, spoonerism, italicize, platonic, chinaware, quixotic.*

27.2 Literary Titles

Capitalize the first word, the last word, and every important word in literary titles.

But What's a Dictionary For (by Bergen Evans)
The Meaning of Ethics (by Philip Wheelwright)
How to Write and Be Read (by Jacques Barzun)

27.3 Rank, Position, Family Relationships

Capitalize the titles of rank, position, and family relationship unless they are preceded by *my, his, their,* or similar possessive pronouns.

Professor J. E. Higgins
I visited Uncle Timothy.
I visited my uncle Timothy.
Dr. Henry W. Sams, Head, Department of English

28. Numbers (Num)

There is a good deal of inconsistency in the rules for handling numbers. We will give you the general rules. Your instructor or your organization may give you others. As in all matters of format, you must satisfy whomever you are working for at the moment. Do, however, be internally consistent within your reports. Do not handle numbers differently from page to page of a report.

28.1 As Words

Generally, you write all numbers ten and under, and rounded-off large numbers, as words.

six generators
about a million dollars

However, when you are writing a series of numbers, do not mix up figures and words. Let the larger numbers determine the form used.

five boys and six girls

But:

It took us 6 months and 25 days to complete the experiment.

28.1.1 Beginning a Sentence

Do not begin sentences with a figure. If you can, write the number as a word. If this would be cumbersome, write the sentence so as to get the figure out of the beginning position.

Fifteen months ago, we saw the new wheat for the first time.
We found 350 deficient steering systems.

28.1.2 Compound Number Adjectives

When you write two numbers together in a compound number adjective, spell out the first one or the shorter one to avoid confusing the reader.

Twenty 10-inch trout
100 twelve-volt batteries

28.1.3 Hyphens

There is much inconsistency in hyphenating numbers. Generally speaking, if a number under 100 is used as a modifier, it is hyphenated. If it stands alone it is not hyphenated.

Eighty-five boxes

But:

Eighty five should be enough.

In technical writing, the common practice is to write any exact number over ten as a figure and avoid the problem altogether.

28.2 Figures

The general rule here is to write all exact numbers over ten as a figure. This rule probably holds more true in technical writing with its heavy reliance on numbers than it does in general writing. However, as we noted in 28.1, rounded-off numbers are commonly written as words. The precise figure could give the reader an impression of exactness that might not be called for.

Certain conventional uses call for figures at all times.

28.2.1 Dates, Exact Sums of Money, Time, Address

1 January 1980 *or* January 1, 1980
$3,422.67 *but* about three thousand dollars
1:57 p.m. *but* two o'clock
660 Fuller Road

28.2.2 Technical Units of Measurement

6 cu ft
4000 rpm

28.2.3 Cross-References

See page 22.
Refer to Figure 2.

28.3 Fractions

When a fraction stands alone, write it as an unhyphenated compound.

two thirds
fifteen thousandths

When a fraction is used as an adjective you may write it as a hyphenated compound. But if either the numerator or the denominator is hyphenated, do not hyphenate the compound. More commonly, fractions used as adjectives are written as figures.

two-thirds engine speed
twenty-five thousandths
¾ rpm

29. Abbreviations (Ab)

Although most people are familiar with the kinds of abbreviations we encounter in everyday conversation and written material, from "mph" to "Mon." to "Dr.," technical abbreviations are something else again. Each scientific and professional field generates hundreds of specialized terms, and many of these terms are often abbreviated for the sake of conciseness and simplicity.

Thus, before deciding to use technical abbreviations in an article or report, you must first consider your audience—laymen or scientists, executives or engineers? Only readers with a technical background will be able to interpret the specialized shorthand for the field in question. When in doubt, then, avoid all but the most common abbreviations. If you must use a technical term, spell it out in full the first time it appears and include both the abbreviation and a definition in parentheses after it. You can then safely use the abbreviation if the term crops up again in your report.

Standard technical and scientific abbreviations include the following:

absolute	abs
acre or acres	(spell out)
alternating current (as adjective)	a-c
atomic weight	at. wt
barometer	bar.
Brinell hardness number	Bhn
British thermal units	Btu or B
meter	m
square meter	m^2
microwatt or microwatts	mu w
miles per hour	mph
National Electric Code	NEC
per	(spell out)
revolutions per minute	rpm
rod	(spell out)
ton	(spell out)

The system implied by these illustrative abbreviations can be described by a brief set of rules.

(1) Use the same (singular) form of abbreviation for both singular and plural terms.

cu ft	either cubic foot or cubic feet
cm	either centimeter or centimeters

But there are some common exceptions.

no.	number
nos.	numbers

p.	page
pp.	pages
ms.	manuscript
mss.	manuscripts

(2) Use small (lowercase) letters except for letters standing for proper nouns or proper adjectives.

abs	*but*	Btu or B
mph	*but*	Bhn

(3) For technical terms, use periods only after abbreviations that spell complete words. For example, *in* is a word, and the abbreviation for inches could be confused with it. Therefore, use a period.

ft	*but*	in.
abs	*but*	bar.
cu ft	*but*	at. wt

(4) Remember the hyphen in the abbreviations a-c and d-c when you use them as adjectives.

This a-c motor can be converted to 28 volts dc.

(5) Spell out many short and common words.

acre   rod   per   ton

(6) In compound abbreviations, use internal spacing only if the first word is represented by more than its first letter.

rpm	*but*	cu ft
mph	*but*	mu w

(7) With few exceptions, form the abbreviations of organization names without periods or spacing.

NEC   ASA

(8) Abbreviate terms of measurement only if they are preceded by an Arabic expression of exact quantity.

55 mph   *and*   20-lb anchor

But:

We will need an engine of greater horsepower.

APPENDIXES

A Student Report

4062 Hoven Street
St. Paul, MN 55108

December 4, 1979

Wyoming Lake Citizens' Committee
Marinette County Court House
Wyoming Lake, WI 54126

Dear Wyoming Lake Citizens' Committee:

Enclosed is the report entitled The Feasibility of Removing Sediment and Muck from Wyoming Lake. It has been submitted to you for your consideration in hopes of initiating the first step in restoring the lake.

The report analyzes the location, depth, and content of muck and sediment deposits. Its main concern is the feasibility of removing such deposits in hopes of rehabilitating the lake and restoring its recreational resources. The study emphasizes the legal, technological, and cost aspects of the proposed project and considers the benefits to property owners and the community. It also considers the problems which may be encountered and the time involved for the completion of the project.

I hope this report will meet with your approval. If you have any questions or problems, please call or write.

Sincerely yours,

Barbara J. Buschatz

Barbara J. Buschatz

enclosure

THE FEASIBILITY OF REMOVING

SEDIMENT AND MUCK

FROM WYOMING LAKE

Prepared for

The Wyoming Lake Citizens' Committee

and Residents of Wyoming Lake and

Marinette County, Wisconsin

by

Barbara J. Buschatz

Abstract

This report presents the results of a study that considered
the feasibility of removing sediment and muck deposits from
Wyoming Lake, Wisconsin.   The report emphasizes the legal
and technological procedures necessary.   It provides cost
estimates of the proposed project and considers the benefits
to property owners and the community.   It also considers the
problems which may be encountered and the time involved for
the completion of the project.

December 4, 1979

TABLE OF CONTENTS

LIST OF ILLUSTRATIONS

THE FEASIBILITY OF REMOVING
SEDIMENT AND MUCK FROM WYOMING LAKE

INTRODUCTION

The purpose of this report is to present the results of a study that evaluated the feasibility of removing sediment and muck from Wyoming Lake.   To be feasible, the removal should be of minimal cost and should significantly restore and increase the lake's recreational resources.

With the installation of sewer systems completed around the entire lake, it seems practical to turn our attention toward the lake's restoration.   Sediment and muck deposits have built up over recent years along most of the shoreline area, making swimming and other recreational activities undesirable and even hazardous.   The report discusses ways in which restoration can be initiated and eventually performed. The main criteria are costs to the property owners, time involved for project completion, and the benefits to the community and to property owners.

FACTUAL SUMMARY

Wyoming Lake was once a widely used recreational area and water body; however, over the past decade the lake has stagnated, clogged by muck and sediments.

There are presently 2,400,000 cubic yards of muck deposits averaging 6 to 11 feet in depth along most of the shoreline of the lake. In order to remove these deposits and rehabilitate the lake, certain steps must be taken. First, sounding tests must be performed, and a permit must be obtained from the Wisconsin Department of Natural Resources, Division of Resource Development, allowing the removal of the deposits. After permission is gained, the lake must be dredged. With prompt action on the part of the Wyoming Lake Citizens' Committee, the project could be designated a pilot project and receive a 50% subsidy.

A hydraulic dredge is the most suitable means for removing the deposits. Three alternatives for accomplishing the dredging were considered:

—Contracting with a commercial firm would cost approximately $720,000 to $1,200,000 for a complete job.

—Local government could buy a dredge and do the job itself. A new dredge and associated equipment cost $390,000. Purchased on a 10-year, 8% mortgage, the dredge and equipment would cost $567,824, including interest charges. The cost of operating the dredge for the two years estimated to be necessary to complete

2

the project would be approximately $304,000. Combining the
total purchase price and the operating costs gives a total of
$871,824. The purchased dredge would be available for future
projects or resale.

—Leasing for the two years needed to complete the project
would cost 108% of the price of a new dredge. Additional costs
would be $180,000 for the purchase of accessory equipment and
$304,000 for operating the dredge. The dredge would not be
available for resale or future use.

Deposit areas where the removed muck and sediment can be used as
fill are available within reasonable distance of the lake. These
areas can actually benefit from this fill because the lake-bottom
material tests about the same as most agricultural soils. Thus, the
deposits would improve subsoils and scalped areas and would be
suitable for filling and reclaiming low marsh lands.

The project will increase property values and tax revenues and
leave the community with a healthy recreational environment in
which to swim, fish, and ski.

Problems that may be encountered, such as payment, poor
communication with the public, and approval of dredging work as it
is accomplished were considered. Long-term payment plans and the
appointment of a project coordinator would seem to solve most such
potential problems.

3

CONCLUSIONS:

1. Wyoming Lake will become completely useless as a recreational resource or water body if its muck and sediment deposits are not removed. (p. 6)

2. Permission for removing the deposits must be obtained from the Department of Natural Resources, Division of Resource Development. (pp. 10-13)

3. With prompt action, state and federal assistance may be available for up to 50% of cost. (p. 16)

4. Leasing of a dredge by the city is not economical. (p. 17)

5. The total cost of purchasing a dredge and operating it is higher than the lowest contract estimate and considerably lower than the highest. (pp. 16-20)

6. The purchased dredge would have value for future projects or resale. (p. 19)

7. Purchasing a dredge seems a sounder plan than either leasing a dredge or contracting the project.

8. The community will benefit from having a healthy recreational environment with benefits to business, industry, and the community residents. Increased property values will bring increased tax revenues. (p. 20-21)

9. Potential problems seem to have practical solutions. (p. 22)

10. Sounding of the lake and prompt application for approval of the project are the next steps to implement the project. (pp. 10-13, 16)

11. The project, given the right decisions and prompt action, is feasible.

4

RECOMMENDATIONS

Local government and the Wyoming Citizens' Committee should
—Perform soundings to initiate the first step toward the lake's rehabilitation.
—Apply for approval for the project from the Wisconsin Department of Natural Resources, Division of Resource Development.
—Take the steps necessary to finance the project through local assessment and state and federal assistance.

If the project can be financed, I recommend that local government carry it out by
—Purchasing a dredge and the associated equipment.
—Appointing a project coordinator.
—Working out a long-term payment plan for property owners.

5

A.   LOCATION AND DEPTH OF MUCK AND SEDIMENT DEPOSITS

The locations of the sediment and muck deposits in
Wyoming Lake are indicated by the shaded areas in Figure 1,
designated A, B, C, etc.   The accompanying table indicátes
the square yardage in each area.   The mapped areas indicate
that a considerable portion of the lake has muck and sediment
deposits.   In over five miles of shoreline there are few
areas that could be used for wading or swimming.   The only
area that does not contain muck and sediment deposits is the
shoreline along the park south end.   However, swimming has
been discontinued at this area also.   (See Figure 1.)

Depths of a selected number of sediment and muck deposits
in Wyoming Lake are indicated in Figure 2.   The readings were
taken at selected locations around the lake to give an overall
picture of the depth of muck deposits beneath the water in
Wyoming Lake.   The depth readings show that muck deposits are
generally 6 to 11 feet deep beneath 3 to 5 feet of water surface.   (See Figure 2.) The muck deposits extend out over a
large surface area, in some cases 300 to 400 yards from the
shoreline.   Removal of 6 feet of the deposits would provide
water depth of 9 to 12 feet deep and eliminate most muck deposits down to the solid base.

Table 1 provides estimates of the volume of muck
removed and the volume of water obtained when removing 3 feet
to 6 feet of the muck deposits. The elimination of the large
quantity of muck deposits, coupled with the addition of 260
to 520 million gallons of water, will substantially improve
the condition of Wyoming Lake.

6

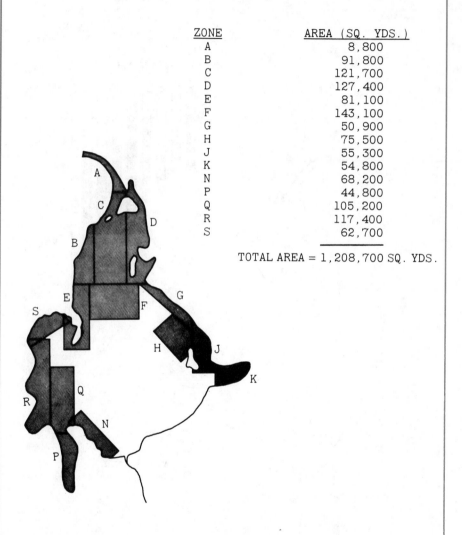

Figure 1.
Muck and Sediment Deposits in Wyoming Lake

ZONE	AREA (SQ. YDS.)
A	8,800
B	91,800
C	121,700
D	127,400
E	81,100
F	143,100
G	50,900
H	75,500
J	55,300
K	54,800
N	68,200
P	44,800
Q	105,200
R	117,400
S	62,700

TOTAL AREA = 1,208,700 SQ. YDS.

Figure based on data from the following unpublished report:
Charles J. Andersen, Muck and Sediment Deposits in Wyom-
ing Lake (A report for the Wyoming Lake Citizens Committee,
Wyoming Lake, Wisconsin, 1973), pp. 6-8.

7

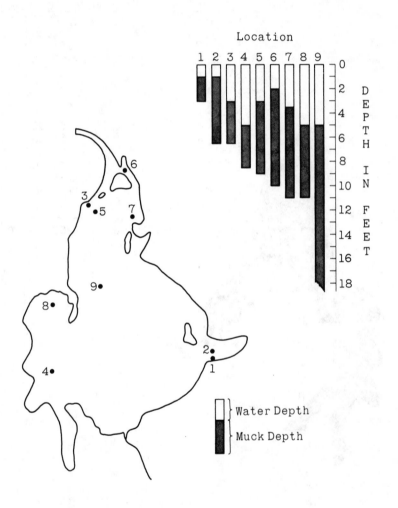

Figure 2.
Sediment and Muck Depths in Wyoming Lake

Location

DEPTH IN FEET

Water Depth

Muck Depth

Figure based on data from the following unpublished report:
Charles J. Andersen, Muck and Sediment Deposits in Wyom-
ing Lake (A report for the Wyoming Lake Citizen's Committee,
Wyoming Lake, Wisconsin, 1973), p. 8.

8

TABLE I. Muck and Sedimentation Volumes in Wyoming Lake

DEPTH OF MUCK REMOVED	VOLUME OF MUCK REMOVED	NEW WATER DEPTH	WATER ADDED TO LAKE BODY
3 feet	1.2 million cubic yards	6 to 8 feet	262 million gallons
6 feet	2.4 million cubic yards	9 to 11 feet	524 million gallons

To determine what the removed sediments consisted of and what possible usefulness they have, three samples were taken to the soils lab at the University of Wisconsin in Madison. The soils specialist was informed as to the source of the sediments and was asked to submit a determination of their possible usefulness.   The submitted samples were compared to average agricultural soils.   The test results (5) indicate that

1.   The lake-bottom sediments test about the same as agricultural soils.

2.   Addition of these sediments would improve subsoils or scalped areas.

The sediments and muck deposits can, therefore, be used to improve subsoils and scalped areas and would be suitable for filling and reclaiming low marsh areas.   If arrangements are made to drain off the water from the deposits, the deposit areas will be ideal for park or recreation areas or agricultural use.

9

B.   METHOD SUITABLE FOR REMOVING THE LAKE DEPOSITS

In establishing a method to remove the lake deposits,
three factors must be considered: needed equipment, legal
steps to be taken, and the disposition of the deposits.

Equipment

Correspondence with two dredging companies and a dredge
manufacturer (2) indicated that the recommended equipment
for the removal of the muck and sediment deposits from Wyoming
Lake is a hydraulic dredge, as illustrated in Figure 3.   A
hydraulic dredge operates by loosening the lake solids with a
large revolving cutterhead (36" diameter and up).   The solids which are then suspended in water are drawn up through a
suction pipe to a pump where they are forced through a
pipeline to the shore.   A dredge advances through the muck by
swinging from side to side using its rear spuds as a
pivot.   The swinging action is obtained by the use of two
swing cables attached to swing anchors.

Choice of the right dredge for a project depends on the
operation and conditions—such as type of material being removed, amount removed, depth of cut, distance to deposit
areas, height to which material must be pumped and time needed
to complete the project.

Dredges are rated by the diameter of their intake pipe,
such as 6", 8", 12", 14", 16" and 20".   Most dredges can be
dismantled into easily reassembled sections and transported
on trucks and rail cars to job locations.

Legal Restrictions and Procedures

The authority and responsibility of approving or reject-
ing lake improvement projects has been delegated to Wisconsin's Department of Natural Resources, Division of Resource

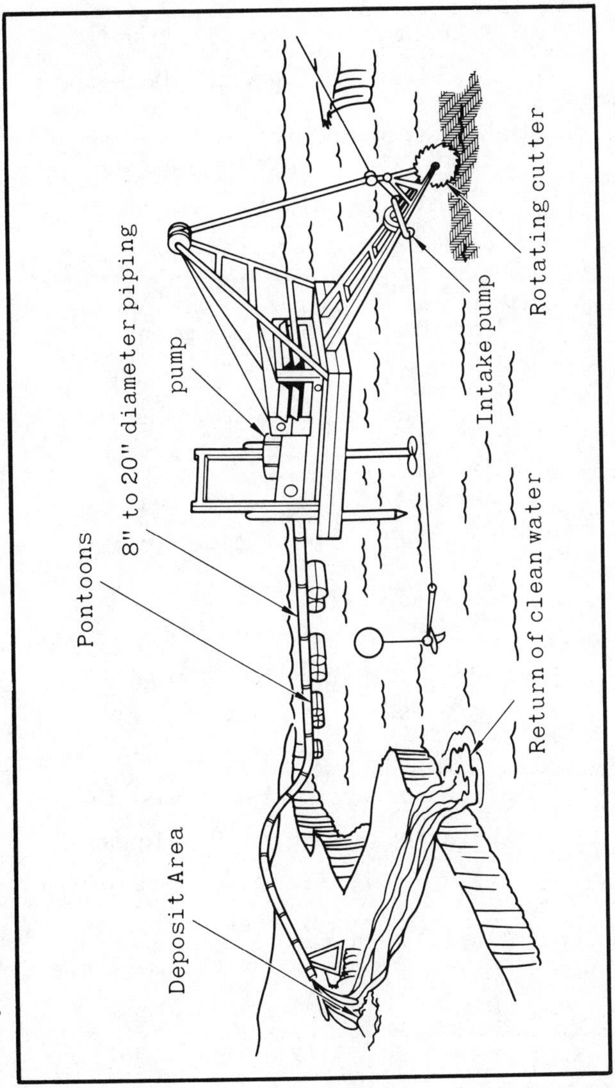

Figure 3.
The Hydraulic Dredge

Development.   (4)   To obtain a permit to remove materials from
Wyoming Lake, applicants must follow this procedure (see also
Figure 4):

1.   Request forms of application from
     the Division of Resource Develop-
     ment.

2.   Submit three copies of the com-
     pleted forms along with three
     copies of the required maps of the
     area to be dredged, showing sound-
     ings (depth measurements) and
     shoreline contour and staking.
     The soundings should indicate the
     depth of water and muck, and should
     be taken at intervals of twenty
     (20) feet, to beyond the limits of
     the proposed dredging.

3.   The Division of Conservation,
     Bureau of Fish Management is then
     informed by the Division of Re-
     source Development of the applica-
     tion, and will in turn advise the
     Division of Resource Development
     if dredging will be consistent with
     conservation interest.

4.   After a review of its findings, the
     Division of Resource Development
     will then notify the applicant
     whether or not permission to pro-
     ceed with the project is granted.

12

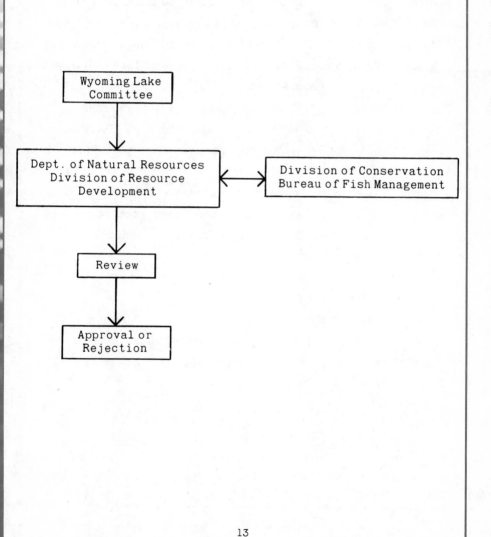

Figure 4.
Application Route for Permission to Remove Muck and Sediment

13

Disposition of Deposits

Suitable deposit areas are indicated in Figure 5.   The
distance to the deposit area is also shown, because of its importance in the selection of suitable equipment.

The most accessible deposit area is the gravel pit located at Tans Drive, approximately 3,000 to 4,000 feet from
the lake body.   It is nearing its depleted stage and is deep
and large in area.   Its elevation above the lake is approximately 80 feet, which would call for the use of larger dredging equipment or a booster pump.

The second most desirable deposit area is the marsh area
adjacent to the intersection of Highways Y and 24.   Although
its depth does not exceed 8 to 10 feet, the area is considerable.   Water will drain away from it toward the south watershed.   This area is closest to the lake and is at a low elevation relative to the lake.   Because new construction for
the relocation of Highway Y is in progress, it would be ad-
vantageous to fill the area after the new highway construction is completed.   The filled marsh area may prove to be an
ideal park or recreation area.

The remaining deposit areas are the Valley Sand and
Gravel Company pits some 4,000 to 5,000 feet northwest of the
lake and the marsh area east of the lake, also in need of fill
to become a useful area.

14

Figure 5.
Proposed Deposit Areas for Muck and Sediment
Removed from Wyoming Lake

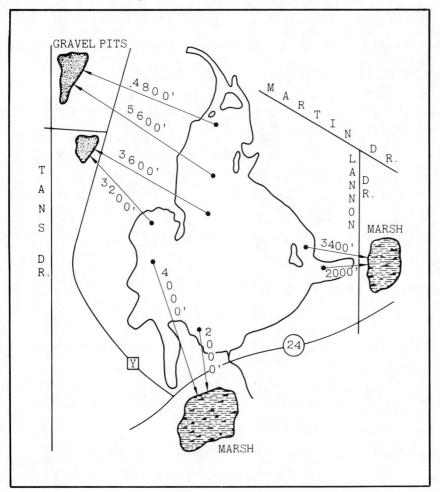

C.   APPLICATION OF CRITERIA

The major criteria bearing on the feasibility of the
project are time and cost and benefits to both property owners
and the community at large.   Cost will depend on which of
three alternatives is chosen: contracting the entire project. leasing the necessary equipment, or buying the necessary equipment.

Estimated Time to Complete the Project

Firms specializing in dredging were asked to provide estimates of the time needed to dredge Wyoming Lake.   An average of their estimates indicates a two-year period, including
four winter months each year when ice on the lake will prevent
operations.   (2)

Estimated Costs of the Project

Under the federal Economic Development Act, the Wisconsin Department of Natural Resources is developing a project
to dredge large lakes in the northern depressed Wisconsin region.   If successful, the state plans to use the knowledge of
methods and costs to advise local development groups.   It is
possible that, through this program, the dredging of Wyoming
Lake could be designated a pilot project and be subsidized at
50% of cost.   (6)

Three plans will be evaluated in this project cost
analysis:

    a.   contract dredging,

    b.   lease of dredging equipment, and

    c.   purchase of dredging equipment.

Contract dredging estimate.   Cost estimates based on
contract dredging are calculated predominantly on the cubic
yardage of the material removed.

16

To calculate the cost of the project on a contracting basis, it was assumed that the removal of 6 feet of muck would be desired in the areas proposed on the map shown in Figure 1.

Therefore, the muck volume removed would be:

2 yards deep x 1.2 million square yards, equaling 2.4 million cubic yards of muck.

Based on a contractor price (2) of $.30 to $.50 per cubic yard--the price span found--the cost would range as follows:

$.30 x 2,400,000 = $720,000

to

$.50 x 2,400,000 = $1,200,000

Based on a shoreline of 5 miles (26,400 feet), the cost per linear foot of shoreline would range from approximately $27.30 to $45.45. If federal and state assistance could be found to cover half the cost, the price per linear foot would be approximately $13.65 to $22.70.

Lease of dredging equipment. Hydraulic dredges are available for lease from the dredge manufacturers. For use of a dredge for a two-year period, the estimated monthly rate is 3% of the selling price of the dredge. If the dredge is used for more than one shift, the monthly rate is 4.5% of the selling price. (2) Rent must be paid during the idle months when ice is on the lake.

Leasing of dredging equipment is not economical when large volumes of materials must be removed. Since Wyoming Lake contains 2,400,000 cubic yards of sediment and muck and will require a two-year dredging period, the leasing of a hydraulic dredge is not recommended. At the 4.5% rate, which will be necessary, the cost of leasing would be 108% the cost of the dredge.

17

The lease also does not include the required accessory equipment which must therefore be purchased.   Accessory equipment for the Wyoming Lake project includes (2):

Booster pump	$80,000
Piping	$50,000
Pontoons and connectors	$50,000
	$180,000

In addition, the operating costs must also be paid by the user of the equipment.

Purchase of dredging equipment.   Estimates of the cost of dredging equipment suitable for Wyoming Lake and of the operating costs are based on the following operational requirements:

Maximum distance pumped .... 6,000 feet
Maximum discharge elevation .... 82 feet
Total material removed .... 2,400,000 cubic yards
Maximum digging depth .... 12 feet
Months per year operating .... 8 months
Hours per day in operation .... 16 to 20 hours
Time to complete the project ...... 2 years
Total personnel required .... 6 workers

Based on the above requirements, the proposed equipment consists of the following (2):

1	14" Hydraulic dredge	$210,000
1	14" Booster pump	$ 80,000
6,000 feet	16" Discharge pipe	$ 50,000
	Pontoons, connectors and joints	$ 50,000
	Total Equipment Cost	$390,000

18

The booster pump is required because of the long distance
to the farthest deposit areas and the necessity of pumping to
an elevation of 82 feet.

The 16" discharge pipes are specified to reduce the fric-
tional loss in the pipe, so that the long distances and high
elevations can be accommodated.

It is assumed that the municipality would purchase the
equipment and bear the cost of ownership.   A well-maintained
dredge will last from 10 to 20 years, and the dredge's re-
sidual value at the end of that time is estimated at 20% of its
original cost.   (2) Purchased on a 10-year, 8% mortgage, the
dredge and associated equipment would cost $567,824, includ-
ing interest costs.   (3)

The operating costs of the city-owned dredge are assumed
to be paid by the property owners on and around the lake
body.   Operating costs are estimated to be as follows (2):

Fuel, oil, lubrication	$13.00/Hour
Labor (3 workers per shift)	$15.00/Hour
Repairs and maintenance (dredge)	$10.00/Hour
Repairs and maintenance (accessories)	$10.00/Hour
Total Operating Costs	$48.00/Hour

With the equipment selected and under the worst conditions (maximum distance and height) the production capacity
of the dredge would be 270 cubic yards per hour.   However, the
dredge has an average hourly production of 300 to 600 cubic
yards per hour, and when the deposit areas are close to the
lake body the cost per cubic yard is significantly reduced:

19

	270 Cu. Yd./Hr.	400 Cu. Yd./Hr.	600 Cu. Yd./Hr.
Dredging Condition	Worst	Average	Best
Operating Cost Per Cubic Yard	$0.18	$0.12	$0.08

Assuming that the project offered all dredging conditions equally (worst, average, and best), and that the total
muck to be removed was 2,400,000 cubic yards, the cost would
be as follows: $144,000 (worst) plus $96,000 (average) plus
$64,000 (best), for a total cost of $304,000.   Based on a
shoreline of 5 miles (26,400 feet), the cost per linear foot
of shoreline would be approximately $11.50.   If federal and
state assistance could be found to cover half the cost, the
price per linear foot would be approximately $5.75.

Since the property owners near the lake benefit most directly by the restoration of the lake, it would be reasonable
to suggest that they bear a portion of the operating costs.

Benefits

If Wyoming Lake is restored by the removal of a large portion of its sediments and muck deposits, there will be many
benefits to the property owners and the community.

Benefits to property owners on the lake.

1.   Property value will be increased.
2.   Recreational activities such as swimming, water ski-
       ing, fishing, and boating will increase.
3.   Health hazards will decrease.
4.   Views and sights will be improved.
5.   Undesirable odors will be eliminated.

<u>Benefits to the community</u>.

1. Lake property valuations will increase.

2. Recreational services for the city residents will im-
    prove.

3. Non-useable wetlands will be converted to tax-
    earning properties.

4. Commercial business to service the recreational
    needs of the citizens and visitors will be attracted.

D.   PROBLEMS WHICH MAY BE ENCOUNTERED

Problems	Method to Solve the Problem
Property owner feels that the project is too costly.	Provide a payment plan which will spread the payments over a five-year period, for those who wish it. Show the property owner that the value of owned property will increase, the recreational utility will be enhanced, and the water will be cleaner and healthier.
Uninformed public	Continuous communication must be established between the dredgers' agencies, local government, and the people. The people should be fully informed at all times, preferably by a newsletter.
Complaints during dredging	Assign a coordinator to manage and supervise the project. All complaints, from the property owners or the dredgers, should be directed to the coordinator for solution.
Problems arising with state agencies	The coordinator should have the authority and responsibility for communicating with the state agencies.
Approval of the work as it proceeds	The completed dredged areas should be approved and signed off by the coordinator.

22

BIBLIOGRAPHY

1. Andersen, Charles J. Muck and Sediment Deposits in Wyoming Lake. A report prepared for the Wyoming Lake Citizens' Committee, Wyoming Lake, Wisconsin, 1973.

2. Correspondence with Madison Dredging Company, Madison, Wisconsin; Capital Dredging Company, St. Paul, Minnesota; and Mississippi Dredges, Inc., St. Louis, Missouri.

3. Interest and Mortgage Tables. Framingham, Mass.: Ottenheimer Publishers, Inc., 1968.

4. Legal Procedures, Policies, and Restrictions. Bulletin 862-74. Wisconsin Department of Natural Resources, 1977.

5. Mason, Roger H. Soil Test Report for Wyoming Lake, Wisconsin. Department of Soil Science, University of Wisconsin, Madison, Wisconsin, 23 September 1979.

6. Telephone conversation with James H. Olson, Wisconsin Department of Natural Resources, Madison, Wisconsin, 20 November 1979.

APPENDIX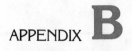

Technical Reference Guides

Prepared by Donald J. Barrett, Chief Reference Librarian, United States Air Force Academy

Book Guides

Information on books in science and technology is generally found in standard guides to book publication and general reviewing tools. A few tools devoted specifically to technical publications do exist, but most cover only a small portion of each year's book production, so the more general tools must be used. Several of the major guides are listed, and will be found in most libraries. Each should be examined for its inclusion policy and limitations the first time you see it.

Book review digest. 1905–
> (Index to selected book reviews, mostly from general periodicals. Gives publication data, brief descriptive notes, exact citations to reviews in about 75 periodicals. Subject and title indexes.)

Books in print and *Subject guide to books in print.* Annual.
> (Guide to book availability (in print) from 3600 American publishers. *Books in print:* Author and title lists—give author, title, usually date of publication, edition, price and publisher. *Subject guide to books in print:* Books entered under Library of Congress subject heading, then by author, with title and other bibliographic data.)

Cumulative book index. 1928–
> (World list of all books published in the English language. Author, title and subject listings arranged in a dictionary sequence. Main entry (fullest information) under author. Give author, title, edition, series, pagination, price, publisher, date, Library of Congress card number.)

Library of Congress catalog—Books: Subjects. 1950–
> (Cumulative list by subject of works represented by Library of Congress printed cards for publications printed 1945 or later.)

New technical books. 1915–
   (Selectively annotated titles by New York Public Library staff. Classed subject arrangement, includes table of contents, annotation for each book.)

Proceedings in print. 1964–
   (Announcement journal for availability of proceedings of conferences. Full citation, price, subject and agency indexes.)

Technical book review index. 1935–
   (Very good guide to reviews appearing in scientific, technical, and trade journals, with bibliographic data and exact references to sources of reviews.)

United Nations documents index. 1950–
   (Contains checklists of documents and publications issued by various U.N. agencies, with subject and author indexes cumulated annually.)

Reference Books

Each subject or academic field frequently has its own literature, often ranging from encyclopedias to indexes, dictionaries, biographical tools, etc. The amount of literature specific to one subject may vary greatly. When first approaching a field, one principal guide, Sheehy's, is available to assist you in becoming familiar with its literature. The current edition of this guide should be found in every major collection.

Guide to reference books. Eugene P. Sheehy, ed. 9th ed. Chicago: American Library Association, 1976. Supplements issued biennially.

Encyclopedias

The encyclopedia, while considered too general by some specialists, is often extremely useful to the researcher in getting under way and learning a field. Several of the encyclopedias for specific subjects are in fact quite detailed and scholarly. The editors and contributors to a well-written special encyclopedia will often be experts in their fields. Of course, the latest developments in a subject would be available only in periodical and report literature. Still, the general works are of great value toward understanding a subject, and often include bibliographic citations to aid in further research.

Dictionary of organic compounds: the constitution and physical, chemical, and other properties of the principal carbon compounds and their derivatives, together with relevant literature references. Ian M. Heilbron, ed. 4th rev. ed. New York: Oxford University Press, 1965. 5 volumes, 2 cumulative supplements, 1965–73, plus annual supplements. 1975–
   (Alphabetical list of compounds, large number of cross-references.)

Encyclopedic dictionary of physics. New York: Pergamon Press, 1961–. 9 volumes, 5 supplements.
   (Scholarly work, alphabetically arranged, articles generally under 3000 words, most with bibliographies. Includes articles on general, nuclear, solid state, molecular, chemical, metal, and vacuum physics. Index, plus a multilingual glossary in six languages.)

Encyclopedia of chemical technology. Raymond Kirk and Donald Othmer, eds. 2d ed. New York: Wiley-Interscience Publishers, 1963–71. 22 volumes, index and supplement. 3d ed. began publication in 1978.
(Main subject is chemical technology; about half the articles deal with chemical substances. There are also articles on industrial processes. A bibliography is included for each product, as well as information on properties, sources, manufacture, and uses.)

Encyclopedia of engineering signs and symbols. New York: Odyssey Press, 1965.
(Compilation of all graphic symbols used on engineering drawings that are most widely accepted in their respective engineering fields. Includes military standards, industry standards established by associations of the particular industries concerned, and engineering standards established by professional societies.)

Encyclopedia of polymer science and technology: plastics, resins, rubbers, fibers. New York: Wiley-Interscience Publishers, 1964–72. 15 volumes plus index, supplement.
(Articles designed to present a balanced account of all aspects of polymer science and technology, with bibliographies included.)

Encyclopedia of the biological sciences. Peter Gray, ed. 2d ed. New York: Van Nostrand Reinhold, 1970.
(800 articles covering the broad field of the biological sciences as viewed by experts in their developmental, ecological, functional, genetic, structural, and taxonomic aspects. Bibliographies, biographical articles, illustrations, and diagrams are helpful features.)

McGraw-Hill encyclopedia of science and technology. 4th ed. New York: McGraw-Hill Book Co., 1977. 15 volumes, annual supplements.
(Main set includes 7600 articles, kept current by annual volumes. Covers the basic subject matter of all the sciences and their major applications in engineering, agriculture, and other technologies. Separate index volume. Has many diagrams and charts, and complicated subjects are treated in clear and readable language. Contributors identified in index volume.)

Van Nostrand's scientific encyclopedia. 5th ed. Princeton: D. Van Nostrand Co., 1976.
(Includes articles on both basic and applied sciences. The encyclopedia defines and explains over 16,500 terms, arranged alphabetically with extensive cross-references.)

Subject Guides
Bibliographers and librarians have gathered research suggestions and bibliographies for many fields into published guides. It should be remembered that the rapidly changing literature in many subjects partially outdates any guide. Therefore, Sheehy's *Guide to reference books* and other tools listing current books and indexes should be consulted to supplement these guides.

Current information sources in mathematics; an annotated guide to books and periodicals, 1960–72. Elie M. Dick. Littleton, Colo.: Libraries Unlimited, 1973.

(Classed arrangement, contents of books noted, with subject and author indexes.)

Geologic reference sources: a subject and regional bibliography to publications and maps in the geological sciences. Dederick Ward and Marjorie Wheeler. 2d ed. Metuchen, N.J.: Scarecrow Press, 1972.

(Subject bibliography, most items not annotated. Subject and geographic indexes. Has a section on geologic maps.)

Guide to basic information sources in engineering. Ellis Mount. New York: Wiley and Sons, 1976.

(Annotated list by type material and subject. Author and title index.)

Guide to the literature of the life sciences. Roger C. Smith and W. M. Reid. 8th ed. Minneapolis: Burgess Publishing Co., 1972.

(Stresses the importance of knowing the literature on a subject and using that knowledge effectively.)

Guide to scientific and technical journals in translation. Carl E. Himmelsbach and Grace Brociner. 2d ed. New York: Special Libraries Association, 1972.

(Lists 278 cover-to-cover English translations of foreign-language journals and their availability.)

Mechanical engineering: the sources of information. Bernard Houghton. Hamden, Conn.: Archon Books, 1970.

(Some British emphasis, but useful to students of the subject.)

Reference sources in science and technology. Earl J. Lasworth. Metuchen, N.J.: Scarecrow Press, 1972.

(Guide to performing literature searches, gives bibliographic references. Author and title indexes.)

Science and engineering literature: a guide to reference sources. Harold R. Malinowsky. 2d ed. Littleton, Colo.: Libraries Unlimited, 1976.

(Selected evaluative list of basic reference sources, arranged by major subjects such as physics, chemistry, astronomy, etc.)

Science and technology: an introduction to the literature. Denis Grogan. 3d ed. Hamden, Conn.: Linnet Books, 1976.

(Student guide to structure of the literature of science and technology.)

Science information sources: a universal and international guide. Herman de Jaeger. 2d ed. Nijmegen, Holland: Association Scientifique et Technique, 1974.

(Selective list by subject of scientific and technical information sources.)

Scientific and technical information sources. Ching-Chih Chen. Cambridge, Mass.: MIT Press, 1977.

(Arranged by type of material, then subject. Annotated entries arranged by title. Author index.)

Technical information sources: a guide to patent specifications, standards, and technical reports literature. Bernard Houghton. 2d ed. Hamden, Conn.: Linnet Books, 1972.

(British emphasis, covers use of patents as source of technical information, use of specifications and reports.)

The use of biological literature. Robert T. Bottle and H. V. Wyatt, eds. 2d ed. Hamden, Conn.: Archon Books, 1971.
(Articles describing indexing and abstracting services to primary sources of biological information.)

The use of earth sciences literature. David Wood, ed. Hamden, Conn.: Archon Books, 1973.
(British author, but with emphasis on international literature sources. Includes bibliographies of basic works and sources of abstracts for each subject area.)

Use of engineering literature. K. W. Mildren, ed. London: Butterworths, 1976.
(Discusses forms of literature and services, with 22 chapters of various engineering fields.)

Use of mathematical literature. A. R. Dorling, Ed. London: Butterworths, 1977.
(Describes general literature and use of particular tools, then critical accounts by subject experts in their fields. Author and subject indexes.)

Use of physics literature. Herbert Coblans, ed. London: Butterworths, 1975.
(Subject chapter bibliographic essays, with selective indexes.)

Using the chemical literature: a practical guide. Henry Woodburn. New York: Marcel Dekker, 1974.
(Practical guide to use of chemical literature, not a comprehensive bibliography of sources.)

Bibliographies

The literature of a field may often be compiled into bibliographies in connection with other publications, and in some cases as an indication of the work of an agency or company. A guide to bibliographies and examples of other types or compilations are given here.

Bibliographic index: a cumulative bibliography of bibliographies. 1937–
(Alphabetical subject list of separately published bibliographies and bibliographies appearing in books, pamphlets, and periodicals.)

Bibliography of agriculture. 1942–
(Classified bibliography of current literature received in the National Agricultural Library, with cumulative annual subject and author indexes.)

Bibliography and index of geology. 1974–
(Index produced by Geological Society of America. Arranged in broad subject categories, with author and subject indexes. Formerly *Bibliography and index of geology exclusive of North America.*)

Bibliography of North American geology. 1919–
(Comprehensive bibliography with detailed indexes, from U.S. Geological Survey.)

Chemical titles. 1960–
(Author and keyword indexes to titles from 700 journals in pure and applied chemistry. A computer-produced bibliography.)

Dissertation abstracts international; abstracts of dissertations available on microfilm or as xerographic reproductions. 1952–
(Compilation of abstracts of doctoral dissertations from most American universities. Since 1966, Part B has been devoted to the sciences and engineering.)

Index of selected publications of the RAND Corporation. 1946–
(Coverage includes unclassified publications of the Corporation. Abstracts, listed by subject and author, describe content and indexes.)

Science citation index. 1961–
(Computer-produced index which provides access to related articles by indicating sources in which a known article by an author has been cited. Not ideally suited to subject searching.)

Translations register-index. 1967–
(Lists new translations received in the Special Libraries Association translations center, plus those available from government and commercial sources.)

Vertical file index. 1935–
(Subject and title index to selected pamphlet material in all subjects of interest to the general library.)

Biographies

Identification of authors and significant figures in the scientific and technical areas is frequently a problem. Some of the most notable biographical sources are commented on here. A check of Sheehy's *Guide to reference books* will reveal many more directories in almost every major subject field.

Biography index. 1946–
(Index to biographical material in books and magazines. Alphabetical, with index by profession and occupation.)

American men and women of science. New York: Jacques Cattell/R. R. Bowker, 13th ed. 1976. 8 volumes.
(Standard biographical set for personages in the sciences.)

Dictionary of scientific biography. New York: Charles Scribner's Sons, 1970–78. 14 volumes and supplement.
(Comprehensive, covers historical and current persons in science fields.)

Who's who in America: a biographical dictionary of notable living men and women. Chicago: Marquis, 1899– . Biennial.
(The standard dictionary of contemporary biographical data. Regional volumes cover persons not of national prominence.)

Who's who in science in Europe: a new reference guide to West European scientists. New York: International Publications Service. 2d ed. 1971–72. 4 volumes.
(More than 40,000 entries including natural and physical sciences.)

Dictionaries

Definition of terms for the student and scholar is a problem in the sciences, as in any field. A few general guides are available in addition to glossaries for a single field.

Chambers dictionary of science and technology. T. G. Collocott, ed. New York: Barnes and Noble, 1972.
(Successor to *Chamber's technical dictionary,* completely revised.)

Compton's dictionary of the natural sciences. Chicago: Compton, 1966, 2 volumes.
(Written in simple language for the nonspecialist at all educational levels.)

A dictionary of physical sciences. John Daintith, ed. New York: Pica Press, 1977.
(Includes some diagrams and cross references.)

Dictionary of technical terms. Frederic S. Crispin, ed. 11th ed. rev. New York: Bruce, 1970.
(Definitions of commonly used terms, first published in 1929.)

McGraw-Hill dictionary of the life sciences. Daniel N. Lapedes, ed. New York: McGraw-Hill Book Co., 1976.
(Provides vocabulary of the biological sciences and related disciplines. Over 20,000 terms, useful appendixes.)

McGraw-Hill dictionary of scientific and technical terms. Daniel N. Lapedes, ed. New York: McGraw-Hill Book Co., 1974.
(Gives almost 100,000 definitions, amplified by 2800 illustrations. Each definition identified with the field of science in which it is primarily used.)

Modern science dictionary. A. Hechtlinger, 2d ed. Palisade, N.J.: Franklin, 1975.
(Briefly defined terms for beginning science students.)

Commercial Guides

Access to materials from companies working on a specific product can be facilitated by the use of product association and company address information. The standards are an example of tools which many industries must use to satisfy a contractor's requirements.

Annual book of ASTM standards, with related material. Philadelphia: American Society for Testing and Materials. Annual. Approximately 50 parts including index.
(Contains 4900 ASTM Standards and Tentatives in effect at the time of publication. An example of an essential reference book in industrial technology.)

MacRae's blue book. Chicago: MacRae's Blue Book. Annual. 5 volumes.
(Buying directory for engineering products from over 50,000 companies.)

Thomas register of American manufacturers and Thomas register catalog. New York: Thomas Publishing Co. Annual. 11 volumes.
(Products lists, alphabetical listing of manufacturers, trade names.)

Periodicals

In almost any current research, the latest developments in a field will be published in the current periodicals and professional journals. Your first task as researcher may be to determine what periodicals are published in a given field. Next, you may want to determine what indexing or abstracting services give access to a specific journal. The guides are the most significant in assisting you to locate this type of information.

Ulrich's international periodicals directory; a classified guide to current periodicals, foreign and domestic. Biennial, with supplement. New York: R. R. Bowker Co.

(Alphabetical subject list of over 55,000 periodicals published in all languages. Alphabetical title index. Indicates coverage of titles in periodical indexes and abstracting services.)

Irregular serials and annuals; an international directory. 4th ed. New York: R. R. Bowker Co., 1976–77.

(Publication information on over 30,000 serials issued annually or irregularly. Lists many yearbooks, proceedings, transactions, etc. Alphabetical subject arrangement, title index.)

Gebbie Press house magazine directory: a public relations and free lance guide to the nation's leading house magazines. New York: Gebbie Press, 1952– Triennial.

(Alphabetical list of magazines published by industrial organizations.)

World list of scientific periodicals published in the years 1900–1960. 4th ed. London: Butterworths, 1963–65, 3 volumes. Supplement, 1960–68. 1970.

(Lists more than 60,000 titles of periodicals concerned with the natural sciences and technology.)

Periodical Indexes

A periodical index provides ready access to articles appearing in professional journals and general periodicals. Each index should be examined for inclusion principles, entry format, and any peculiarities unique to that index. The principal indexes for your consideration cover both general and specific fields, and the primary newspaper index to the *New York Times* is also listed.

Agricultural index, subject index to a selected list of agricultural periodicals and bulletins. 1916–64.

(Detailed alphabetical subject index. Continued as *Biological and agricultural index.*)

Air University Library index to military periodicals. 1949–

(Subject index to significant articles in approximately 70 military and aeronautical periodicals not covered in readily available commercial indexes.)

Applied science and technology index. 1958–

(Subject index to periodicals in aeronautics, automation, chemistry, construction, electricity and electrical communication, engineering, geology and metallurgy, industrial and mechanical arts, machinery, physics, transportation and related subjects. Formerly part of the *Industrial arts index.*)

Art Index. 1929–

(Author and subject index to fine arts periodicals, but includes coverage of architecture, graphic arts, industrial design, planning, and landscape design.)

Biological and agricultural index. 1964–
   (Detailed subject index to approximately 190 English language periodicals. Reports, bulletins, and other agricultural agency publications formerly covered in the *Agricultural index* are no longer covered.)

British technology index. 1962–
   (Current subject guide to articles in British technical journals, with author index.)

Business periodicals index. 1958–
   (Subject index to business, financial, and management periodicals, and specific industry and trade journals. Formerly part of the *Industrial arts index.*)

General science index, 1978–
   (Subject index to 90 periodicals in fields including astronomy, atmospheric sciences, biological sciences, earth sciences, environment and conservation, genetics, oceanography, physics, physiology, and zoology.)

Index to U.S. Government periodicals. 1974–
   (Computer-generated guide to 139 selected titles by author and subject.)

Industrial arts index. 1913–1957.
   (Subject index, split into *Applied science and technology index* and the *Business periodicals index.*)

New York Times index. 1851–
   (Subject index with precise reference to date, page and column for each article. Well cross-referenced, brief synopses of articles. Can serve as a guide to locating articles in other unindexed papers.)

Public affairs information service bulletin (PAIS). 1915–
   (Very useful index to government, economics, sociology, etc., covering books, periodicals, documents, and reports. Includes selective indexing of over 1000 periodicals.)

Readers' guide to periodical literature. 1900–
   (Best known periodical index. Covers U.S. periodicals of a broad, general nature in all subjects and scientific fields.)

Selected references on environment quality as it relates to health. 1971–
   (Indexing service from the National Library of Medicine. Subject and author sections. Covers U.S. and foreign titles.)

Social sciences index. 1974–
   (Author and subject index to periodicals in fields including economics, environmental science, psychology, public administration, and related fields. Covers 263 titles on a more scholarly level than the *Readers' guide.* Formerly part of the *Social sciences and humanities index* and the *International index to periodicals.*)

Abstract Services

   The abstracting journals assist access to periodical, book, and report literature as does an index. The significant difference is the abstract itself, which frequently gives a better indication of article content, and hence makes the abstracting journal significantly more useful than the periodical index.

A guide to the world's abstracting and indexing services in science and technology. Washington: National Federation of Science Abstracting and Indexing Services, 1963.

(List of 1855 titles originating in 40 countries. Covers the pure and applied sciences including medicine, agriculture, etc. Includes country and subject indexes.)

Abstracts and indexes in science and technology: a descriptive guide. Dolores B. Owen. Metuchen, N.J.: Scarecrow Press, 1974.

(Describes about 125 abstract and indexing services by subject coverage, arrangement, indexes, abstracts, and other features.)

Abstracts of North American geology. 1966–71.

(Abstracts of books, technical papers, maps on the geology of North America. Complements the *Bibliography of North American Geology.*)

Air pollution abstracts. 1970–76.

(Produced by the Air Pollution Technical Information Center of the Environmental Protection Agency. Broad subject arrangement with specific author-subject indexes. Covers periodicals, books, proceedings, legislation, and standards.)

ASM review of metal literature. 1944–67.

(American Society for Metals abstracting journal of the world's literature concerned with the production, properties, fabrication, and application of metals, their alloys and compounds. Combined to form part of *Metal abstracts.*)

Biological abstracts. 1926–   .

(Broad subject coverage of periodicals, books, and papers in all biological fields. Author, keyword, and systematic indexes.)

Ceramic abstracts. 1907–   .

(Published as a section of the *American Ceramic Society journal.*)

Chemical abstracts. 1918–   .

(Covers chemical periodicals in all languages. Arranged in 80 subject sections; entry includes title, author, publication date, and abstract. Index sections by chemical substance, formula, numbered patent, patent concordance, author, and keyword.)

Energy research abstracts. 1976–   .

(Covers all scientific and technical reports, journal articles, conference papers and proceedings, books, patents, theses originated by the Department of Energy, its laboratories and contractors. Classed subject arrangement, with author, title, subject, report, and contract indexes. Succeeds *Nuclear science abstracts.*)

Engineering Index. 1906–   .

(Abstracting journal includes coverage of serial publications, papers of conferences and symposia, separates and nonserial publications, some books. Excludes patents. Entries arranged by subject. Separate author index.)

Geophysical Abstracts. 1929–71.

(Abstracts of current literature pertaining to the physics of the solid earth and to geophysical exploration. Annual author and subject indexes.)

Information science abstracts. 1966– .
(Classified subject arrangement listing books and periodicals, international in scope. Formerly *Documentation abstracts.*)

International aerospace abstracts. 1961– .
(Covers published literature in aeronautics and space science and technology. Companion publication to *Scientific and technical aerospace reports.* Arranged in 75 subject categories, each entry gives accession number, title, author, source (book, conference, periodical, etc.), date, pagination, and abstract. Indexed by specific subject, personal author, contract number, meeting paper, accession number.)

Mathematical reviews. 1940– .
(Subject index to mathematical periodicals and books. Arranged by broad subject, with abstract for most entries. Separate author index, subject classification.)

Metallurgical abstracts. 1934–1967.
(British publication, combined to form part of *Metals abstracts.*)

Metals abstracts. 1968– .
(Covers all aspects of the science and practice of metallurgy and related fields. Classed subject arrangement with author index. Publication is a merger of *Metallurgical abstracts* and the *ASM review of metal literature.*)

Meteorological and geoastrophysical abstracts. 1950– .
(Includes foreign publications, arranged by Universal Decimal Classification subjects. Has separate author, subject, and geographical location indexes.)

Mineralogical abstracts. 1920– .
(Classified list of abstracts covering current international literature, including books, periodicals, pamphlets, reports.)

Nuclear science abstracts. 1947–1976.
(Covers reports of the U.S. Energy Research and Development Administration [formerly the Atomic Energy Commission], government agencies, universities, industrial and independent research organizations, and worldwide book, journal, and patent literature dealing with nuclear science and technology. Arranged by subject field. Indexes by corporate author, personal author, subject, and report number. Continued by *Energy Research Abstracts.*)

Oceanic abstracts. 1964– .
(Covers the worldwide book, periodical, and report literature on the oceans, including pollution, engineering, geology, and oceanography.)

Pollution abstracts. 1970– .
(Classed subject listing of books, periodicals, reports, documents.)

Psychological abstracts. 1927– .
(Covers books, periodicals, and reports, arranged by subject, with full author and subject indexes, cumulated indexes for 1927–74.)

Science abstracts. Section A–Physics abstracts. 1898–   . *Section B–Electrical and electronic abstracts.* 1898–   . *Section C–Computer and control abstracts.* 1966–   .
(Covers books, periodicals, papers in all languages; sections do not overlap. All sections arranged by subjects. Separate author and conference indexes.)

Selected water resources abstracts. 1968–   .
(Covers water in respect to quality, resources, engineering, and related aspects from books, journals, and reports. Subject and author indexes.)

Report Literature

The publication of reports by academic, industrial, and government agencies has been a major development since World War II. Many contracts funded by government agencies have required publication of such reports. The report is now frequently the most recent information on a subject. The bibliographical guides to this type of literature are only recently being developed.

Government reports announcements and index. 1946–   .
(Formerly titled *Government reports announcements* (1971–75), *U.S. government research and development reports* (1965–71), and *U.S. government research reports* (1954–64). Covers new reports of U.S. government-sponsored research and development released by the Department of Defense and other federal agencies. Arranged by broad subject areas, each entry gives complete bibliographical citation, descriptors and availability data, and usually an abstract.)

Government reports index. 1965–75.
(Continued as part of *Government reports announcements and index.* Previously issued under other titles. Indexes reports by subject, author, corporate source, and report number.)

Scientific and technical aerospace reports. 1963–   .
(Comprehensive abstracting journal covering worldwide report literature on the science and technology of space and aeronautics. Companion publication to *International aerospace abstracts.* Arranged in 74 subject categories. Indexed by subject, corporate source, individual author, contract number, report number, and accession number.)

Technical abstract bulletin. 1953–   .
(Primary guide to report literature until the mid-1960s, published by Defense Documentation Center. Covers technical reports of U.S. and foreign governments and commercial contractors as submitted to DDC. Now covers material with distribution limitations; abstract section issues before 1978 are security classified until 6 years after issue. All index sections are declassified. Both sections are unclassified after July 1978.)

Use of reports literature. Charles P. Auger, ed. Hamden, Conn.: Archon Books, 1975.
(Discusses report literature nature, control, and value. Evaluates sources.)

Dictionary of report series codes. Lois Godfrey and Helen Redman, eds. 2nd ed. New York: Special Libraries Association, 1973.
(Explains and identifies letter and number codes used in issuing report literature. Reference notes explain some of the systems used by various agencies in assigning report codes. Alphabetical list of designations related to issuing agency, alphabetical agency list to related series codes.)

Atlases and Statistical Guides

Basic data of interest to the technical researcher on many subjects is found in reliable and frequently updated standard guides. The quality atlas generally contains much more than maps giving geographical locations. Statistical guides are of great reference importance to original research in economic, industrial, and social questions.

Commercial atlas and marketing guide. Annual. Chicago: Rand McNally & Co.
(In addition to maps, contains much statistical data on trade, manufacturing, business, population, and transportation.)

The National atlas of the United States of America. Washington: U.S. Department of the Interior Geological Survey, 1970.
(Outstanding collection of 765 maps, many colored. Covers general reference, special subjects including landforms, geophysical forces, geology, marine features, soils, climate, water, history, economic, socio-cultural, administrative, with maps, data tables and diagrams.)

Statistical abstract of the United States. 1878– . Annual. Washington: Bureau of the Census. (Official government standard summary of statistics on the social, political and economic organization of the U.S. Excellent source citations, index.)

The World almanac and book of facts. 1868– . Annual. New York: Newspaper Enterprise Association.
(The most comprehensive and generally useful of the almanacs of miscellaneous information. Excellent statistical and news summary coverage.)

Computerized Information Retrieval

In the 1970s, a growing number of bibliographic files have been made available for on-line interactive searching and information retrieval. Normal access is from a local computer terminal to a firm offering access to data base or a system of bases. Charges are calculated on the number of minutes a file is used and the number of citations received, usually by off-line printing and airmail delivery. Over 10,000,000 citations beginning in about 1970 are now available for searching. Three commercial services offer access to a wide spectrum of data bases. They are ORBIT II, from Systems Development Corporation, Santa Monica, California; DIALOG, Lockheed Information Retrieval Service, Palo Alto, California; and BRS, Bibliographic Retrieval Services, Scotia, New York. An average search can be expected to cost a minimum of $20–40, with a key to the cost being good preplanning of search terms and search strategy by an experienced operator. Many major colleges, universities, and information centers offer these services through service bureaus or libraries. Individual access directly through originating firms is also possible. Custom searches of over 600,000 reports from federal agencies and federally sponsored research are available from National Technical Information Service, U.S. Department of Commerce, Springfield, Virginia 22161. Cost is about $100 for up to 100 abstracts. Some of the current subjects available in the scientific and technical areas are given.

Agriculture: AGRICOLA, developed by the National Library of Agriculture.

Biological sciences: BIOSIS, prepared by Biological Sciences Information Service.

Business: ABI/INFORM, produced by Data Courier, Inc.

Chemistry: CA CONDENSATES/CASIA, produced by Chemical Abstracts Service of the American Chemical Society.

Education: ERIC (Educational Resources Information Center), developed by the National Institute of Education.

Engineering: COMPENDEX, produced by Engineering Index, Inc.

Environmental studies: ENVIROLINE, prepared by Environment Information Center, Inc.

Geosciences: GEO-REF, produced by the American Geological Institute.

Government research and development reports: NTIS, produced by the National Technical Information Service of the Department of Commerce.

Mechanical engineering: ISMEC, prepared by Data Courier, Inc.

Physics: SPIN, Searchable Physics Information Notices, by American Institute of Physics.

Pollution and environment: POLLUTION, produced by the publishers of *Pollution abstracts.*

Psychology: Psychological Abstracts, by the American Psychological Association.

Science abstracts: INSPEC, produced by the Institution of Electrical Engineers.

Science and technology: SCISEARCH, produced by the Institute for Scientific Information.

Many other data bases are available for searching, but some have special use restrictions, or are limited to specific industrial group members.

U.S. Government Publications

Access to government publications, releases, and directives is possible only from a series of federally produced and commercially supplemented catalogs and indexes. The nature of the subject indexing is generally less specific than with nongovernmental periodical indexes and abstracting services. Therefore, more ingenuity on the part of the researcher is generally needed to find pertinent resources.

Monthly catalog of United States Government publications, 1895–   .

(List by agency of publications, printed and processed, issued each month. Subject index only until 1973, then added author and title indexes cumulated annually. Entry gives title, author, publication data, price or availability indication, Superintendent of Documents classification number.)

Cumulative subject index to the Monthly catalog of United States Government publications, 1900–1971. Washington: Carrollton Press, 1973–75. 15 volumes.
(Covers about 800,000 publications. Be sure to read the introduction for exclusions and entry policies.)

CIS/Index to publications of the United States Congress. Washington: Congressional Information Service, 1970– .
(Part one: Abstracts of Congressional publications and legislative histories; Part two: Index of Congressional publications and public laws. Commercial index giving significant insight into contents of Congressional publications, with a detailed subject index.)

United States Government manual. 1935– . Annual. Washington: U.S. Government Printing Office.
(The official handbook of the federal government, describing the purposes and programs of most official and quasi-official agencies, with lists of current officials.)

Federal register. 1936– .
(Daily publication of executive orders, presidential proclamations, and announcements of important rules and regulations of the federal government. Indexed by agency and significant subjects.)

Weekly compilation of presidential documents. 1965– .
(Makes available transcripts of the president's news conferences, messages to Congress, public speeches and statements and other presidential materials. Indexed.)

Code of federal regulations. 1949– .
(Contains codifications of general and permanent administrative rules and regulations of general applicability and future effect.)

Monthy checklist of state publications. 1910– .
(Records those documents and publications issued by the various states and received in the Library of Congress.)

Subject bibliographies. 1975– .
(Lists of publications available in specified subject areas from the Government Printing Office.)

American statistics index and abstracts, annual and retrospective edition. A comprehensive guide and index to the statistical publications of the U.S. Government. Washington: Congressional Information Service.
(Aims to be a master guide and index to all federally produced statistical data. Does not contain the data, but describes data and identifies sources.)

Bureau of the Census catalog of publications, 1790–1972.
(An example of a departmental catalog, kept up-to-date by frequent supplements. Many agencies issue such retrospective catalogs.)

APPENDIX C

Brief History of Measurement Systems* With a Chart of the Modernized Metric System

Weights and measures may be ranked among the necessaries of life to every individual of human society. They enter into the economical arrangements and daily concerns of every family. They are necessary to every occupation of human industry; to the distribution and security of every species of property; to every transaction of trade and commerce; to the labors of the husbandman; to the ingenuity of the artificer; to the studies of the philosopher; to the researches of the antiquarian, to the navigation of the mariner, and the marches of the soldier; to all the exchanges of peace, and all the operations of war. The knowledge of them, as in established use, is among the first elements of education, and is often learned by those who learn nothing else, not even to read and write. This knowledge is riveted in the memory by the habitual application of it to the employments of men throughout life.

John Quincy Adams
Report to the Congress, 1821

Weights and measures were among the earliest tools invented by man. Primitive societies needed rudimentary measures for many tasks: constructing dwellings of an appropriate size and shape, fashioning clothing, or bartering food or raw materials.

Man understandably turned first to parts of his body and his natural surroundings for measuring instruments. Early Babylonian and Egyptian records and the Bible indicate that length was first measured with the forearm, hand, or finger and that time was measured by the periods of the

---

*Special Publications 304A, U.S. Department of Commerce, National Bureau of Standards (Washington, D.C.: U.S. Government Printing Office, 1974).

sun, moon, and other heavenly bodies. When it was necessary to compare the capacities of containers such as gourds or clay or metal vessels, they were filled with plant seeds which were then counted to measure the volumes. When means for weighing were invented, seeds and stones served as standards. For instance, the "carat," still used as a unit for gems, was derived from the carob seed.

As societies evolved, weights and measures became more complex. The invention of numbering systems and the science of mathematics made it possible to create whole systems of weights and measures suited to trade and commerce, land division, taxation, or scientific research. For these more sophisticated uses it was necessary not only to weigh and measure more complex things—it was also necessary to do it accurately time after time and in different places. However, with limited international exchange of goods and communication of ideas, it is not surprising that different systems for the same purpose developed and became established in different parts of the world—even in different parts of a single continent.

The English System

The measurement system commonly used in the United States today is nearly the same as that brought by the colonists from England. These measures had their origins in a variety of cultures—Babylonian, Egyptian, Roman, Anglo-Saxon, and Norman French. The ancient "digit," "palm," "span," and "cubit" units evolved into the "inch," "foot," and "yard" through a complicated transformation not yet fully understood.

Roman contributions include the use of the number 12 as a base (our foot is divided into 12 inches) and words from which we derive many of our present weights and measures names. For example, the 12 divisions of the roman "pes," or foot, were called *unciae*. Our words "inch" and "ounce" are both derived from that Latin word.

The "yard" as a measure of length can be traced back to the early Saxon kings. They wore a sash or girdle around the waist—that could be removed and used as a convenient measuring device. Thus the word "yard" comes from the Saxon word "gird" meaning the circumference of a person's waist.

Standardization of the various units and their combinations into a loosely related system of weights and measures sometimes occurred in fascinating ways. Tradition holds that King Henry I decreed that the yard should be the distance from the tip of his nose to the end of his thumb. The length of a furlong (or furrow-long) was established by early Tudor rulers as 220 yards. This led Queen Elizabeth I to declare, in the sixteenth century, that henceforth the traditional Roman mile of 5,000 feet would be replaced by one of 5,280 feet, making the mile exactly 8 furlongs and providing a convenient relationship between two previously ill-related measures.

Thus, through royal edicts, England by the eighteenth century had achieved a greater degree of standardization than the continental countries. The English units were well suited to commerce and trade because they had been developed and refined to meet commercial needs. Through colonization and the dominance of world commerce during the seventeenth, eighteenth, and nineteenth centuries, the English system of weights and meaures was spread to and established in many parts of the world, including the American colonies.

However, standards still differed to an extent undesirable for commerce among the 13 colonies. The need for greater uniformity led to clauses in the Articles of Confederation (ratified by the original colonies in 1781) and the Constitution of the United States (ratified in 1790) giving power to the Congress to fix uniform standards for weights and measures. Today, standards supplied to all the States by the National Bureau of Standards assure uniformity throughout the country.

The Metric System

The need for a single worldwide coordinated measurement system was recognized over 300 years ago. Gabriel Mouton, Vicar of St. Paul in Lyons, proposed in 1670 a comprehensive decimal measurement system based on the length of one minute of arc of a great circle of the earth. In 1671 Jean Picard, a French astronomer, proposed the length of a pendulum beating seconds as the unit of length. (Such a pendulum would have been fairly easily reproducible, thus facilitating the widespread distribution of uniform standards.) Other proposals were made, but over a century elapsed before any action was taken.

In 1790, in the midst of the French Revolution, the National Assembly of France requested the French Academy of Sciences to "deduce an invariable standard for all the measures and all the weights." The Commission appointed by the Academy created a system that was, at once, simple and scientific. The unit of length was to be a portion of the earth's circumference. Measures for capacity (volume) and mass (weight) were to be derived from the unit of length, thus relating the basic units of the system to each other and to nature. Furthermore, the larger and smaller versions of each unit were to be created by multiplying or dividing the basic units by 10 and its multiples. This feature provided a great convenience to users of the system, by eliminating the need for such calculations as dividing by 16 (to convert ounces to pounds) or by 12 (to convert inches to feet). Similar calculations in the metric system could be performed simply by shifting the decimal point. Thus the metric system is a "base-10" or "decimal" system.

The Commission assigned the name *metre* (which we also spell meter) to the unit of length. This name was derived from the Greek word *metron*, meaning "a measure." The physical standard representing the meter was to be constructed so that it would equal one ten-millionth of the distance from the north pole to the equator along the meridian of the earth running near Dunkirk in France and Barcelona in Spain.

The metric unit of mass, called the "gram," was defined as the mass of one cubic centimeter (a cube that is 1/100 of a meter on each side) of water at its temperature of maximum density. The cubic decimeter (a cube 1/10 of a meter on each side) was chosen as the unit of fluid capacity. The measure was given the name "liter."

Although the metric system was not accepted with enthusiasm at first, adoption by other nations occurred steadily after France made its use compulsory in 1840. The standardized character and decimal features of the metric system made it well suited to scientific and engineering work. Consequently, it is not surprising that the rapid spread of the system coincided with an age of rapid technological development. In the United States, by Act of Congress in 1866, it was made "lawful throughout the United States of America to employ the weights and measures of the metric system in all contracts, dealings or court proceedings."

By the late 1860s, even better metric standards were needed to keep pace with scientific advances. In 1875, an international treaty, the "Treaty of the Meter," set up well-defined metric standards for length and mass, and established permanent machinery to recommend and adopt further refinements in the metric system. This treaty, known as the Metric Convention, was signed by 17 countries, including the United States.

As a result of the Treaty, metric standards were constructed and distributed to each nation that ratified the Convention. Since 1893, the internationally agreed-to metric standards have served as the fundamental weights and measures standards of the United States.

By 1900 a total of 35 nations—including the major nations of continental Europe and most of South America—had officially accepted the metric system. Today, with the exception of the United States and a few small countries, the entire world is using predominantly the metric system or is committed to such use. In 1971 the Secretary of Commerce, in transmitting to Congress the results of a three-year study authorized by the Metric Study Act of 1968, recommended that the U.S. change to predominant use of the metric system through a coordinated national program. The Congress is now considering this recommendation.

The International Bureau of Weights and Measures located at Sevres, France, serves as a permanent secretariat for the Metric Convention, coordinating the exchange of information about the use and refinement of the metric system. As measurement science develops more precise and easily reproducible ways of defining the measurement units, the General Conference of Weights and Measures—the diplomatic organization made up of adherents to the Convention—meets periodically to ratify improvements in the system and the standards.

In 1960, the General Conference adopted an extensive revision and simplification of the system. The name Le Système International d'Unités (International System of Units), with the international abbreviation SI, was adopted for this modernized metric system. Further improvements in and additions to SI were made by the General Conference in 1964, 1968, and 1971.

COMPARING THE COMMONEST MEASUREMENT UNITS

Approximate conversions from customary to metric and vice versa.

	When you know:	You can find:	If you multiply by:
LENGTH	inches	millimeters	25
	feet	centimeters	30
	yards	meters	0.9
	miles	kilometers	1.6
	millimeters	inches	0.04
	centimeters	inches	0.4
	meters	yards	1.1
	kilometers	miles	0.6
AREA	square inches	square centimeters	6.5
	square feet	square meters	0.09
	square yards	square meters	0.8
	square miles	square kilometers	2.6
	acres	square hectometers (hectares)	0.4
	square centimeters	square inches	0.16
	square meters	square yards	1.2
	square kilometers	square miles	0.4
	square hectometers (hectares)	acres	2.5
MASS	ounces	grams	28
	pounds	kilograms	0.45
	short tons	megagrams (metric tons)	0.9
	grams	ounces	0.035
	kilograms	pounds	2.2
	megagrams (metric tons)	short tons	1.1
LIQUID VOLUME	ounces	milliliters	30
	pints	liters	0.47
	quarts	liters	0.95
	gallons	liters	3.8
	milliliters	ounces	0.034
	liters	pints	2.1
	liters	quarts	1.06
	liters	gallons	0.26
TEMPERATURE	degrees Fahrenheit	degrees Celsius	5/9 (after subtracting 32)
	degrees Celsius	degrees Fahrenheit	9/5 (then add 32)

NAMES AND SYMBOLS FOR METRIC PREFIXES

Prefix	means
tera (10^{12})	one trillion times
giga (10^9)	one billion times
mega (10^6)	one million times
kilo (10^3)	one thousand times
hecto (10^2)	one hundred times
deca (10)	ten times
deci (10^{-1})	one tenth of
centi (10^{-2})	one hundredth of
milli (10^{-3})	one thousandth of
micro (10^{-6})	one millionth of
nano (10^{-9})	one billionth of
pico (10^{-12})	one trillionth of

SOME COMMON UNITS

Length	Mass	Volume	Temperature	Electric Current	Time
METRIC					
meter	kilogram	liter	Celsius	ampere	second
CUSTOMARY					
inch	ounce	fluid ounce	Fahrenheit	ampere	second
foot	pound	teaspoon			
yard	ton	tablespoon			
fathom	grain	cup			
rod	dram	pint			
mile		quart			
		gallon			
		barrel			
		peck			
		bushel			

APPENDIX 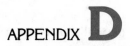 **D**

A Selected Bibliography

Technical writing

Coleman, Peter, and Brambleby, Ken. *The Technologist as Writer.* New York: McGraw-Hill Book Company, 1971.

Damerst, William A. *Clear Technical Reports.* New York: Harcourt Brace Jovanovich, Inc., 1972.

Fear, David E. *Technical Communication.* Glenview, Ill.: Scott, Foresman & Company, 1977.

Herbert, A. J. *The Structure of Technical English.* New York: Longman, Inc., 1975.

Jones, W. Paul. *Writing Scientific Papers and Reports.* 7th ed. Dubuque, Iowa: William C. Brown Company, Publishers, 1976.

Mathes, J. C., and Stevenson, Dwight. *Designing Technical Reports.* Indianapolis: The Bobbs-Merrill Company, Inc., 1976.

Mills, Gordon H., and Walter, John A. *Technical Writing.* 4th ed. New York: Holt, Rinehart & Winston, 1978.

Pearsall, Thomas E., and Cunningham, Donald H. *How to Write for the World of Work.* New York: Holt, Rinehart & Winston, 1978.

Pickett, Nell A., and Laster, Ann A. *Technical English.* San Francisco: Canfield Press, 1975.

Souther, James W., and White, Myron L. *Technical Report Writing.* 2nd ed. New York: John Wiley & Sons, 1977.

Business communication

Barry, Robert E. *Business English for the 70's.* 2d ed. Englewood Cliffs, N.J.: Prentice-Hall, Inc., 1975.

Janis, Harold J. *Writing and Communicating in Business.* 3d ed. New York: Macmillan Publishing Company, Inc., 1978.

Lindauer, J. S. *Communicating in Business.* New York: Macmillan Publishing Company, Inc., 1974.

Menning, Jack H., and Wilkinson, C. W. *Communicating Through Letters and Reports.* 6th ed. Homewood, Ill.: Richard D. Irwin, Inc., 1976.

Writing Better Letters, Reports and Memos. New York: American Management Association, Inc., 1976.

Writing in general

Flesch, Rudolf. *Art of Readable Writing.* Rev. ed. New York: Harper & Row Publishers, Inc. 1974.

———. *Say What You Mean.* New York: Harper & Row Publishers, Inc., 1972.

Hall, Donald, ed. *Modern Stylists: Writers on the Art of Writing.* New York: Free Press, 1968.

Lambuth, David, *et al. The Golden Book on Writing.* New York: Penguin Books, Inc., 1976.

Quiller-Couch, Sir Arthur. *On the Art of Writing.* New York: G. P. Putnam's Sons, 1916.

Trimble, John H. *Writing with Style.* Englewood Cliffs, N.J.: Prentice-Hall, Inc., 1975.

Style and usage

Bernstein, Theodore. *Miss Thistlebottom's Hobgoblins.* New York: Farrar, Straus & Giroux, Inc., 1971.

CBE Style Manual Committee. *Council of Biology Editors Style Manual.* 4th ed. Arlington, Va.: Council of Biology Editors, 1978.

Fowler, Henry W. *A Dictionary of Modern English Usage.* 2d ed. Ernest Gowers, ed. New York: Oxford University Press, 1965.

Fowler, H. W., and Fowler, F. G. *The King's English.* 3d ed. New York: Oxford University Press, 1974.

Jordon, Lewis. *The New York Times Manual of Style and Usage.* New York: Quadrangle, 1975.

A Manual of Style. 12th ed. Chicago: The University of Chicago Press, 1969.

Perrin, Porter G., and Ebbitt, Wilma R. *Writer's Guide and Index to English.* 5th ed. Glenview, Ill.: Scott, Foresman & Company, 1972.

Speech

Bryant, Donald C., and Wallace, Karl R. *Oral Communication.* 4th ed. Englewood Cliffs, N.J.: Prentice-Hall, Inc., 1976.

Monroe, Alan H., and Ehninger, Douglas. *Principles of Speech Communication.* 7th ed. Glenview, Ill.: Scott, Foresman & Company, 1975.

Semantics

Condon, John C., Jr. *Semantics and Communication.* 2d ed. New York: Macmillan Publishing Company, Inc., 1975.

Hayakawa, S. I. *Language in Thought and Action.* 4th ed. New York: Harcourt Brace Jovanovich, Inc., 1978.

Logic

Cohen, Morris R., and Nagel, Ernest. *Introduction to Logic.* New York: Harcourt Brace Jovanovich, Inc., 1962.

Copi, Irving M. *Introduction to Logic.* 5th ed. New York: Macmillan Publishing Company, Inc., 1978.

Graphics

Levens, A. S. *Graphics.* New York: John Wiley & Sons, Inc., 1968.

Morris, George E. *Technical Illustrating.* Englewood Cliffs, N.J.: Prentice-Hall, Inc., 1975.

Pocket Pal: A Graphic Arts Production Handbook. 11th ed. New York: International Paper Company, 1974.

Library research

Barzun, Jacques, and Graff, Henry F. *The Modern Researcher.* 3d ed. New York: Harcourt Brace Jovanovich, Inc., 1977.

Downs, Robert B., and Keller, Clara D. *How to Do Library Research.* 2d ed. Urbana, Ill.: University of Illinois Press, 1975.

Jackson, Ellen. *Subject Guide to Major United States Government Publications.* Chicago: American Library Association, 1968.

McCormick, Mona. *The New York Times Guide to Reference Materials.* New York: Popular Library, 1978.

Schmeckebier, Laurence F., and Eastin, Roy B. *Government Publications and Their Use.* Rev. ed. Washington, D.C.: Brookings Institution, 1969.

Index

MARKING SYMBOLS

This list of marking symbols refers you to either Part 3, the "Handbook," or to other parts of the book where comments about style and form can be found. The list furnishes you with a caption and a page reference.